ac circuit analysis with illustrative problems

ac circuit analysis

with illustrative problems

Phillip Cutler
Education Research Associates
and
Orange Coast
College
Costa Mesa
California

McGraw-Hill Book Company
New York
St. Louis
San Francisco
Düsseldorf
Johannesburg
Kuala Lumpur
London
Mexico
Montreal
New Delhi
Panama
Rio de Janeiro
Singapore
Sydney
Toronto

Other Problems Books by Phillip Cutler:

DC Circuit Analysis with Illustrative Problems
Solid-state Device Theory with Illustrative Problems
Linear Electronic Circuits with Illustrative Problems

Other Texts:

Electronic Circuit Analysis: Passive Networks

The editors for this book were Alan W. Lowe and Cynthia Newby, the designer was Marsha Cohen, and its production was supervised by Patricia Ollague. It was set in Times Roman by Holmes Typography, Inc. It was printed and bound by Edwards Brothers Incorporated.

ac circuit analysis with illustrative problems

07–014996–8

1 2 3 4 5 6 7 8 9 0 E B E B 7 9 8 7 6 5 4

contents

To Professor E. L. (Roy) France.
The salt of the earth.

preface

This is the second text in the four-volume circuit analysis series designed to upgrade electronics technology courses. These volumes provide a completely integrated sequence of instruction which commences at the dc level in the first, the ac level in the second, solid-state devices in the third, and circuit theory and applications in the fourth.

By introducing specific theoretical concepts at appropriate points and ensuring comprehension by adequate drill in practice problems of a meaningful nature, a strong thread of continuity is preserved throughout the complete series. This makes possible a considerable saving in time in the advanced courses by avoiding repetition of material that should have been mastered in prior courses. More significantly, it allows for a broader and deeper presentation of modern electronic devices and circuits in the latter courses—a presentation which also provides the foundation for mastering specialized courses. For example, the student is so thoroughly familiarized with Thévenin's theorem in the first volume that this powerful tool of circuit analysis does not have to be reviewed in detail when he is evaluating the equivalent circuit seen at the output terminals of a transformer in the second volume, an FET amplifier in the third volume, or a voltage regulator in the fourth volume.

As another example, we have Bode plots introduced in this, the second volume, followed by sufficient exercise problems to ensure their comprehension. This analytic tool is then used to investigate the frequency response characteristics of devices and basic amplifier configurations in the third and fourth volumes. It also affords an insight—one that is unfortunately not even considered in most courses of a similar nature—into the stability of feedback amplifiers in the fourth volume.

Chapter 1 of this text introduces the sinusoidal waveform. In particular it emphasizes the significance and theory behind the average and effective value concepts so that the reader can apply them to nonsinusoidal waveforms at some future time. It is not at all necessary to know integral calculus to follow the text material. Chapter 2 explains how the voltage and current notation developed in the dc volume may be applied to ac sources. Chapter 3 rather extensively explores the concept of power in the resistive circuit with sinusoidal excitation. This material is most useful in the future analysis of power amplifiers.

Chapters 4 and 5 introduce, in a practical manner, the concepts of capacitance and inductance. In particular, the circuit properties of these elements with respect to a time-varying voltage or current are stressed so that the reader acquires an intuitive feel for the effect of these parameters. The concepts of reactance and susceptance are introduced and applied to the analysis of series and parallel configurations. A few practical problems pertaining to distributed L or C are also injected so that the reader may acquire some insight into why a few inches of wire may mean the difference between the success or failure of a circuit in the real world.

Chapter 6 introduces the impedance concept and investigates the analysis of *RLC* circuits, while Chapter 7 introduces the concept of Q and series-parallel conversions which are most useful in circuit simplification and impedance transformation.

Chapter 8 presents a fairly detailed but practical discussion of the resonance phenomenon and its implications in both the series and parallel (antiresonant) cases. The relation between Q and bandwidth is emphasized, so that an insight into the properties of tuned circuits is established. Chapter 9 explores, under sinusoidal excitation, the properties of the four-terminal networks introduced in the first volume.

Chapter 10 develops the analysis of inductively coupled circuits with a minimum of magnetic circuit theory. Although the physical behavior of transformer action is considered, the discussion soon turns to regarding the transformer as a four-terminal network from which suitable equivalent circuits may be derived and applied. The transition from the nonideal to ideal transformer is carefully made so that the student knows under what conditions a transformer may be considered ideal.

Chapter 11 introduces the transfer function concept. While this may sound formidable, it requires only that the student be able to utilize the j operator. The reader is then shown how to construct Bode plots so that the frequency response of various *RC*, *RL*, and *RLC* networks may be displayed. The contents of this chapter provide the foundation for examining the frequency response of various solid-state devices and amplifier configurations in the third volume, and in the fourth volume Bode plots are again used to study op amps and stability in feedback systems.

Audiovisual materials to supplement the contents of this and other volumes in the series have been prepared by the author and are available through Education Research Associates, 3290 Turlock Drive, Costa Mesa, California, 92626.

The author wishes to express his thanks to Professor Brinton Mitchell of East Los Angeles College, to Professor Robert C. Carter, Southern Technical Institute at Marietta, Georgia, and to Robert Decker for reviewing the manuscript and making helpful suggestions. In particular, I am once again indebted to Mr. Decker for his uncanny ability to ferret out errors with amazing speed.

My thanks also to Professor Edward L. (Roy) France whose dedication and administrative skill contributed mightily to the success of the annual workshops at Utah State University, where much of the material in these volumes has been tested. I am also most grateful to Professor Don James and his magnificent crew at Utah Technical College at Provo for their loyalty to the philosophy of these volumes and for visibly demonstrating that the practical and theoretical content of electronics technology can be interwoven into an outstanding curriculum. And again my appreciation to Ted Hoffman at Orange Coast College for adhering to technical excellence and providing me with invaluable feedback.

Last but not least, I thank Bertha Rapaport for her friendship and beautiful spirit; my beloved parents, to whom I owe so much; and Betty, my dear wife, who not only typed this manuscript but makes it all worthwhile.

Phillip Cutler

ac circuit analysis with illustrative problems

1 the sinusoid

It is advisable to begin this text on ac circuit analysis with an introductory comment regarding our primary goal in this course of study. Essentially it is to be able to perform ac circuit analysis in a manner and at a level comparable to our study of dc circuit analysis in the preceding text of this series.[1] The notation established at that time for describing current and voltage, Kirchhoff's laws, and all the network theorems is still applicable. You will find, however, that whereas fairly simple algebra provided the means of analytically exploring dc circuits, more advanced mathematical concepts are required to analyze ac circuits.

It turns out that the reason we require more sophisticated mathematics is because the currents and voltages in ac circuits vary with time, and this causes additional circuit parameters besides resistance to manifest themselves. When the driving voltage and/or current sources were fixed in magnitude and polarity, as in dc circuits, the only opposition offered to the flow of current was due to the circuit resistance (R).

However, if the driving voltage and/or current should vary with time, we would find that any electric circuit, be it so humble as a loop of wire, exhibits the following to some degree:

1. An opposition to the flow of current (resistance, R)
2. An opposition to a *change* in current (inductance, L)
3. An opposition to a *change* in voltage (capacitance, C)

The opposition to current flow in the first case is simply that due to ohmic resistance (R), and it manifests itself whether or not the current or voltage is changing.

The opposition presented to a *change* in current (case 2) is due to a parameter called inductance, which is given the symbol L.

The opposition presented to a change in voltage (case 3) is due to a parameter called capacitance, which is given the symbol C.

These circuit parameters will be examined more thoroughly below. For the moment just file away the fact that not only resistance but inductance and capacitance must be considered whenever there is a time variation in voltage or current.

1-1 Waveform parameters

A plot of voltage or current versus time is called a waveform. Figure 1-1*a*, for example, is a plot of the dc voltage e versus time t. As you can see, e is of constant value. If the polarity of this dc voltage source is reversed, we have the result shown in Fig. 1-1*b*.

[1] Phillip Cutler, "DC Circuit Analysis with Illustrative Problems," McGraw-Hill Book Company, New York, 1974.

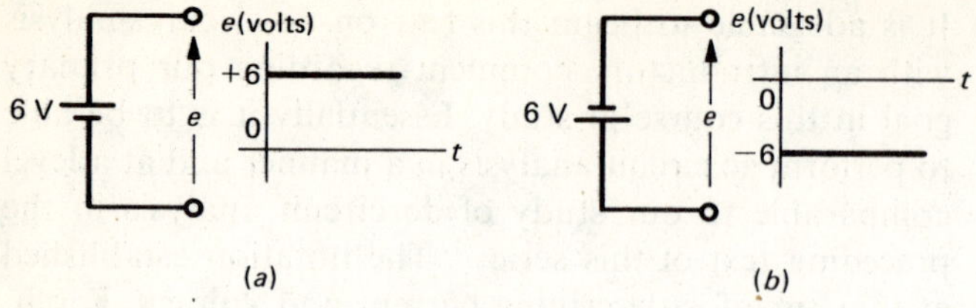

FIGURE 1-1

If the waveform periodically reverses polarity, it is referred to as an alternating current or alternating voltage. It is common practice to use the abbreviation AC, or ac, for alternating current and also to use it as an adjective, so that we speak of an ac current or an ac voltage. The circuit of Fig. 1-2*a* would, for example, generate the ac voltage waveform of Fig. 1-2*b* if the switch were periodically shifted between positions *A* and *B*.

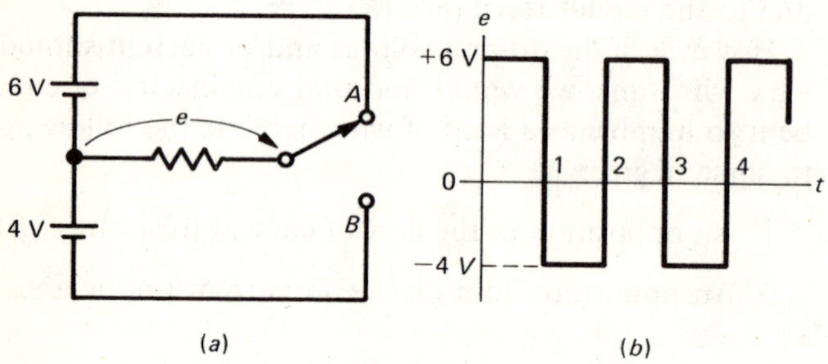

FIGURE 1-2

An enormous variety of waveforms may be generated by suitable circuitry; some of the more commonly encountered ones are the pulse, square, rectangular, triangular, sweep, and sine waves of Fig. 1-3.

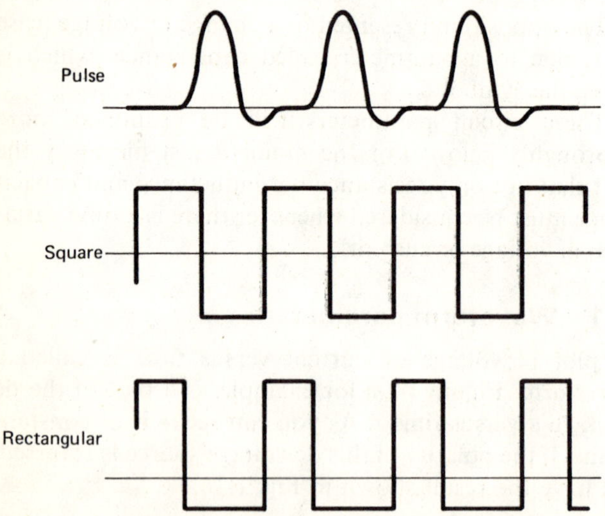

FIGURE 1-3

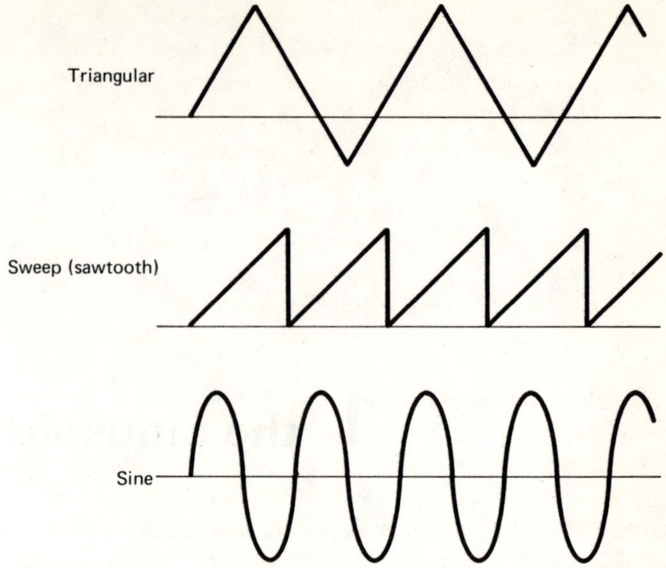

FIGURE 1-3 (*Continued*)

In general our interest in any particular current or voltage waveform extends to parameters such as

1. Positive peak amplitude
2. Negative peak amplitude
3. Peak-to-peak amplitude
4. Period
5. Frequency
6. Average value (dc component)
7. Root mean square (rms) value

We will assume that the waveforms under discussion are periodic, which means that they repeat themselves in time.

To understand the significance of these waveform parameters, we will consider the illustrative example shown in Fig. 1-4. The positive peak amplitude is the

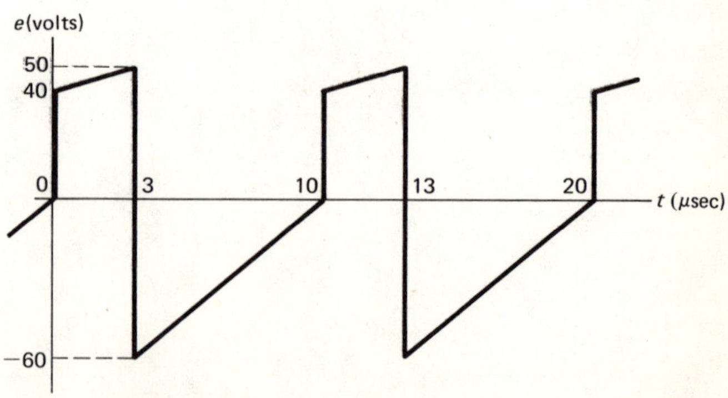

FIGURE 1-4

most positive value of the waveform, +50 volts in this case. It is possible for the peak positive amplitude of some waveform to be a negative value if the complete waveform lies below the zero axis.

The negative peak amplitude is the most negative value of the waveform, −60 volts in Fig. 1-4. If some waveform is of such a nature that all of it lies above the zero axis, the peak negative amplitude will be a positive number.

The peak-to-peak amplitude is the absolute value of the amplitude between the positive and negative peak amplitudes. Thus for Fig. 1-4 the peak-to-peak (abbreviated p-p or pk-pk) amplitude is measured from −60 volts to +50 volts for a total of 110 volts.

The period (given the symbol T) of a waveform is the length of time between corresponding points on it, or, in other words, the time it takes to repeat itself. Thus in Fig. 1-4 the period $T = 10\,\mu$sec.

The frequency (given the symbol f) represents the number of times the waveform repeats itself in a given time interval. If that time interval is equal to the period, we say the waveform has completed one full cycle, or one cycle for short. Thus frequency and period are reciprocal quantities related by

$$f = \frac{1}{T} \tag{1-1a}$$

$$T = \frac{1}{f} \tag{1-1b}$$

Frequency is measured in a unit called the Hertz (Hz), where

$$1\text{ Hz} = \frac{1\text{ cycle}}{\text{sec}} \tag{1-2}$$

Therefore, instead of describing a frequency by saying 20,000 cycles/sec or 20 kilocycles/sec (20 kc/s), we say 20,000 Hertz or 20 kilohertz (20 kHz).

1-2 The average value

The average value of some periodic waveform is its average value as determined over a time interval equal to the period T. The average value is the dc component of a waveform and it represents what a dc voltmeter would read if it were driven by a voltage waveform.

To find the average value of any function, say $y = f(x)$, over some interval, say $x_2 - x_1$ as shown in Fig. 1-5, it is necessary first to determine the area bounded by the curve (shaded area) during this interval and then to divide this area by the length of the interval $(x_2 - x_1)$. This yields the average value. In Fig. 1-5 some of the area is positive and some is negative. To find the net area we must subtract the negative area from the positive. Should the negative area exceed the positive area, there will result a negative average value.[2]

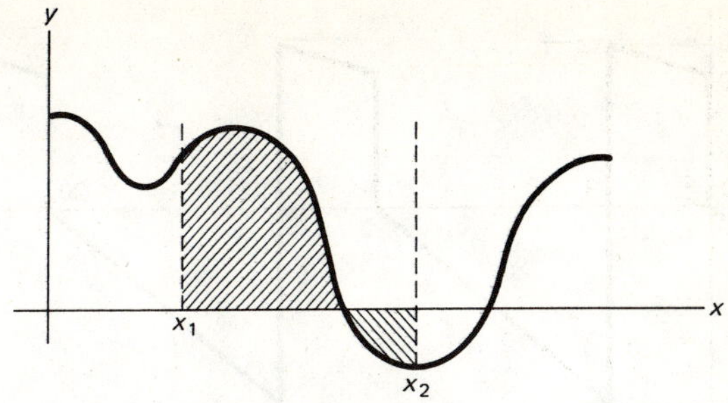

FIGURE 1-5

Mathematically the average value of a waveform is expressed as

$$\text{Average value} = \frac{1}{t_2 - t_1}\int_{t_1}^{t_2} f(t)\,dt \tag{1-3}$$

where $(t_2 - t_1)$ is the time interval over which we seek the average value, usually equal to the period, and

$$\int_{t_1}^{t_2} f(t)\,dt$$

is a mathematical symbolism used to designate the area under a curve (in this case a plot of some $f(t)$ meaning function of time) over an interval $t_2 - t_1$.

Don't panic if the integral sign ($\int$) and Eq. (1-3) seem strange. You may instead bear in mind

$$\text{Average value} = \frac{\text{net area between the waveform and the time } (t) \text{ axis during a given time interval}}{\text{the given time interval}} \tag{1-4}$$

which is really identical to Eq. (1-3).

Thus to compute the average voltage for the waveform of Fig. 1-4 (repeated as Fig. 1-6*a*), we determine the area under the curve over the period (10 μsec). First notice that the positive area under the curve during the interval from 0 to 3 μsec may be found by resolving this area into a rectangle and triangle, as shown in Fig. 1-6*b*. The area of the rectangle (A_1) is the base × height, whereas that of the triangle (A_2) is (base × height)/2. Thus the net positive area A^+ is 135 V μsec.

The negative area as determined from Fig. 1-6*c* is $A^- = -210\text{ V }\mu\text{sec}$. Therefore the net area under a full period is $A = A^+ + A^- = 135\text{ V }\mu\text{sec} + (-210\text{ V }\mu\text{sec}) = -75\text{ V }\mu\text{sec}$.

The average or dc voltage of this waveform is

$$E_{av} = \frac{A}{T} = \frac{-75\text{ V }\mu\text{sec}}{10\,\mu\text{sec}} = -7.5\text{ V}$$

[2]For a more detailed but nevertheless elementary discussion of the average value concept, see Sec. 2-25 in Phillip Cutler, "Electronic Circuit Analysis—Passive Networks," McGraw-Hill Book Company, New York, 1967.

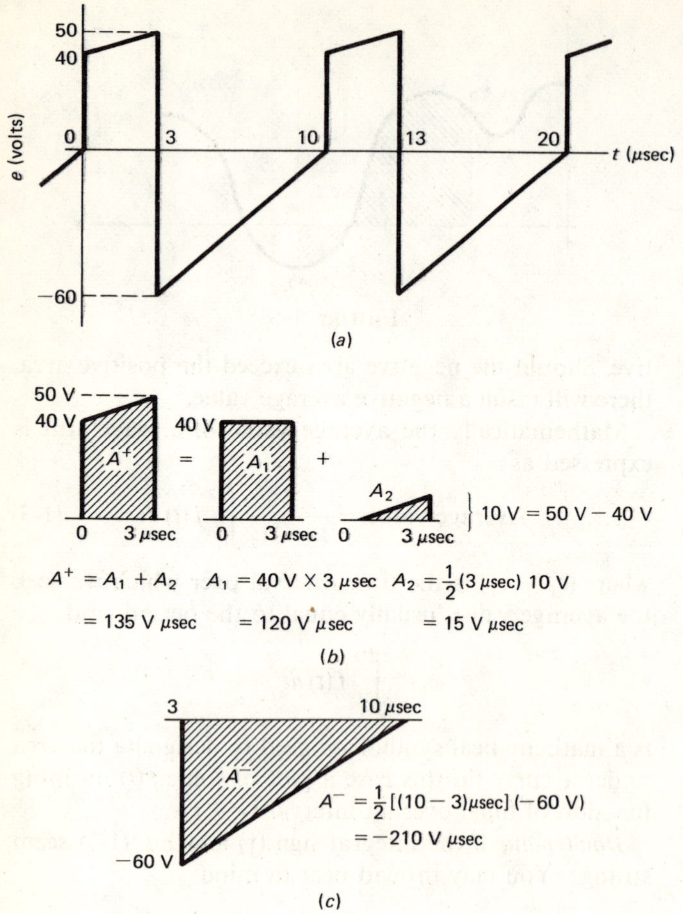

FIGURE 1-6

1-3 The effective or root-mean-square (rms) value

The root-mean-square or effective value of a waveform pertains to the power it can develop. For example, suppose we impress some periodic voltage waveform across a resistor. The resistor will get warm and dissipate heat. It will do this even if the average value of voltage (or current) is zero, because it does not matter in which direction the resistor is passing current. The mere fact that current goes through it causes energy to be lost.

Now by definition the effective or rms value of a voltage (or current) waveform is that value which causes just as much average power to be developed in the resistor as an equivalent dc voltage. Or, in other words, that dc voltage (current), which produces just as much heat in a resistor as some arbitrary waveform, is equal to the effective voltage (current) of the waveform.

To determine the rms value of a waveform, say voltage, we may reason as follows. The instantaneous power at any moment in time is given by

1) $$p = \frac{e^2}{R} \tag{1-5}$$

where e is the instantaneous voltage across the resistor R.

Note that lower-case symbols are used for instantaneous values. To find the average power (P) over some time interval ($t_2 - t_1$), we must generate a curve of p versus t during this interval and then determine the average value (P) of the p curve during this same interval. This yields the average waveform power over the interval in question (usually the period T). Thus

2) $$P_{av} = \frac{\text{area under } p \text{ curve}}{t_2 - t_1} \tag{1-6}$$

Remember, the area under the p curve is evaluated over the $t_2 - t_1$ interval. Substituting (1) into (2),

3) $$P_{av} = \frac{\text{area under } \dfrac{e^2}{R} \text{ curve}}{t_2 - t_1}$$

or

4) $$P_{av} = \frac{1}{R} \times \frac{\text{area under } e^2 \text{ curve}}{t_2 - t_1}$$

Equating the average waveform power in R with its equivalent dc power,

5) $$P_{dc} = P_{av}$$

or

6) $$\frac{E_{dc}^{\ 2}}{R} = \frac{1}{R} \times \frac{\text{area under } e^2 \text{ curve}}{t_2 - t_1}$$

Multiplying both sides by R and solving for E_{dc}, which by our definition is the effective value E_{rms}, we obtain

7) $$E_{rms} = \sqrt{\frac{\text{area under } e^2 \text{ curve}}{t_2 - t_1}} \tag{1-7}$$

For the more mathematically inclined, Eq. (7) may be expressed as

8) $$E_{rms} = \sqrt{\frac{1}{t_2 - t_1}\int_{t_1}^{t_2} e^2\, dt} \tag{1-8}$$

If current is used in place of voltage in Eqs. (1-7) or (1-8), we obtain I_{rms}.

The significance of the rms value is that once it is determined for any periodic current or voltage waveform, the corresponding power developed in a resistor can be determined from.

$$\left.\begin{aligned} P &= E_{rms}^{\ 2}/R \\ P &= I_{rms}^{\ 2}\, R \end{aligned}\right\} \text{ always valid} \qquad \begin{aligned}(1\text{-}9a)\\(1\text{-}9b)\end{aligned}$$

$$P = E_{rms} I_{rms} \text{ sometimes valid} \tag{1-9c}$$

The exception taken with Eq. (1-9c) is due to the fact that in general the product of two average values (and the rms value is a special kind of average) is not the same as the average of their products. As an extreme example, consider an open switch with some voltage waveform across it. This voltage waveform will have some rms value, but clearly no power is dissipated in the open switch since the current is zero. Now if the switch is closed, some current waveform which has some rms value will pass through it. However, the switch power dissipation is again zero since there is no voltage developed across a closed switch. Hence the dissipated power in the switch must be zero. However, blindly plugging into Eq. (1-9*c*) would yield a definite value of power.

1-4 Power

This is a good time to inject some comments regarding power calculations. There are two kinds of power we are usually interested in—instantaneous power (p) and average power (P). The instantaneous power for any waveform of voltage and any corresponding waveform of current is given by

$$p = ei \tag{1-10}$$

where p = instantaneous power
e = instantaneous voltage
i = instantaneous current

The current and voltage waveforms need *not* be the same, and in many electronic circuits this is often true.[3]

To determine the average power in any device and with any e and i waveforms, the safest procedure to use is to generate the p curve by Eq. (1-10) and then to determine the average value of the p curve in a manner similar to finding the average voltage or current previously discussed. Thus for a waveform of period T,

$$P = \frac{\text{area under the } p \text{ curve during the interval } T}{T} \tag{1-11}$$

Or for those familiar with calculus,

$$P = \frac{1}{T}\int_0^T p\,dt = \frac{1}{T}\int_0^T (ei)\,dt \tag{1-12}$$

We will resume the subject of power in succeeding chapters.

In general the average and rms values of a waveform will not be the same. Thus one must be sure to use rms rather than average voltage or current values when computing power.

Now, for illustrative reasons, let us determine the effective value of the waveform of Fig. 1-4, repeated here as Fig. 1-7*a*. If we choose some e values in Fig. 1-7*a*

[3]For example, the plate of a pentode tube might be driven with a rectangular pulse of constant voltage while the grid is driven with a sawtooth voltage, causing a sawtooth of plate current.

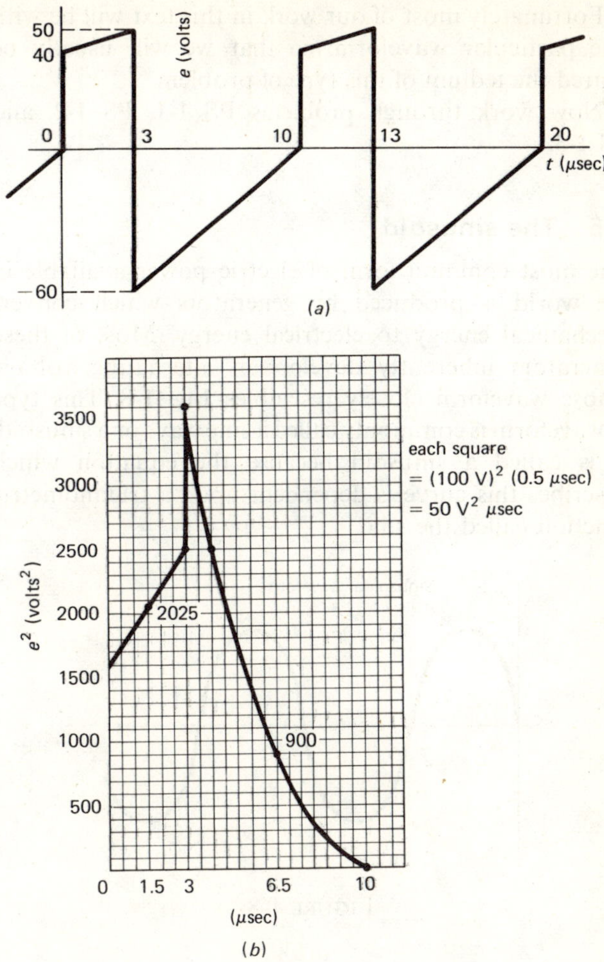

FIGURE 1-7

and square them, we obtain Fig. 1-7*b*. As you can see, all the area under the e^2 curve is positive since the square of a negative number is positive. Also, the e^2 curve bears an uncomfortable lack of resemblance to the e curve, but let the chips fall where they may. Next, by any means, we must approximate the area under the curve. If you count squares or make some geometric approximations to the area, you will obtain values ranging from perhaps 12,000 V^2 μsec to 18,000 V^2 μsec depending upon your accuracy. The rms value is then

$$e_{\text{rms}} = \sqrt{\frac{A}{T}} \qquad T = 10\,\mu\text{sec}$$

and for $A = 12{,}000\ \text{V}^2\ \mu\text{sec},\ e_{\text{rms}} = 34.6\ \text{V}$
$A = 18{,}000\ \text{V}^2\ \mu\text{sec},\ e_{\text{rms}} = 42.4\ \text{V}$

If the calculus is employed to achieve an exact solution (left as an exercise for the reader familiar with calculus), we obtain $e_{\text{rms}} = 38.1$ V as the correct value. Thus with reasonable care we can expect 10 percent or better accuracy in most graphic approximations.

Fortunately most of our work in this text will be with one particular waveform, so that we will usually be spared the tedium of this type of problem.

Now work through problems PS 1-1, PS 1-2, and PS 1-3.

1-5 The sinusoid

The most common form of electric power available in the world is produced by generators which convert mechanical energy to electrical energy. Most of these generators inherently develop an alternating voltage whose waveform closely resembles Fig. 1-8. This type of waveform is commonly called a sine wave or a sinusoid. It is called a sinusoid because the equation which describes this curve is dependent upon a trigonometric function called the sine.

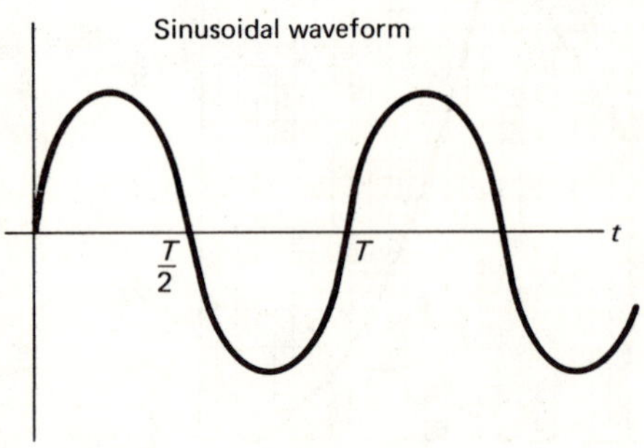

FIGURE 1-8

Suffice it to say that the sinusoid is of extreme interest to us, not only because it is the waveform of the power line voltage (usually 117 volts, 60 Hz), but also because it is inherently produced in electronic circuits called oscillators which form the basis of most forms of wireless communication (radio, TV). Furthermore, the sinusoid is related to considerable natural phenomena and it has certain mathematical properties which make it very useful.[4] For example, it can be shown by a mathematical technique called Fourier analysis that nonsinusoidal waveforms such as we previously discussed can be resolved into a combination of specifically related sine waves.

To better understand the sinusoid, consider Fig. 1-9. Here we have a rectangular coordinate system defined by the *xy* axis and divided into the usual four quadrants, I, II, III, and IV. We also have a rotating radius vector (or phasor, as it is sometimes called) *r* which generates the angle θ as it rotates in the counterclockwise (CCW) direction. The vector *r*, irrespective of what quadrant it lies in, is always a positive quantity, and the CCW

[4] For example, it can be differentiated or integrated and still yield a sinusoidal result. Also, the superposition of sinusoids of the same frequency results in a sinusoid of identical frequency.

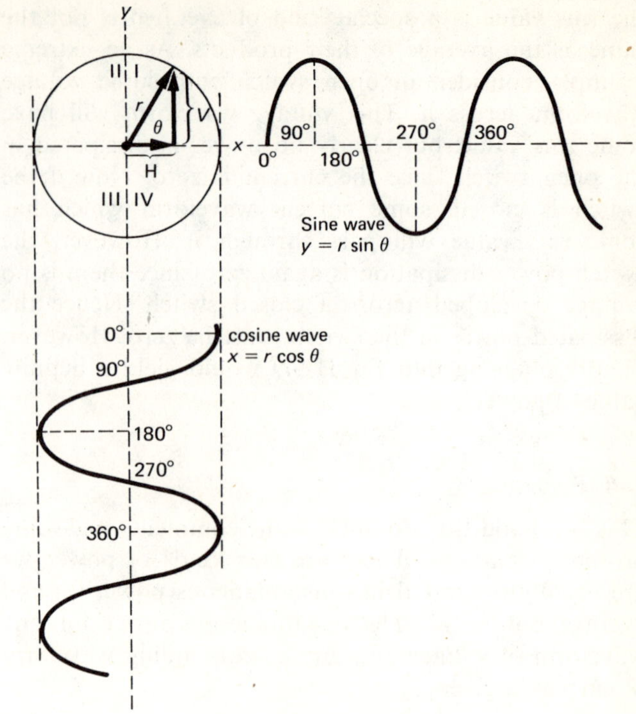

FIGURE 1-9

direction of rotation is taken as the positive reference direction. Thus, if *r* rotates CCW, it generates positive values of θ. If *r* rotates clockwise (CW), it generates negative angles. In Fig. 1-9, *r* rotates CCW. The vector *r* may be resolved into a horizontal component *H*, which always lies on the *x* axis, and a vertical component *V*, which lies on the *y* axis. Thus sine $\theta = V/r = y/r$, and

$$1) \qquad y = V = r \sin \theta \qquad (1\text{-}13)$$

Thus (1) represents the vertical (*y* axis) projection of *r* for various values of the angle θ. A plot of (1) yields the sine wave shown in Fig. 1-9. Note that the waveform repeats itself every 360° (corresponding to one revolution of *r*), that it is also symmetrical, and that the peak amplitude is *r*.

Since $\cos \theta = H/r = x/r$, we have

$$2) \qquad x = H = r \cos \theta \qquad (1\text{-}14)$$

We see that Eq. (2) represents the horizontal (*x* axis) projection of *r*. A plot of Eq. (2) yields the cosine wave of Fig. 1-9. Note that the cosine wave also repeats itself every 360° with a peak amplitude equal to *r*. However, the cosine wave amplitude is a maximum at 0° when the sine wave amplitude is zero. In essence, a cosine wave is a sine wave displaced 90°.

Although the sine wave amplitude of Fig. 1-9 is plotted as a function of θ in degrees, we could have expressed θ in radians. You will recall that 2π rad = 360°, or π rad = 180°. Thus in converting degrees to radians, we multiply the number of degrees by the conversion factor π rad/180°. Conversely, to change radians to degrees, we multiply the radians by the conversion factor 180°/π rad. For example

$$78° = 78° \times \frac{\pi \text{ rad}}{180°} = 1.361 \text{ rad}$$

and

$$0.32 \text{ rad} = 0.32 \text{ rad} \times \frac{180°}{\pi \text{ rad}} = 18.33°$$

Now work through problem PS 1-4.

The sine or cosine functions could also be plotted as functions of time (t) instead of θ. The relation between θ and t may be developed as follows. If the frequency (f) of the sine wave is the number of cycles (complete revolutions of r) per second and there are 2π rad/cycle, it follows that the number of radians r that sweep out per second is $2\pi f$. This defines the angular velocity (ω) of the rotating vector. Thus

$$\omega \text{ rad/sec} = (2\pi \text{ rad})(f \text{ Hz}) = 2\pi f \qquad (1\text{-}15)$$

where f is in Hz. Even though we write cycles as if it were a dimension for convenience, it is actually a dimensionless quantity, so that frequency has the dimensions of 1/sec ($f = 1/\text{sec} = \text{sec}^{-1}$).

Since frequency is the number of cycles per unit time and the period (T) is the time per cycle, f and T are reciprocally related by

$$f = \frac{1}{T} \qquad (1\text{-}16a)$$

$$T = \frac{1}{f} \qquad (1\text{-}16b)$$

Thus Eq. (1-15) could be written as

$$\omega = 2\pi f = \frac{2\pi}{T} \qquad (1\text{-}17)$$

Hence the angle swept through in a time t is given by

$$\theta = (\omega \text{ rad/sec})(t \text{ sec}) = \omega t = 2\pi f t = \frac{2\pi}{T} t \qquad (1\text{-}18)$$

If there is an initial angular displacement ϕ, called the phase angle at time zero (or zero degrees), the net angle θ at time t is given by

$$\theta = \omega t + \phi \qquad (1\text{-}19)$$

Thus the cosine wave which leads the sine wave by a phase angle of 90° is in fact equivalent to a sine wave displaced 90° or $\pi/2$ rad. For example, Fig. 1-10 illustrates two CCW rotating vectors frozen in motion at time zero. Note that vector r_b leads r_a (in the positive reference direction of rotation) by 90°. Consequently, at time zero, r_a makes a zero angle with our starting

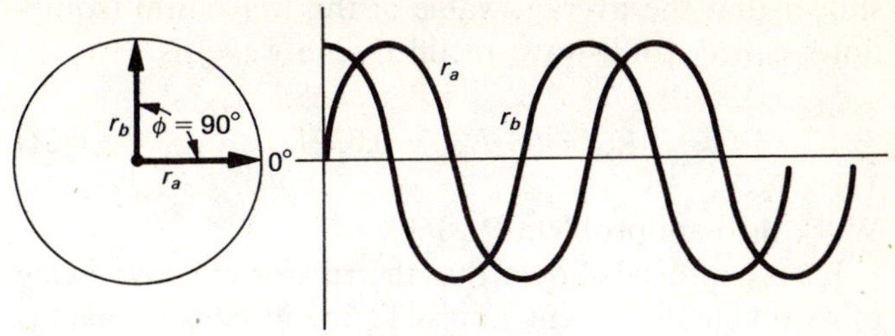

FIGURE 1-10

line on the x axis, whereas r_b makes a 90° angle. If we were to start these vectors rotating and plotted the sine of their respective angles, we would obtain the curves of Fig. 1-10. Clearly the r_b waveform is identical with the cosine wave of Fig. 1-9, which proves that a cosine wave is a sine wave with a 90° phase lead. Any two periodic waveforms which are 90° out of phase are said to be in quadrature.

Now work through problems PS 1-5, PS 1-6, PS 1-7, PS 1-8, and PS 1-9.

Since a sine or cosine wave is symmetrical with respect to its horizontal axis, the positive and negative peak values are the same. Hence it is common to speak of just the peak or maximum value, a voltage in this case. The peak-to-peak voltage in Fig. 1-11 is then

$$E_{pk-pk} = 2E_{pk} \qquad (1\text{-}20)$$

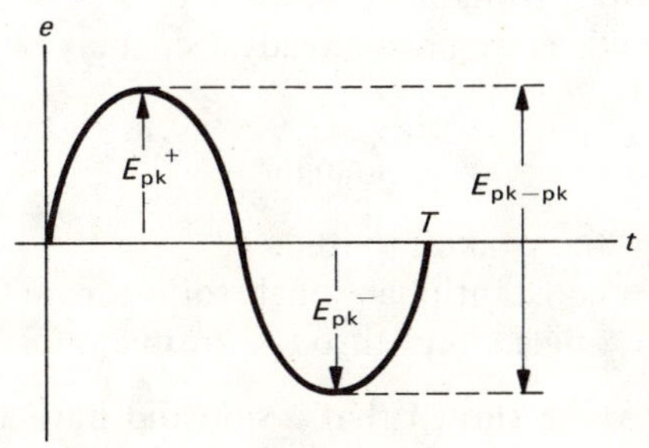

FIGURE 1-11

Clearly, because of its symmetry about the zero axis, a sine or cosine wave has a zero average value. However, it is common practice to speak of the average value of a sine wave with its negative alternation flipped above the zero axis, as shown in Fig. 1-12. It may be

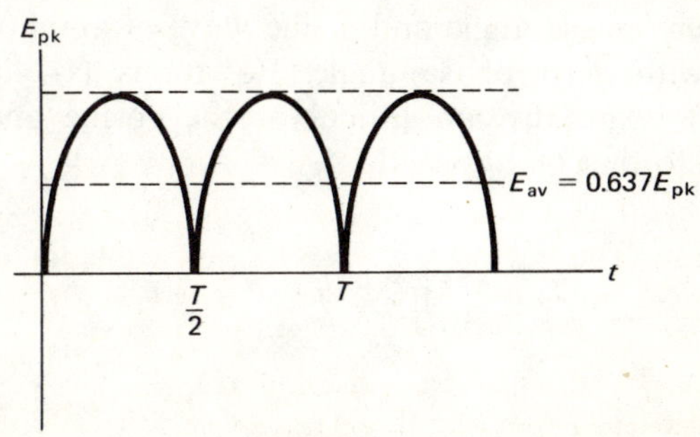

FIGURE 1-12

shown that the average value of this waveform (sometimes called a full-wave rectified sine wave) is[5]

$$E_{av} = \frac{2}{\pi} E_{pk} = 0.637E_{pk} \tag{1-21}$$

Work through problem PS 1-10.

It may also be shown that the rms or effective value of a voltage or current sinusoid (sine or cosine wave) is given by[6]

$$E_{rms} = E = \frac{E_{pk}}{\sqrt{2}} = 0.707E_{pk} \tag{1-22}$$

as shown in Fig. 1-13.

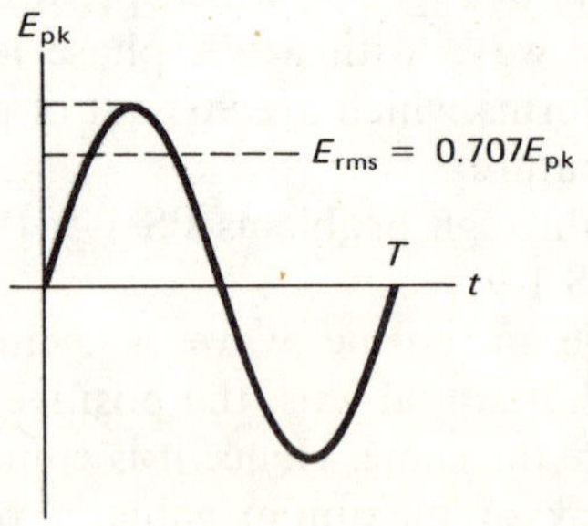

FIGURE 1-13

Now work through problem PS 1-11.

A sinusoid, as we have already seen, may be expressed in the form

1) $$y = A \sin(\omega t + \phi) \tag{1-23}$$

where A = the peak amplitude
ϕ = constant phase angle (offset from 0°)
ωt = angle swept through in time interval t

Now it may be shown that a sinusoid may also be expressed in the form

2) $$y = B \cos \omega t + C \sin \omega t \tag{1-24a}$$
where $$B = A \sin \phi \tag{1-24b}$$
$$C = A \cos \phi \tag{1-24c}$$

Equation (1-24*a*) tells us the sinusoid of Eq. (1-23), which has an amplitude A and phase angle ϕ, is actually equivalent to the sum of a cosine wave of amplitude B and zero phase angle and a sine wave of amplitude C, also with zero phase angle. Equations (1-24*b*) and (1-24*c*) relate the amplitudes of the cosine and sine waves to that of amplitude A.

The equivalence of Eqs. (1-23) and (1-24) is often useful, and since the proof is very simple and will help your understanding, let's do it. There is a trigonometric identity which states that

3) $$\sin(x + y) = \sin x \cos y + \cos x \sin y \tag{1-25}$$

Thus Eq. (1) may be written as

4) $$y = A[\sin \omega t \cos \phi + \cos \omega t \sin \phi]$$

Multiplying through by A and recognizing that ϕ is a constant enables us to rewrite (4) as

5) $$y = \underbrace{A \sin \phi}_{B} \cos \omega t + \underbrace{A \cos \phi}_{C} \sin \omega t$$

Letting

6) $$B = A \sin \phi$$
7) $$C = A \cos \phi$$

we can rewrite (5) as

8) $$y = B \cos \omega t + C \sin \omega t$$

which completes the proof.

To demonstrate the equivalence of Eqs. (1-23) and (1-24) more convincingly, let us use a numerical example and also work the problem backwards. That is, let us assume we have a cosine wave of peak amplitude $B = 5$ and a sine wave of amplitude $C = 8$, both with an angular velocity ω. Thus the cosine and sine waves would plot as curves b and c in Fig. 1-14, where

$$b = B \cos \omega t = 5 \cos \omega t$$

$$c = C \sin \omega t = 8 \sin \omega t$$

Now given the two waveforms, let us add them together graphically, point by point with a pair of dividers to obtain the resultant waveform shown as curve a.[7] Curve a seems to show a peak value $A \cong 9.4$ and $\phi \cong 32°$. This at least demonstrates that the sum of an identical frequency cosine wave and sine wave still yields a similar frequency sinusoid, but with some phase angle.

To go a step further and verify how B and C are related to A, let us take the peak amplitudes of curves a and b, as shown over to the right of Fig. 1-14 and construct two concentric circles having radii B and C equal to these peak amplitudes. Forget about the circle of radius vector A for the moment. Now at time zero the cosine wave radius vector B will be vertical whereas the sine-wave radius vector C is horizontal. So here we

5. $$E_{av} = \frac{2}{2\pi} \int_0^{\pi} E_{pk} \sin \theta \, d\theta = \frac{2}{\pi} E_{pk}$$

6. $$E_{rms} = \sqrt{\frac{2}{2\pi} \int_0^{\pi} (E_{pk} \sin \theta)^2 \, d\theta} = \frac{E_{pk}}{\sqrt{2}}$$

[7] If the reader is familiar with the calculus, he may do all sorts of interesting things to sketch curve a from the function $y = 5 \cos \theta + 8 \sin \theta$ where $\theta = \omega t$. For example, this equation may be differentiated to determine θ_{pk} at the peak value of curve a, and then substitution of θ_{pk} into $f(\theta)$ yields $A = a_{pk}$. Also, setting $y = 0$ would yield ϕ.

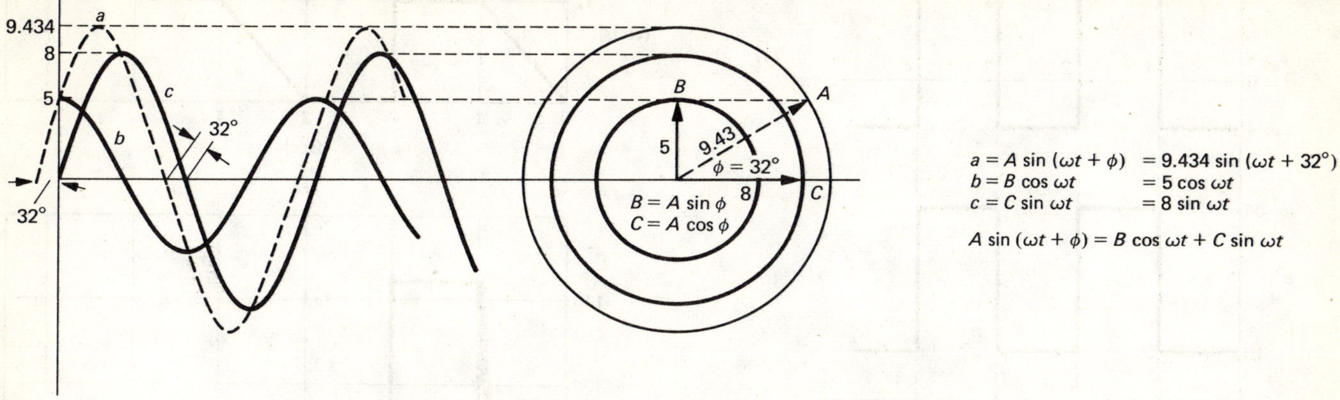

FIGURE 1-14

have a flash picture of our rotating vectors frozen in time at $t = 0$. If we add vectors B and C by vector addition, the result is A, which graphically appears to be about 9.4 units in length. Now using the length of A as a radius, construct the third concentric circle and project this back to curve a with some dashed lines in a manner similar to that shown for curves b and c. Note that, happily, the radius vector A has a peak amplitude similar to that of curve a.

We may, analytically rather than graphically, determine A and ϕ by observing our vector triangle and noting that

$$A = \sqrt{B^2 + C^2} = \sqrt{5^2 + 8^2} = 9.43$$

which checks (allowing for errors in graphic methods), and

$$\phi = \tan^{-1}\frac{B}{C} = \tan^{-1} 5/8 = 32°$$

So what we have shown to this point is how to convert the summation of a cosine and sine wave to a resultant sinusoid. That is, we showed

$$5 \cos \omega t + 8 \sin \omega t = 9.43 \sin (\omega t + 32°)$$

Now let us consider the inverse problem, which is of equal interest, and our original goal. Let us assume that we were given the sinusoid

$$a = A \sin (\omega t + \phi) = 9.43\, (\omega t + 32°)$$

and wished to determine the corresponding cosine and sine waves it is equivalent to. We have already seen from Fig. 1-14 that resultant curve a is the sum of the cosine and sine waves. The problem is to express them analytically. Clearly, inspection of the radius vector A in Fig. 1-14 shows that it can be resolved into these horizontal and vertical components respectively:

$$C = A \cos \phi = 9.43 \cos 32° = 8$$

$$B = A \sin \phi = 9.43 \sin 32° = 5$$

which checks and also indicates the physical significance of Eqs. (1-24*b*) and (1-24*c*).

Now study problems PS 1-12 and PS 1-13.

The following table (Table 1-1) represents a few trigonometric identities which are often useful in deriving other trigonometric relationships such as that between Eqs. (1-23) and (1-24). As another example, work through PS 1-14.

Table 1-1 Useful identities

(1) $\tan x = \dfrac{\sin x}{\cos x}$

(2) $\sin^2 x + \cos^2 x = 1$

(3) $\sin (x \pm y) = \sin x \cos y \pm \cos x \sin y$

(4) $\cos (x \pm y) = \cos x \cos y \mp \sin x \sin y$

(5) $\tan (x + y) = \dfrac{\tan x \pm \tan y}{1 \mp \tan x \tan y}$

PROBLEMS WITH SOLUTIONS

PS 1-1 Determine the peak positive, peak negative, and peak-to-peak amplitude of the waveform shown in Fig. PS 1-1*a*. Also, find the period, frequency, dc, and rms values.

SOLUTION

peak positive voltage = +8 V
peak negative voltage = −4 V
peak-peak voltage = 12 V

$$T = 2 \text{ sec}$$

$$f = \frac{1}{T} 0.5 \text{ Hz}$$

To determine the dc component or average voltage,

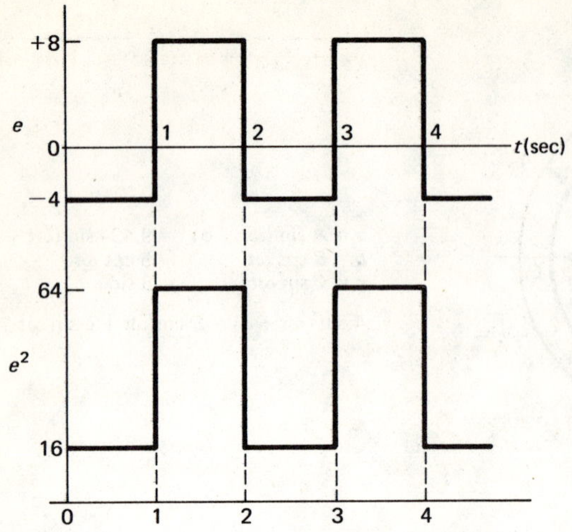

FIGURE PS 1-1

we note in Fig. PS 1-1*a* that the positive area in the 2-sec period is

$$A^+ = 8\text{ V}(1\text{ sec}) = 8\text{ V sec}$$

and the negative area is

$$A^- = -4\text{ V}(1\text{ sec}) = -4\text{ V sec}$$

Therefore

$$A = A^+ + A^- = 8 + (-4) = 4\text{ V sec}$$

and the average voltage is

$$E_{dc} = \frac{A}{T} = \frac{4\text{ V sec}}{2\text{ sec}} = 2\text{ V}$$

To determine the rms voltage, generate the e^2 waveform of Fig. PS 1-1*a* by squaring every e value (see Fig. PS 1-1*b*) and noting that the area under the e^2 curve during the 2-sec period is

$$A = 16\text{ V}^2\,(1\text{ sec}) + 64\text{ V}^2\,(1\text{ sec}) = 80\text{ V}^2\text{ sec}$$

The effective value is then

$$E_{rms} = \sqrt{\frac{A}{T}} = \sqrt{\frac{80\text{ V}^2\text{ sec}}{2\text{ sec}}} = 6.23\text{ V}$$

PS 1-2 Determine the peak amplitudes, period, average, and rms values of the waveform shown in Fig. PS 1-2*a*.

SOLUTION The waveform of Fig. PS 1-2*a* has a positive peak amplitude of 12 volts, a negative peak amplitude of zero volts, and a peak-to-peak amplitude of 12 volts. The period is 3 sec.

To find the average or dc voltage, we find the area, A, under the e curve as follows:

$$A = \tfrac{1}{2}(1\text{ sec})\,12\text{ V} + (2\text{ sec} - 1\text{ sec})\,12\text{ V} + 0 = 18\text{ V sec}$$

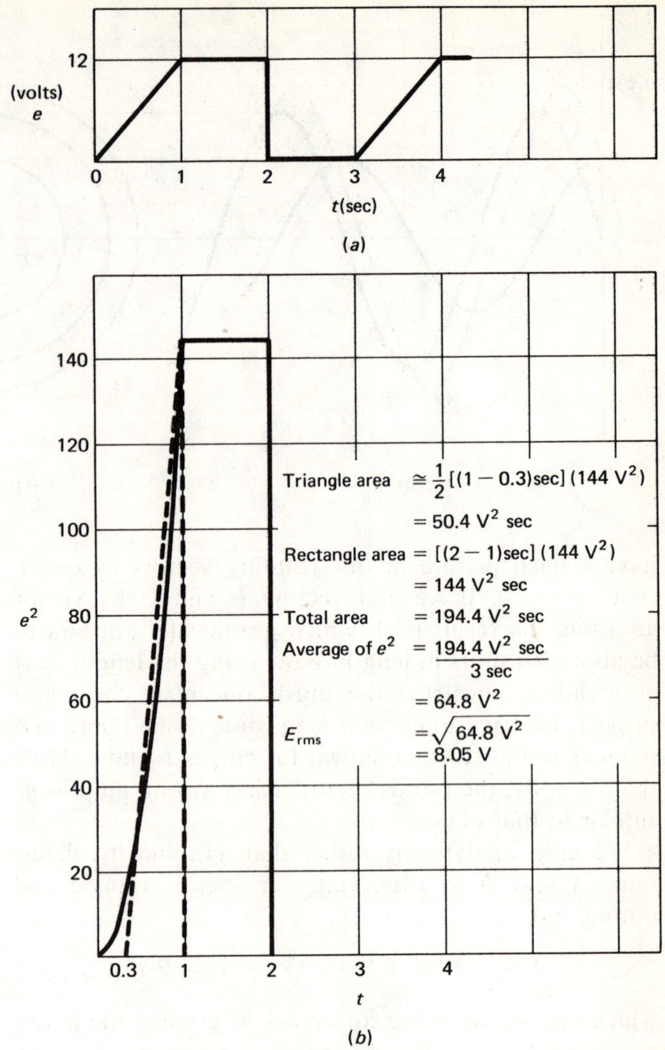

FIGURE PS 1-2

and obtain the average value as

$$E_{av} = E_{dc} = \frac{A}{T} = \frac{18\text{ V sec}}{3\text{ sec}} = 6\ V$$

To obtain the effective value, we generate the e^2 curve as shown in Fig. PS 1-2*b*. Eyeballing the e^2 curve area, we see that it can be resolved into a rectangle and an approximate triangle. The slope of the triangle portion is so drawn that we try to include just as much excess area above the point where the dashed chord intersects the curve as below it, where we lose area. The area of the triangle is about 50.4 V^2 sec, the rectangle is 144 V^2 sec, and the total area is about 194.4 V^2 sec. Dividing the total area by the period yields the average value of e^2 as 64.8 volts, and taking the square root yields $E_{rms} =$ 8.05 volts. The exact value as determined by the methods of calculus (integration) is exactly 8 volts.

PS 1-3 A four-terminal network (Fig. PS 1-3*a*) has the following characteristic: If a triangular waveform of

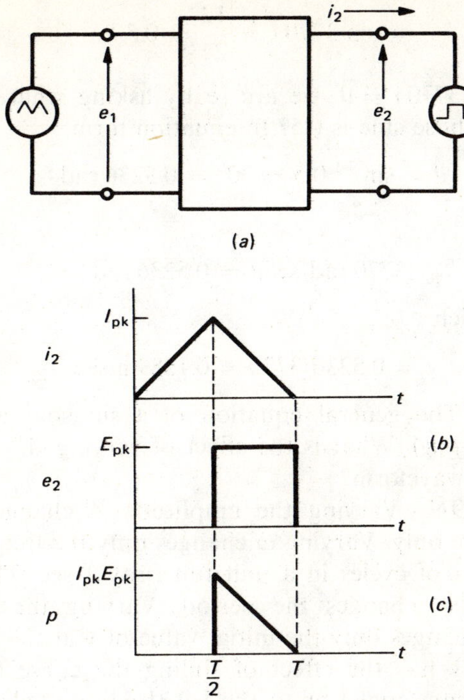

FIGURE PS 1-3

voltage e_1 is applied at the input (1-1) terminals, a triangular waveform of current i_2 flows out of the output (2-2) terminals. If a rectangle of voltage e_2 is simultaneously impressed so that the instantaneous i_2 and e_2 waveforms appear as shown in Fig. PS 1-3*b*, what general expression yields the power dissipated looking into the 2-2 terminals?

SOLUTION To derive an expression for the power, we first multiply the i_2 and e_2 waveforms point by point to obtain Fig. PS 1-3*c*. Since e_2 is zero from $t = 0$ to $t = T/2$, $p = 0$. At $t = T/2$ the instantaneous power jumps to a peak value of $p = I_{pk}E_{pk}$. From $T/2$ to T the current ramp is multiplied by a constant voltage so the shape of the p waveform is still a ramp which equals zero at $t = T$.

The average power is the area under the p waveform divided by the period. Since the area of a triangle is 1/2 (base × altitude), we have

$$P = \frac{A}{T} = \frac{1}{T}\left[\frac{1}{2}\left(T - \frac{T}{2}\right)I_{pk}E_{pk}\right]$$

$$P = \frac{I_{pk}E_{pk}}{4}$$

PS 1-4 Determine the cos 2300° and the tan 2300 rad.

SOLUTION

$$2300° = 2300° \times \frac{1 \text{ cycle}}{360°}$$

$$= 6.389 \text{ cycles or complete revolutions}$$

Thus the rotating radius vector has completed 6 plus 0.389 revolutions. Since the trigonometric functions repeat themselves every 360°, we do not care about the 6 complete revolutions but only the 0.389 parts of a revolution. Thus

$$2300° \text{ is equivalent to } 0.389\,(360°) = 140°$$

and 140° is equivalent to 180° − 140° = 40° in the second quadrant (see Fig. PS 1-4). Since the cosine is negative in the second quadrant we have

$$\cos 2300° = \cos 140° = -\cos 40° = -0.766$$

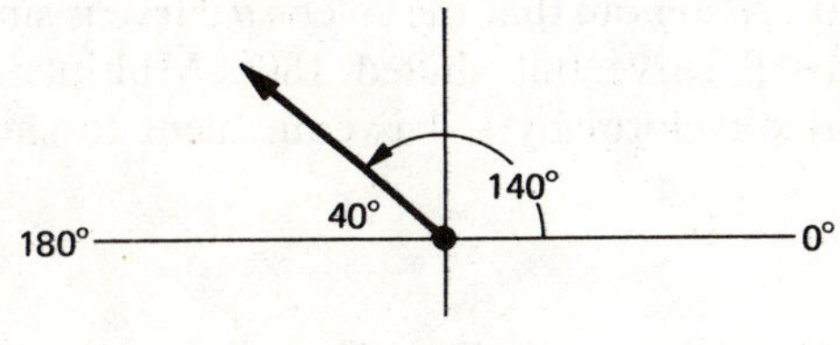

FIGURE PS 1-4

To find tan 2300 rad, let us first convert to degrees.

$$2300 \text{ rad} = 2300 \text{ rad}\left(\frac{180°}{\pi \text{ rad}}\right) = 131{,}780°$$

Dividing by 360 to determine the number of complete revolutions and any fractional remainder, we have

$$\frac{131{,}780°}{360°} = 366.056 \text{ cycles or revolutions}$$

and

$$0.056 \times 360° = 20.160°$$

The tangent in the first quadrant is positive. Hence

$$\tan 2300 \text{ rad} = \tan 131{,}780° = \tan 20.160° = 0.36977$$

PS 1-5 Express 30°, 45°, 60°, 90°, and 180° in radians.

SOLUTION

$$30° = 30° \times \frac{\pi \text{ rad}}{180°} = \frac{\pi}{6} = 0.524 \text{ rad}$$

$$45° = 45° \times \frac{\pi}{180°} = \frac{\pi}{4} = 0.785 \text{ rad}$$

$$60° = 60° \times \frac{\pi}{180°} = \frac{\pi}{3} = 1.047 \text{ rad}$$

$$90° = 90° \times \frac{\pi}{180°} = \frac{\pi}{2} = 1.57 \text{ rad}$$

$$180° = 180° \times \frac{\pi}{180°} = \pi = 3.14 \text{ rad}$$

PS 1-6 Plot $y = \sin\theta$, $y = \sin(\theta + 90°)$, and $y = \pm\cos\theta$ for 30° increments from 0° to 720°.

SOLUTION

θ	0°	30°	60°	90°	120°	150°	180°
$\sin\theta$	0	0.5	0.866	1	0.866	0.5	0
$\cos\theta$	1	0.866	0.5	0	−0.5	−0.866	−1
$\theta + 90°$	90°	120	150	180	210	240	270°
$\sin(\theta + 90°)$	1	0.866	0.5	0	−0.5	−0.866	−1

See Fig. PS 1-6 for the resultant curves. Note that the $y = \cos\theta$ and $y = \sin(\theta + 90°)$ curves are identical, which verifies that a cosine curve is a sine curve advanced 90°. Also note that the $-\cos\theta$ curve is similar to the $+\cos\theta$ curve but shifted 180°. Multiplying any sinusoidal waveform by -1 is equivalent to shifting it 180°.

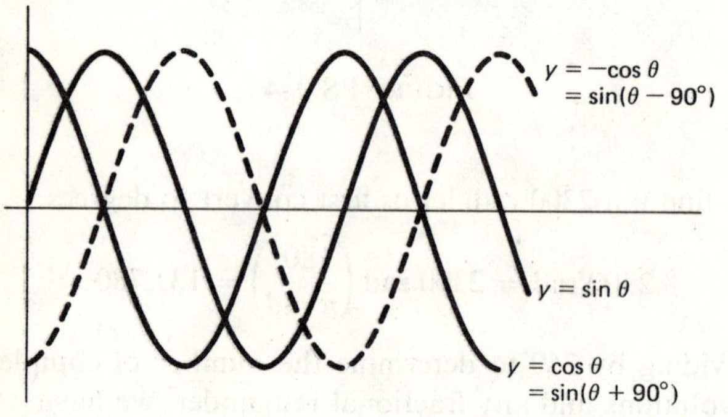

FIGURE PS 1-6

PS 1-7 Given the function $y = 24 \sin 3770\,t$, determine the peak amplitude, frequency, period, and value of t when $y = 12$.
SOLUTION The peak amplitude is 24. To determine the frequency f, which is the number of cycles per second, we may directly compare the function

1) $$y = 24 \sin 3770\,t$$

with the general form

2) $$y = A \sin \omega t = A \sin 2\pi f t$$

A comparison of 2 with 1 indicates

3) $$\omega = 2\pi f = 3770 \text{ rad}$$

or

4) $$f = \frac{3770}{2\pi} = 600 \text{ Hz} = 600 \text{ Hz}$$

Thus

5) $$T = \frac{1}{f} = \frac{1}{600} = 1.667 \times 10^{-3} \text{ sec}$$

To determine t for $y = 12$, we obtain from 1

6) $$\sin 3770\,t = \frac{12}{24} = 0.5$$

If we let $3770\,t = \theta$, we are really asking what is the angle θ whose sine is 0.5? In equation form,

7) $$\theta = \sin^{-1} 0.5 = 30° = 0.5236 \text{ rad}$$

or

8) $$(3770 \text{ rad/sec})t = 0.5236 \text{ rad}$$

From which

9) $$t = 0.5236/3770 = 0.1389 \text{ msec}$$

PS 1-8 The general equation of a sinusoid is $y = A \sin(\omega t \pm \phi)$. What is the effect of varying A, ω, and ϕ on the waveform?
SOLUTION Varying the amplitude A changes the peak value only. Varying ω changes only the frequency or number of cycles in a unit time, say 1 sec. This, of course, also changes the period. Varying the phase angle ϕ changes only the initial value of y at $t = 0$ and graphically has the effect of sliding the curve to the right (lagging angle) or to the left (leading angle). The sign of ϕ is negative for lagging angles and positive for leading angles.

PS 1-9 The input source (e_{in}) of Fig. PS 1-9, which is a sinusoid expressed as $e_{in} = 12 \sin 5 \times 10^6 t$, is used to drive two separate networks. The first is a phase shifter which retards the phase of its input signal so that the output e_1 lags e_{in} by 60°. The second network is a delay line which has the property of causing the output e_2 to lag its input e_{in} by 0.1 μsec for each inch of delay line. How many inches of line are required to guarantee that e_1 and e_2 are in phase?

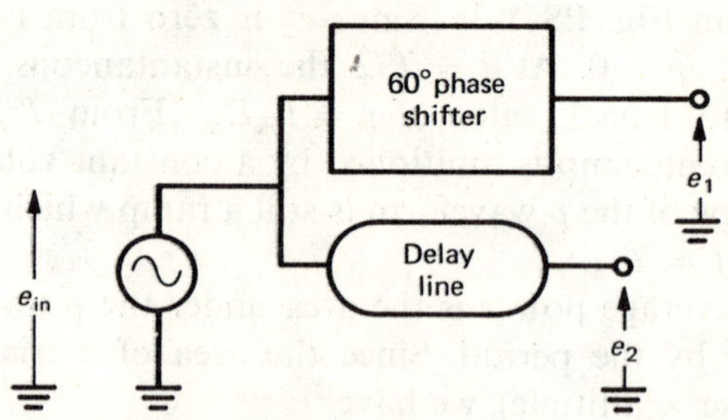

FIGURE PS 1-9

SOLUTION First find the time required to shift the signal 60°. Converting 60° to radians, we have $60° \times (\pi \text{ rad}/180°) = 1.0472$ rad. Since $\omega = 5 \times 10^6$, we can set the general expression for the angle generated equal to the 60° shift in radians as follows:

$$\omega t = (5 \times 10^6 \text{ rad/sec})t = 1.0472 \text{ rad}$$

or

$$t = 0.2094\ \mu\text{sec}$$

The required line length is then

$$\text{line length} = \frac{1\ \text{in.}}{0.1\ \mu\text{sec}} \times 0.2094\ \mu\text{sec} = 2.094\ \text{in.}$$

PS 1-10 Given the sinusoidal function $y = 120 \cos (157t - 30°)$, determine the following: amplitude, period, phase angle, and sketch of the waveform.

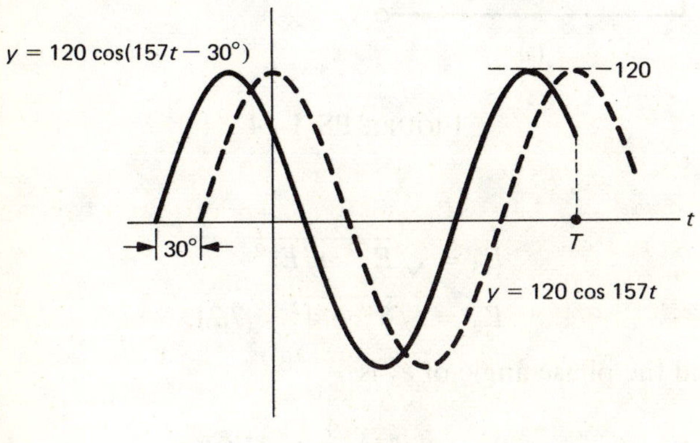

FIGURE PS 1-10

SOLUTION Amplitude $A = 120$.
To find the period T, we may write

$$\omega\ \text{rad/sec} = 2\pi f = \frac{2\pi\ \text{rad}}{T} = 157\ \text{rad/sec}$$

or

$$T = \frac{2\pi\ \text{rad}}{157\ \text{rad/sec}} = 0.04\ \text{sec}$$

The phase angle $\phi = -30°$

Note that in the expression $y = 120 \cos (157t - 30°)$, the total angle involved is $157t - 30°$. It should be borne in mind that the $157t$ or ωt term is in radians whereas the 30° is in degrees, and hence the units are not consistent. Although it would be better to write 30° as $\pi/6$ rad, it is common practice to mix degrees and radians in this type of expression, and this is a point to bear in mind.

PS 1-11 Estimate by graphic methods the average value of a full-wave rectified sine wave.

SOLUTION Construct one-half or even one-fourth of a sine wave. In figure PS 1-11 we show a full cycle of a unit amplitude full-wave rectified sine wave, but we will determine the area of only one-fourth, from 0 to 90°, and multiply by 4 for the area over a full cycle. A variety of ways may be used to estimate the area of interest. The approach we will use this time is to approximate the area as the sum of rectangles having equal bases (180°) and judiciously chosen heights with the vertical scale divided into 0.1 unit per division. Thus the area for $\frac{1}{4}$ cycle is

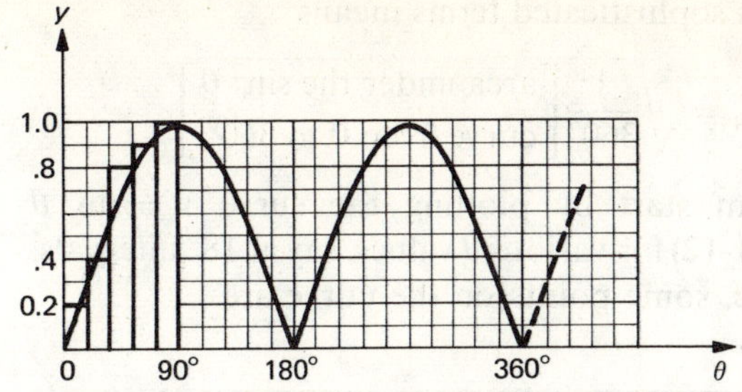

Each vertical division = 0.1 unit
Each horizontal division = 18°
A over 90° = 18°(0.2 + 0.4 + 0.8 + 0.9 + 1) =
= 18°(3.3 units) = 59.4° units

A over 360° = 4(59.4° units) = 237.6° units

$$\text{Average value} = \frac{237.6°\ \text{units}}{360°} = 0.66\ \text{units}$$

FIGURE PS 1-11

$$A_{(90°)} = 18° (0.2) + 18° (0.4) + 18° (0.8) + 18° (0.9) + 18° (1) = 59.4°\ \text{units}$$

And over a full cycle

$$A_{(360°)} = 4\,(59.4°\ \text{units}) = 237.6°\ \text{units}$$

The average height is therefore

$$Y_{av} = \frac{237.6°\ \text{units}}{360°} = 0.66\ \text{units}$$

which is reasonably close to the correct value of 0.6366.

PS 1-12 Graphically estimate the rms value of a sinusoid.

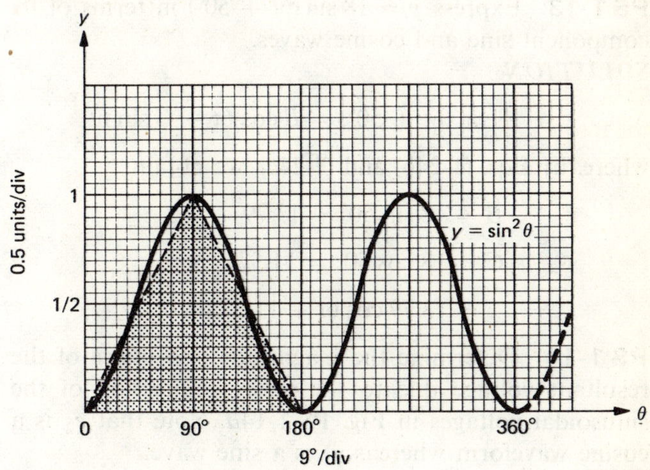

FIGURE PS 1-12

SOLUTION Let us choose a unit sinusoid, say $y = \sin \theta$. The rms value is

$$Y_{rms} = \sqrt{\frac{1}{2\pi} \int_0^{2\pi} \sin^2 \theta\, d\theta}$$

which in less sophisticated terms means

$$Y_{\text{rms}} = \sqrt{\frac{1}{360°}\left[\begin{array}{l}\text{area under the } \sin^2\theta \\ \text{curve from 0 to } 360°\end{array}\right]}$$

Thus we can start by plotting the curve $y = \sin^2\theta$ (see Fig. PS 1-12) for various θ values, say at 18° intervals. For example, some points on the curve are

θ	0	36°	72°	90°	180°
$\sin\theta$	0	0.5878	0.9511	1	0
$\sin^2\theta$	0	0.3455	0.9045	1	0

Constructing the $\sin^2\theta$ curve and examining it with an eye toward finding the area under it indicates that a straight line from the origin to the peak at 90° will lose just as much area inside the curve between $y = \frac{1}{2}$ and $y = 1$ as it gains outside the curve between $y = 0$ and $y = \frac{1}{2}$. Therefore the area enclosed by the shaded portion of the curve is

$$A = \tfrac{1}{2}(\text{base})(\text{height}) = \tfrac{1}{2}(180°)(1) = 90° \text{ units}$$

and over a full period,

$$A = 2(90° \text{ units}) = 180° \text{ units}$$

Thus

$$Y_{\text{rms}} = \sqrt{\frac{1}{360°}(180° \text{ units})} = \sqrt{\frac{1}{2}} = \frac{1}{\sqrt{2}} = 0.707$$

which miraculously is the correct value.

PS 1-13 Express $y = 18 \sin(6t + 50°)$ in terms of its component sine and cosine waves.
SOLUTION

$$y = 18 \sin(6t + 50°) = B \cos 6t + C \sin 6t$$

where, by Eqs. (1-24*b*) and (1-24*c*), we obtain

$$B = 18 \sin 50° = 13.79$$

$$C = 18 \cos 50° = 11.57$$

$$y = 13.79 \cos 6t + 11.57 \sin 6t$$

PS 1-14 Determine the magnitude and phase of the resultant voltage due to the series connection of the sinusoidal voltages in Fig. PS 1-14*a*. Note that e_2 is a cosine waveform whereas e_1 is a sine wave.
SOLUTION Since the sum of two sinusoids is in general a sinusoid of similar frequency but with some different amplitude and phase, we can write

$$e_3 = E_3 \sin(\omega t + \phi)$$

Sketching a vector diagram of the peak amplitudes at, say, time zero, we have Fig. PS 1-14*b*, from which we obtain

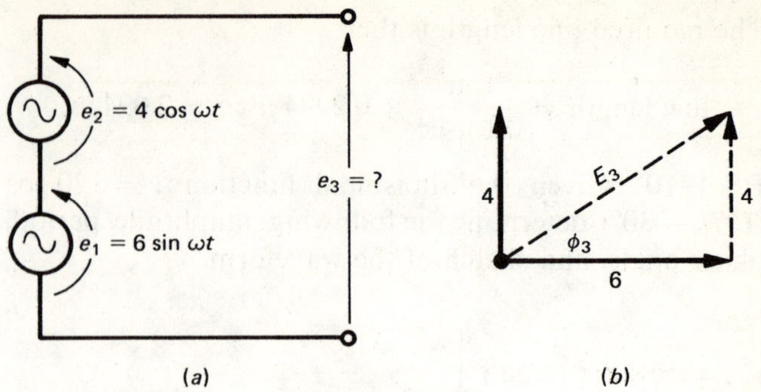

FIGURE PS 1-14

$$E_3 = \sqrt{E_1{}^2 + E_2{}^2}$$

$$E_3 = \sqrt{6^2 + 4^2} = 7.21$$

and the phase angle of e_3 is

$$\theta_3 = \tan^{-1}\frac{4}{6} = 33.69°$$

Thus the resultant voltage is

$$e_3 = 7.21 \sin(\omega t + 33.69°)$$

PS 1-15 Prove that

$$\sin^2\theta = \tfrac{1}{2} - \tfrac{1}{2}\cos 2\theta$$

SOLUTION If in (4) of Table 1-1 we let $x = y = \theta$ and use the sum form, we may write

1) $$\cos(x + y) = \cos x \cos y - \sin x \sin y$$

or

2) $$\cos 2\theta = \cos^2\theta - \sin^2\theta$$

From (2) of Table 1-1, we obtain

3) $$\cos^2\theta = 1 - \sin^2\theta$$

Substituting (3) into (2) and simplifying, we first obtain

4) $$\cos 2\theta = 1 - 2\sin^2\theta$$

Solving (4) for $\sin^2\theta$ yields

5) $$\sin^2\theta = \frac{1 - \cos 2\theta}{2} = \frac{1}{2} - \frac{1}{2}\cos 2\theta$$

which completes the proof.

This is a very useful result. Apparently the $\sin^2\theta$ waveform consists of a dc or average value component equal to $\frac{1}{2}$ (the first term), from which is subtracted a cosine wave of amplitude (peak value) $\frac{1}{2}$ and frequency twice that of the original sinusoid. If we plot these individual waveform components and then add them as shown in Fig. PS 1-15*a*, we see that the resultant waveform is a double-frequency sinusoid which swings about an average value of $\frac{1}{2}$. Figure PS 1-15*b* compares $y =$

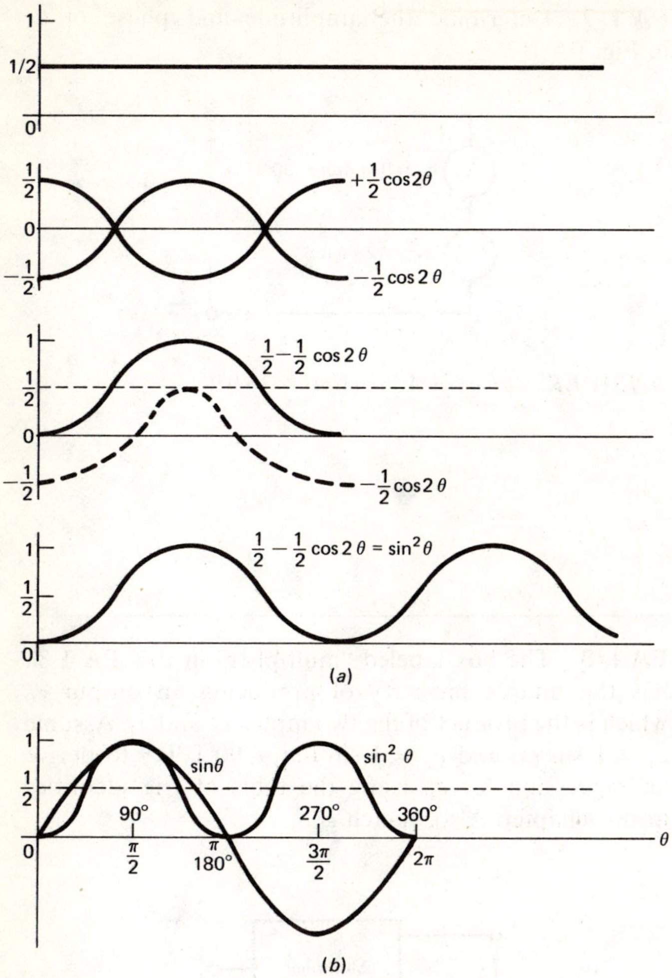

FIGURE PS 1-15

sin θ versus $y = \sin^2 \theta$ graphically to help you remember this.

PS 1-16 Using the results of problem PS 1-15, prove that the effective value of a sinusoid is 0.707 times the peak value.

SOLUTION If we assume a sinusoid of general form

1) $$y = A \sin \theta$$

the rms value would be given by

2) $$Y_{\text{rms}} = \sqrt{\frac{1}{2\pi} \int_0^{2\pi} y^2 \, d\theta}$$

which really means the same as

3) $$Y_{\text{rms}} = \sqrt{\frac{1}{2\pi} \left[\begin{array}{c} \text{area under a full} \\ \text{cycle of the } y^2 \text{ curve} \end{array} \right]}$$

Now in general the area under a full cycle of a periodic curve is equal to its average value (Y_{av}) multiplied by T or 2π. Inspection of Fig. PS 1-15*b* clearly indicates that Y'_{av} for the $\sin^2 \theta$ curve is $\frac{1}{2}$. Thus if we allow for the square of the amplitude A, scaling the $\sin^2 \theta$ curve by A^2, we may write

4) $$Y_{\text{rms}} = \sqrt{\frac{1}{2\pi} \left[A^2 Y'_{\text{av}}(2\pi) \right]}$$

$$= A \sqrt{Y'_{\text{av}}}$$

$$= A \sqrt{\frac{1}{2}} = \frac{A}{\sqrt{2}} = 0.707\,A$$

PROBLEMS WITH ANSWERS

PA 1-1 Determine the positive and negative peak voltages, the peak-to-peak voltage, the period, the frequency, and the dc and rms voltages for the waveform of Fig. PA 1-1.

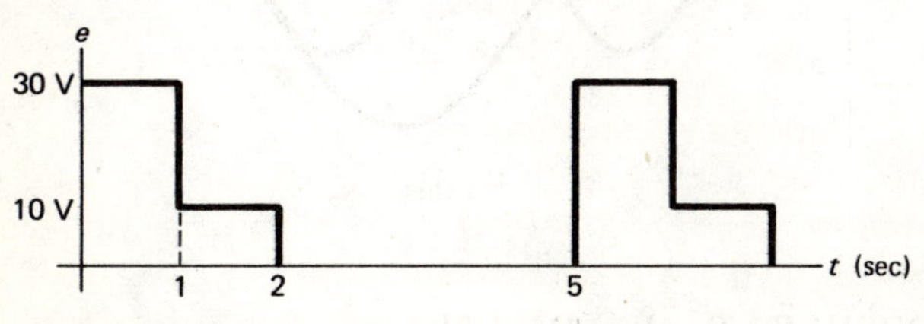

ANSWER $E_{\text{dc}} = 8$ V, $E_{\text{rms}} = 14.14$ V

$E_{\text{pk}}{}^{+} = 30$ V, $E_{\text{pk}}{}^{-} = 0$, $E_{\text{pk-pk}} = 30$ V

$f = \frac{1}{5}$ sec, $T = 5$ sec

PA 1-2 The voltage across and the current through some device is given by the waveforms of Fig. PA 1-2. Determine the power dissipated in the device.

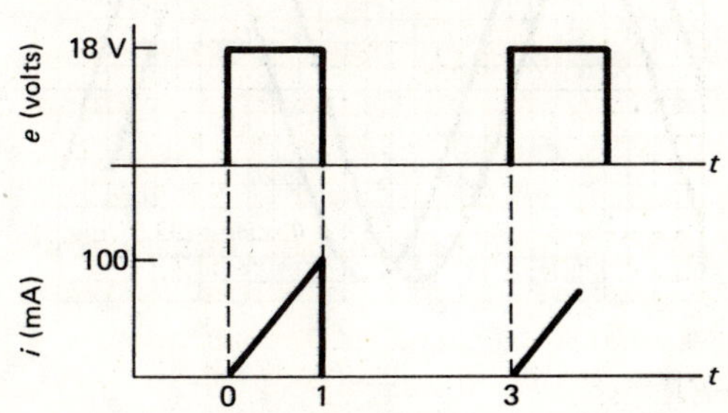

ANSWER $P = 0.3$ W

PA 1-3 What is the frequency of a shaft which rotates 120π rad/sec?

ANSWER 60 cycles/sec = 60 Hz

PA 1-4 Determine the sine of 540° and cosine of 12 rad.

ANSWER $\sin 540° = 0$, $\cos 12 \text{ rad} = 0.844$

PA 1-5 Sketch the function $y = 5 \sin 20t$.

ANSWER See Fig. PA 1-5.

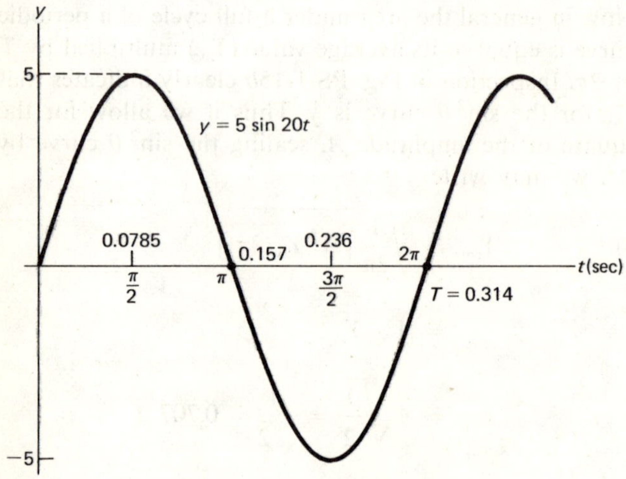

PA 1-6 What functional relationship do you feel is expressed by the waveforms of Fig. PA 1-6?

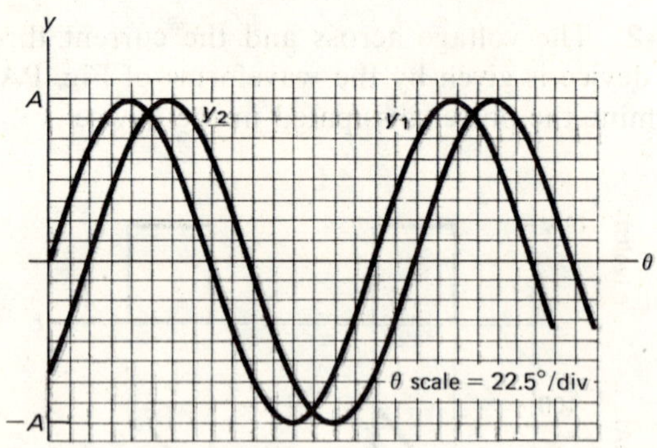

ANSWER

$$y_1 = A \sin \theta, \; y_2 = A \sin (\theta - 45°)$$

PA 1-7 Determine the amplitude and phase of e_0 in Fig. PA 1-7.

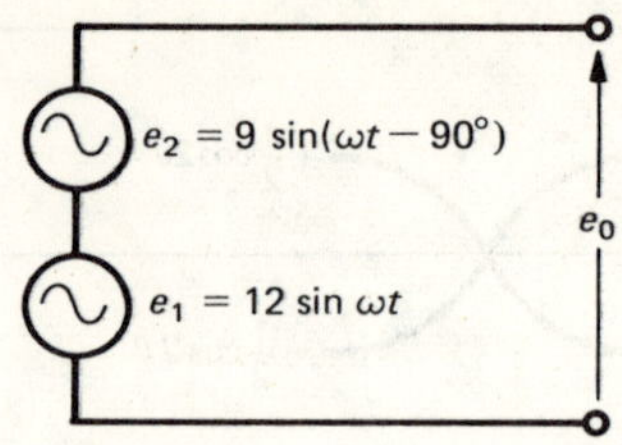

ANSWER $e_0 = 15 \sin (\omega t - 36.9°)$

PA 1-8 The box labeled "multiplier" in Fig. PA 1-8*a* has the unique property of producing an output e_0 which is the product of the two inputs e_1 and e_2. Assume $e_1 = 1 \sin \omega t$ and $e_2 = 1 \sin (\omega t + 90°)$. Try to derive an expression for e_0 using the table of trig identities in this chapter. Also, sketch e_0.

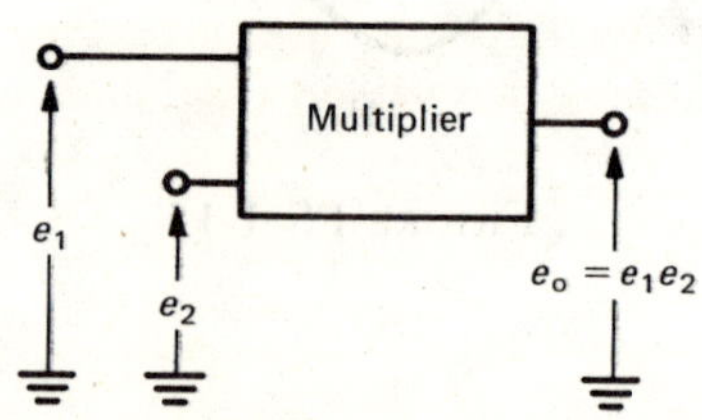

(*a*)

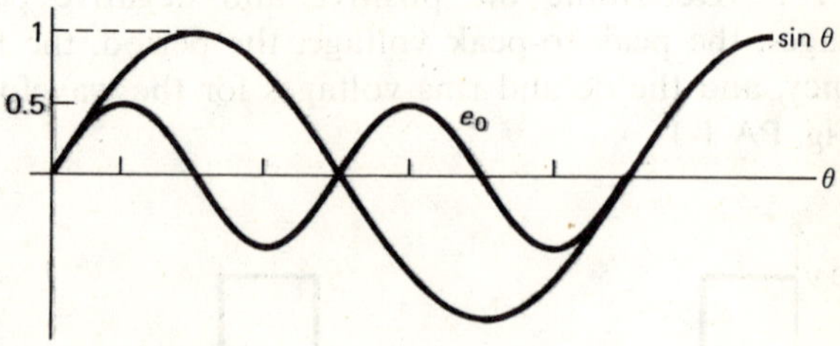

(*b*)

ANSWER See Fig. PA 1-8*b*.

PROBLEMS WITHOUT ANSWERS

P 1-1 Determine the average and rms value of the waveform shown in Fig. P 1-1.

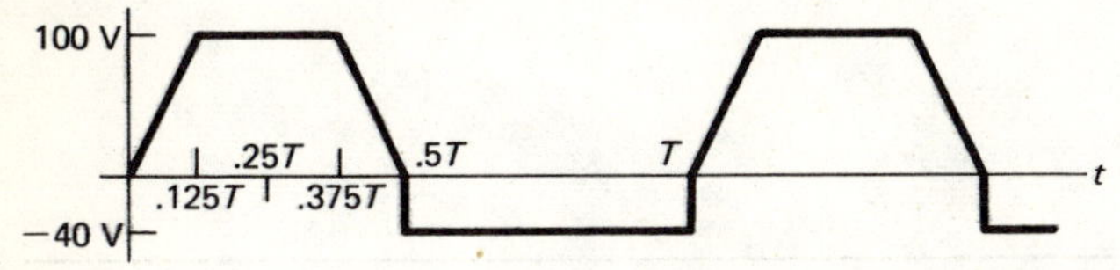

P 1-2 Determine the power in a device if the voltage across and current through it exhibit the waveforms shown in Fig. P 1-2.

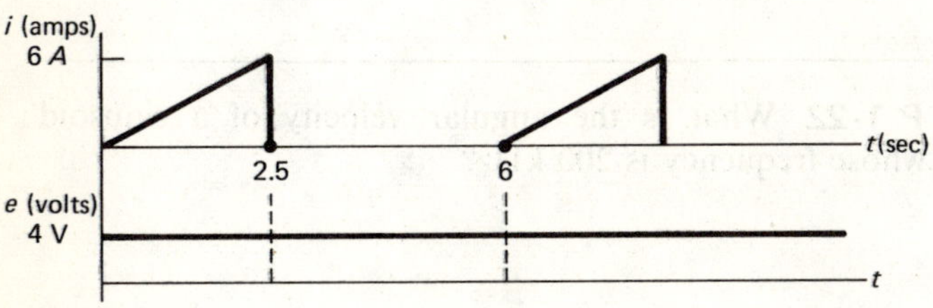

P 1-3 What is the frequency in cycles per second, Hertz, and radians per second of a shaft which rotates at 3800 rpm?

P 1-4 What is the amplitude and frequency in Hertz of the sinusoid $y = 0.05 \sin 4000t$?

P 1-5 Sketch the sinusoids $y = \sin \theta$, $y = -\sin \theta$, $y = \cos \theta$, and $y = -\cos \theta$.

P 1-6 Sketch the sinusoids $y = \sin(\theta + 90°)$, $y = \sin(\theta - 90°)$.

P 1-7 What is the amplitude, frequency, period, and phase angle of the sinusoid $y = 7 \sin(48t - 60°)$?

P 1-8 Evaluate y in problem P 1-7 at $t = 0.1$ sec.

P 1-9 What is the amplitude, frequency, period, and phase angle of the sinusoid $y = -7 \sin(48t - 60°)$?

P 1-10 What is the average and rms value of the sinusoid $y = 32 \cos(18t + 30°)$?

P 1-11 The effective voltage of a sinusoid is 117 volts. What is the peak value?

P 1-12 Resolve the sinusoid $y = 19 \sin(\omega t + 55°)$ into its equivalent sine and cosine waves.

P 1-13 The box labeled "multiplier" in Fig. P 1-13 has the unique property that the output e_0 is the product of the two inputs e_1 and e_2. If $e_1 = 3 \sin \omega t$ and $e_2 = 4$ volts, determine the dc voltage across the load resistor and the power dissipated in it.

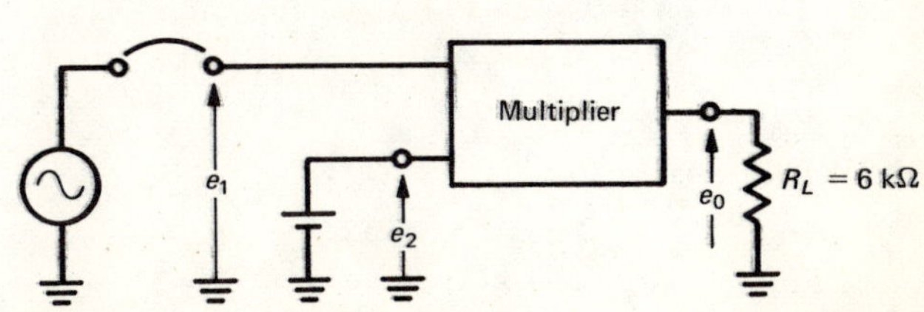

P 1-14 The peak to peak voltage of a sine wave is 296 volts. What is the effective value? What is the average value?

P 1-15 An 8 volt sinusoidal source is connected in series with a 14 volt battery. Sketch the resultant waveform and determine the peak to peak value.

P 1-16 The peak voltage of a 60 Hz sinusoid is 36 volts. Assuming zero phase angle determine the instantaneous value at 30°, 45°, 60°, 90°, 140°, 250°, and 300°.

P 1-17 What is the period of a 27 MHz sinusoid?

P 1-18 A certain electronic circuit can change its operating mode in 0.8 manoseconds. How far could a beam of light travel in the same interval if the speed of light is 186,000 miles per second?

P 1-19 Measurement of a sinusoidal voltage with an oscilloscope indicates a period of 2.4 microseconds and a peak to peak amplitude of 18 volts. Determine the equation of the sinusoid and its effective value.

P 1-20 A sine wave of current reaches its peak amplitude of 89 milliamperes in 0.4 milliseconds. What is the frequency and effective value of this current waveform? Assume zero phase angle.

P 1-21 A sinusoid has an angular velocity of 35×10^6 radians/second. Determine the frequency.

P 1-22 What is the angular velocity of a sinusoid whose frequency is 200 kHz?

P 1-23 What is the effective value and frequency of a sinusoid whose equation is $e = 24 \sin (159 \times 10^6)t$?

P 1-24 A particular sinusoid is given by $i = -12 \sin \omega t$. Another sinusoid is given by $e = 24 \cos \omega t$. What is the phase angle of the current waveform with respect to the voltage waveform?

2 voltage and current notation

2-1 Voltage notation

Because an ac voltage is constantly changing in polarity, the student may be confused as to how a voltage may be described. You will recall in the previous text, "DC Circuit Analysis with Illustrative Problems," that a voltage was described by the symbol E or V and an arrow or double subscripts as shown in Fig. 2-1. It was also agreed that the tail of the arrow, or the first subscript, would always be the reference point to which the potential at the head of the arrow is compared. Or, in other words, we start at the reference point and follow some path of known or expressible potentials till we reach the head of the arrow or the second subscript. In following this path, we keep track of all the ups and downs in potential, and the net voltage in question is then the algebraic sum of all these ups and downs in potential. You will recall that if in following the path from tail to head (or first to second subscript) you enter the negative end of some circuit element and then emerge at the positive end, you have gone up in potential, and this is denoted with a positive sign. Conversely, if you first enter the positive end of a circuit element and emerge at the negative end, you have gone down in potential, and this is denoted with a negative sign. Thus

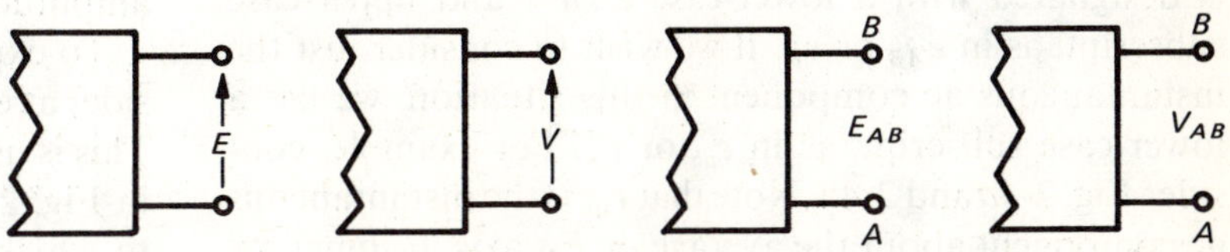

FIGURE 2-1

in Fig. 2-2, $E = -6\text{ V} + 4\text{ V} = -2\text{ V}$. This is identical with the method used in the preceding text of this series and should be familiar to you.

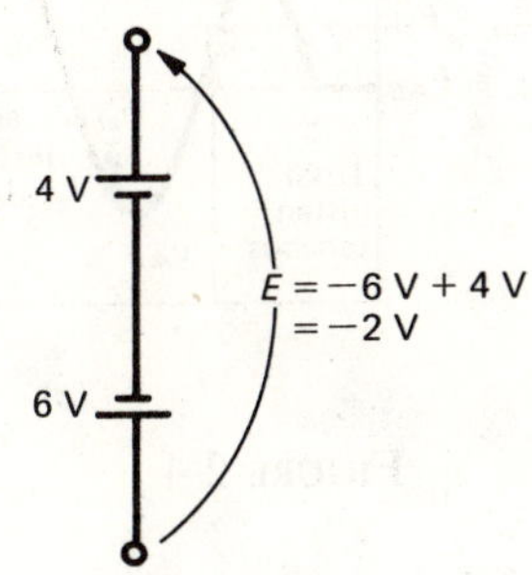

FIGURE 2-2

Now, if we are talking about an ac voltage source as shown in Fig. 2-3, it is clear that V must be changing in polarity every half period and hence may be either positive or negative at any given moment. This may prove disturbing, and rightfully so. Hence, in working

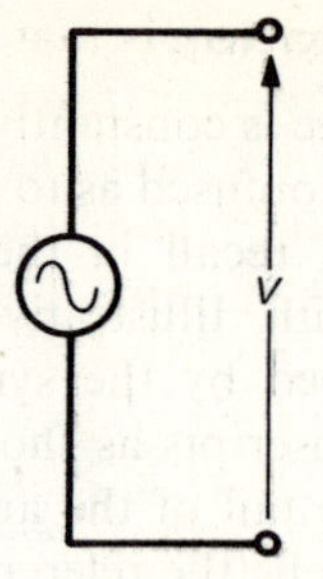

FIGURE 2-3

with strictly sinusoidal ac circuits, we will also agree that an upper-case V or E represents the effective (rms) value which in itself is independent of polarity.

If alphabetic subscripts are used, as is necessary when dealing with mixed sources, the effective value is designated with a lower-case subscript to differentiate it from a dc voltage component, which is also designated with an upper-case subscript. Thus E_a is an rms voltage, whereas E_A is a dc voltage. This differentiation is necessary since an upper-case E or V in itself implies a dc or rms voltage.

If a dc and ac component (or any other combination of sources) are superimposed, the total instantaneous value is designated with a lower-case e or v and upper-case subscript, as in e_{AB} or v_F. If we wish to consider just the instantaneous ac component in this situation, we use a lower-case subscript, as in e_{ab} or v_f. For example, consider Fig. 2-4*a* and 2-4*b*. Note that e_{ab} is the instantaneous ac component about the average or E_{dc} axis. If somehow this dc component were removed so that the waveform swings about the zero voltage level, e_{AB} and e_{ab} would be identical.

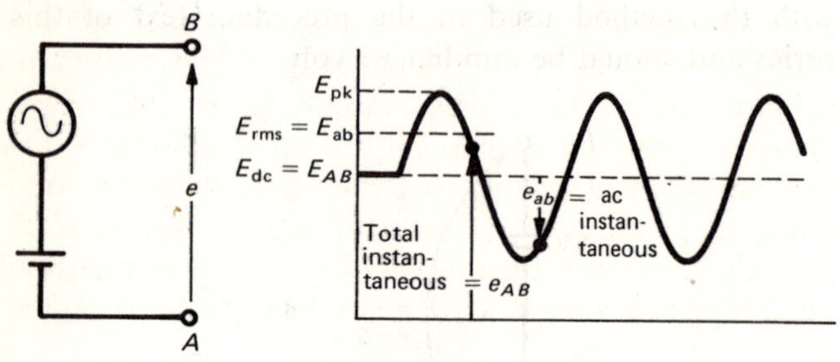

FIGURE 2-4

If there should be some problem in differentiating between the peak value of the total waveform or just the peak value of the ac component, we will use E_{pk} and $E_{pk(ac)}$ respectively.

Now work through problems PS 2-1 and PS 2-2.

To further complicate the matter of voltage notation in ac circuits, we must consider the possibility of sinusoidal sources in series which may differ or be similar in frequency. If they differ in frequency, we have an awful mess, and the resultant waveform is in general nonsinusoidal. This case will be considered at some future time.

If they are of similar frequency, as is usually true, we must consider the possibility of phase differences between the sources. For example, at, say, time zero, the rotating vectors (phasors) that in effect generate the sinusoidal waveforms may not only have different amplitudes but may differ in phase as shown in Fig. 2-5. Here E_2 leads E_1 by 90° and hence E_2 and E_1 generate the cosine and sine waves respectively.

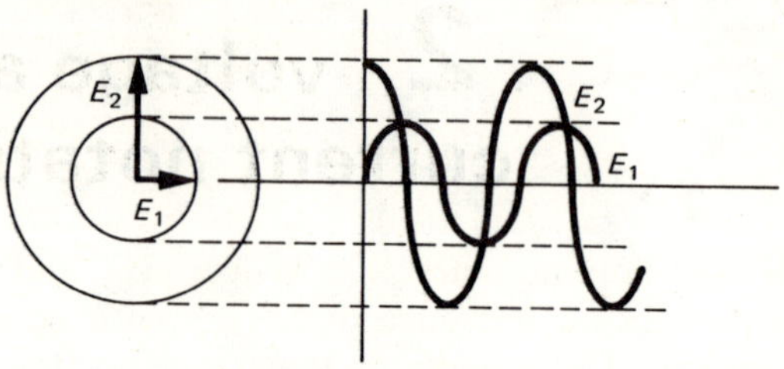

FIGURE 2-5

What if the phase angle is other than 90°? And how does one go about finding the resultant of two different amplitude, out-of-phase sinusoids?

To explore these questions it is first necessary to consider a reference vector to which the others are compared. This is usually a unit vector at zero degrees. For example, in Fig. 2-5 we could consider E_1 as the reference vector to which E_2 is compared. A convenient way to do this is to use the polar form of vector notation so that $\mathbf{E}_1 = E_1\underline{/0°}$ and $\mathbf{E}_2 = E_2\underline{/90°}$. Or, if complex number algebra (the j operator notation) is used, $\mathbf{E}_1 = E_1 + j0$ and $\mathbf{E}_2 = 0 + jE_2$. Remember, unless otherwise indicated, all voltages are effective values.[1] Now all the tools of vector analysis including complex number algebra may be utilized. Thus, vectors may be graphically added or subtracted by completing the parallelogram or using the polygon head-to-tail method. Or, better still, resolving them into their horizontal (real) and vertical (imaginary) components. Multiplication and division of vectors is best performed in the polar or exponential form. The details of these techniques should be familiar to you at this time.[2]

Now work through problem PS 2-3.

So far we have used arrows or double subscripts to indicate the points between which we wish to denote voltage, and the tail of the arrow or first subscript is the

[1] When necessary vector (phasor) quantities are indicated with boldface type, dot, or arrowhead.

[2] Most any algebra or electronics math text such as N. M. Cooke and H. R. F. Adams, "Basic Math for Electronics," 3d ed., McGraw-Hill Book Company, New York, 1970, has a treatment of elementary plane vectors and vector algebra.

reference point from which we start reckoning voltage. Occasionally it is desirable in the interest of clarity to use both an arrow and double subscripts simultaneously. If this procedure is followed, it is absolutely imperative that the tail of the arrow rest on the point corresponding to the first subscript or else the two forms of notation will be in conflict. See Fig. 2-6, for example.

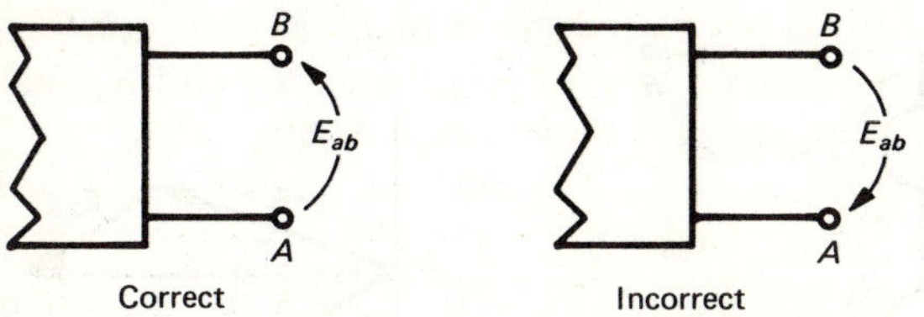

FIGURE 2-6

It is also common practice to use plus and minus marks to indicate reference directions with ac as well as dc sources. This usually bugs the student because he had difficulty in separating the concept of notation from the actual polarity of the ac source which is changing with time. When you see plus and minus marks around an ac source, just pretend the minus and plus terminals respectively correspond to the tail and head of a voltage arrow (see Fig. 2-7).

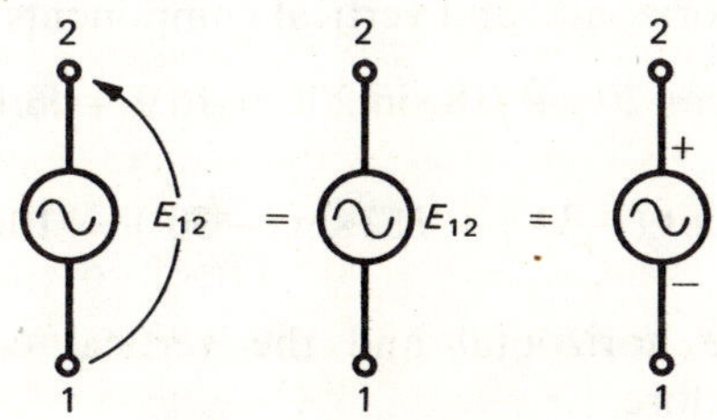

Equivalent forms of voltage notation

FIGURE 2-7

As our final point, you should recall that whenever you move through a circuit element in a direction opposite to the indicated reference direction, the resultant voltage will be the negative of that obtained by going in the indicated reference direction. In other words, if you obtain -4 volts or, say, E_{xy} volts in following some path from tail to head of an arrow, you would obtain $+4$ volts or $-E_{xy}$ going in the opposite direction.

Now work through problems PS 2-4 and PS 2-5.

2-2 Current notation

Since sinusoidal current as well as voltage varies in magnitude and direction in an ac circuit, we have a notation problem somewhat similar to that for voltage notation.

The first point to clarify is that current in this text, as in the preceding and other texts of this series, implies electron flow rather than so-called conventional current flow. This means when you see an I arrow or some other reference to a current, it implies the transfer of negative rather than positive charge in the arrow direction. Should you have a need to work with conventional current flow, just change the arrow direction and all else will work out. The previous voltage notation is still valid.

The symbol I or i and appropriate subscripts if needed will be used to designate current. If an arrow is drawn in alongside the current symbol, this arrow represents the *assumed positive reference direction* of current (electron) flow. We say "assumed" because we do not really know which way the current is flowing if the circuit is at all messy. Hence, we take a guess at it. We say "positive reference direction" to give us some kind of handle in discussing the current or writing circuit equations. If a detailed circuit analysis reveals the current to be a positive quantity (although it may have magnitude and direction in ac circuitry), it simply means our assumed positive reference direction was correct. For example, $I = 6\angle 30^\circ$ amps in the arrow direction indicates that the current I does flow in assumed positive reference direction. However, $I = -6\angle 30^\circ$ means that the current actually flows opposite to the assumed positive reference direction of I. The fact that multiplication by -1 is equivalent to a phase rotation of 180° should cause no confusion insofar as assuming reference directions or analyzing a circuit is concerned. In other words, you can interpret, say, $I = -6\angle 30^\circ$ as $I = 6\angle 30^\circ \pm 180^\circ = 6\angle 210^\circ$, or as $I = 6\angle -150^\circ$ with no change in meaning.

As in voltage notation, the upper-case symbol standing alone or with a numerical subscript in general implies the effective value. However, if it is necessary to differentiate between dc, rms, instantaneous, etc., values, we use double subscripts. With double subscripts, the first subscript is analogous to the tail and the second subscript to the head of the arrow in arrow notation for currents. And in a manner similar to voltage notation (assuming a dc component may be superimposed on the ac),

I_{AB} = dc current component if any from A to B

I_{ab} = rms value

$I_{AB(\text{pk})}$ = total peak value

$I_{AB(\text{pk})(\text{ac})}$ or $I_{ab(\text{pk})}$ = peak of ac component

i_{AB} = total instantaneous value

i_{ab} = instantaneous value of ac component only

As a final comment, we might consider that it may be necessary to differentiate between a dc and an ac current source on a schematic diagram. Usually the current designation, say I_{AB}, and i_{cd} or I_{cd}, would be sufficiently indicative of which is the dc and ac source. However, if there should arise a situation where further clarification is needed, let us agree to use the schematic representations of Fig. 2-8.

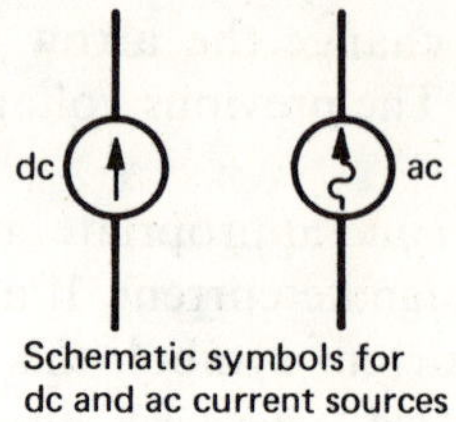

FIGURE 2-8

Since currents as well as voltages are, in general, vector (phasor) quantities in ac circuit analysis, we also find a dot, boldface type, or a small arrow ($\vec{E}$) to indicate a vector if necessary. Usually the context is sufficiently clear to avoid such notation.

Now work through problems PS 2-6 and PS 2-7.

PROBLEMS WITH SOLUTIONS

PS 2-1 How would you designate the dc, rms, and instantaneous voltage of some source, say *B*?

SOLUTION Since it is voltage we wish to specify, we use *E* or *V*, and upper-case *B* for the dc voltage (E_B). The total instantaneous value is e_B. The rms value is written as E_b.

PS 2-2 Designate the rms and determine the dc, positive peak, negative peak, peak-to-peak, peak ac, and the instantaneous and ac instantaneous voltages of some source *A* at time t_1 if the waveform appears as shown in Fig. PS 2-2. Do not try to calculate the rms value. Just designate it.

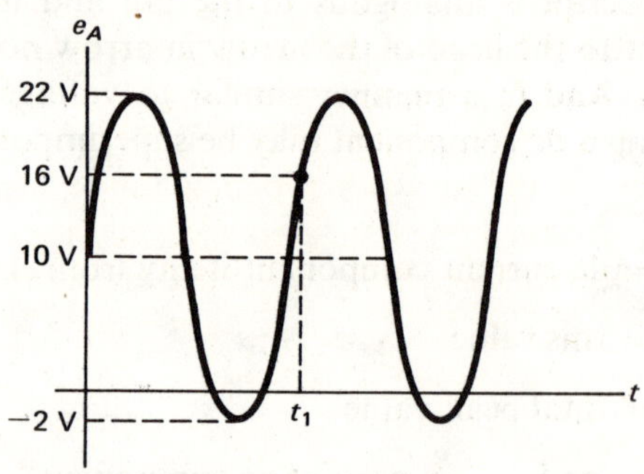

FIGURE PS 2-2

SOLUTION

$E_{rms} = E_a$

$E_{A(pk-pk)} = 22 - (-2) = -24\ V \quad e_A = 16\ V$

$E_{A(pk}{}^{+} = 22\ V$

$E_{A(pk)ac} = 22 - 10 = 12\ V \qquad e_a = 16 - 10 = 6\ V$

$E_{A(pk)}{}^{-} = -2\ V$

PS 2-3 Graphically determine $\mathbf{E}_{ab}$ if $\mathbf{E}_1 = 18\angle 20°$ and $\mathbf{E}_2 = 12\angle -30°$ in Fig. PS 2-3*a*. After you complete the graphic solution, verify it analytically.

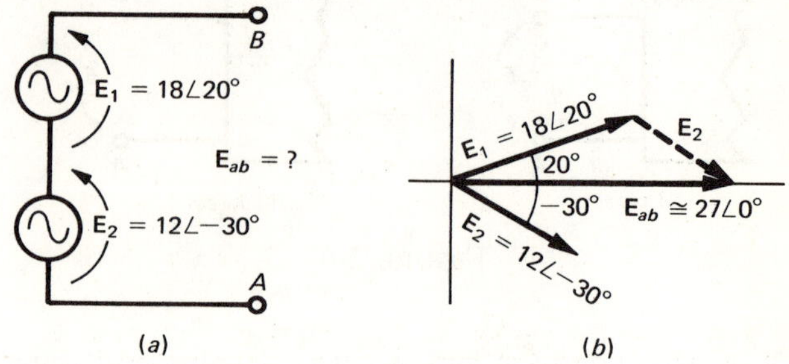

FIGURE PS 2-3

SOLUTION First construct the $\mathbf{E}_1$ and $\mathbf{E}_2$ vectors on a vector diagram as shown in Fig. PS 2-3*b*. Then graphically add them by translating $\mathbf{E}_2$ so that its tail sits on the head of $\mathbf{E}_1$. Construct the resultant vector from the tail of $\mathbf{E}_1$ to the new location of the head of $\mathbf{E}_2$ to obtain the vector sum $\mathbf{E}_{ab} \cong 27\angle 0°$.

For an analytic solution we will resolve $\mathbf{E}_1$ and $\mathbf{E}_2$ into their horizontal and vertical components. Thus

$$\mathbf{E}_1 = 18 \cos 20° + j18 \sin 20° = 16.9 + j6.16$$

$$\mathbf{E}_2 = 12 \cos(-30°) + j12 \sin(-30°) = 10.4 - j6.00$$

Adding the horizontal and the vertical components yields, therefore,

$$\mathbf{E}_{ab} = 27.3 + j0.16$$

or, in polar form,

$$\mathbf{E}_{ab} = \sqrt{(27.3)^2 + (0.16)^2}\ \angle \tan^{-1}\frac{0.16}{27.3} = 27.3\angle 0.336°$$

which is in close agreement with the graphic solution.

PS 2-4 Analytically determine E_{ab} in Fig. PS 2-4.

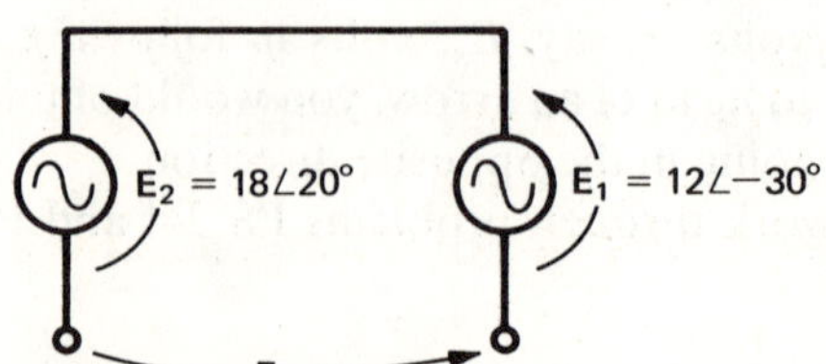

FIGURE PS 2-4

SOLUTION 1) $\mathbf{E}_{ab} = \mathbf{E}_2 - \mathbf{E}_1$

Note that $\mathbf{E}_1$ is preceded with a negative sign because, in following the path from the tail of the $\mathbf{E}_{ab}$ arrow to its head, we went through the $\mathbf{E}_1$ arrow backwards—that is, from its head to its tail. Substituting values into 1,

2) $\mathbf{E}_{ab} = 18\underline{/20^\circ} - 12\underline{/-30^\circ}$

$= (18\cos 20^\circ + j18\sin 20^\circ) - [12\cos(-30^\circ) + j12\sin(-30^\circ)]$

$= 16.91 + j6.156 - [10.39 - j6]$

3) $\mathbf{E}_{ab} = 6.52 + j12.16 = 13.8\underline{/61.8^\circ}$ V

PS 2-5 Determine $\mathbf{E}_{ab}$ in Fig. PS 2-5.

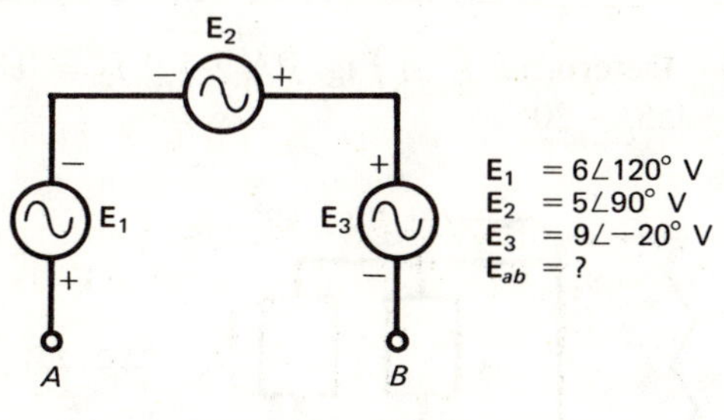

FIGURE PS 2-5

SOLUTION

1) $\mathbf{E}_{ab} = -\mathbf{E}_1 + \mathbf{E}_2 - \mathbf{E}_3$

$= -(6\underline{/120^\circ}) + (5\underline{/90^\circ}) - (9\underline{/-20^\circ})$

$= -[-3 + j5.2] + [0 + j5] - [8.46 - j3.08]$

$= 3 - j5.2 + j5 - 8.46 + j3.08$

$\mathbf{E}_{ab} = -5.46 - j2.88 = 6.17\underline{/-152^\circ}$

or alternatively we have the same result expressed as

$\mathbf{E}_{ab} = -(5.46 + j2.88) = 6.17\underline{/27.8^\circ + 180^\circ}$

$= 6.17\underline{/208^\circ}$

PS 2-6 Express Kirchhoff's current law if all currents flowing into node X in Fig. PS 2-6 are assumed positive while those leaving are considered negative.

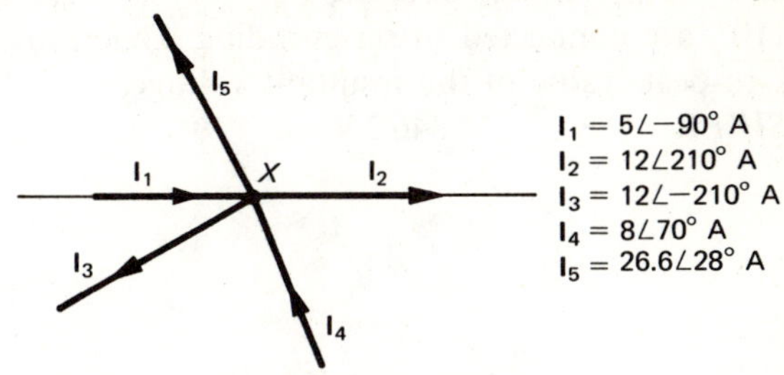

FIGURE PS 2-6

SOLUTION

$0 = \mathbf{I}_1 - \mathbf{I}_2 - \mathbf{I}_3 + \mathbf{I}_4 - \mathbf{I}_5$

$0 = -5\underline{/-90^\circ} - 12\underline{/210^\circ} - 12\underline{/-210^\circ} + 8\underline{/70^\circ} - 26.6\underline{/28^\circ}$

$0 = -(-j5) - (-10.4 - j6) - (-10.4 + j6) + (2.74 + j7.52) - (23.5 + j12.5)$

$0 = 0$

PS 2-7 The current gain of some four-terminal current amplifying network as shown in Fig. PS 2-7 is $A_i = I_2/I_1$. Suppose $I_1 = 1 + j0$ and $I_2 = -(40 - j8)$ at a frequency of 100 kHz. What conclusions can we draw?

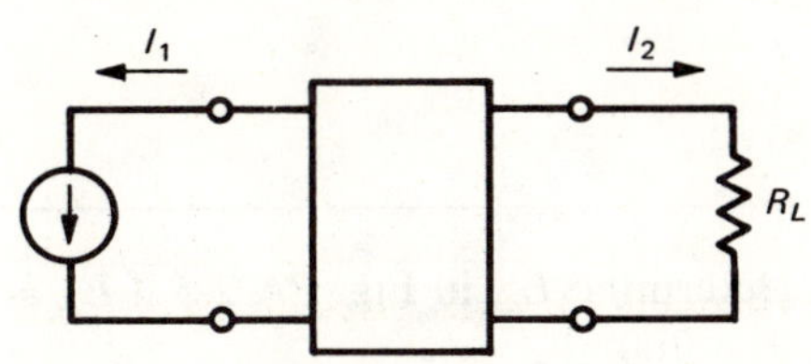

FIGURE PS 2-7

SOLUTION

$$A_i = \frac{I_2}{I_1} = \frac{-(40 - j8)}{1 + j0} = \frac{-(40.8\underline{/-11.3^\circ})}{1\underline{/0^\circ}}$$

$$A_i = -(40.8\underline{/11.3^\circ})$$

Inspection of the above expression for A_i leads us to conclude that I_2, as indicated by the negative sign, actually flows through R_L in a direction opposite to that shown in Fig. PS 2-7*a*, and that it lags I_1 by 11.3°. Alternatively, we could have said

$$A_i = -40.8\underline{/-11.3^\circ} = 40.8\underline{/-11.3^\circ \pm 180^\circ}$$

PROBLEMS WITH ANSWERS

PA 2-1 Two voltage sources, $E_a = 50\underline{/70^\circ}$ and $E_b = 80\underline{/110^\circ}$, are connected in series-aiding. Determine the peak-to-peak value of the resultant voltage.

ANSWER 346.2 V

PA 2-2 Determine E_3 in Fig. PA 2-2 if $E_1 = 12\underline{/270^\circ}$ and $E_2 = 15\underline{/-180^\circ}$.

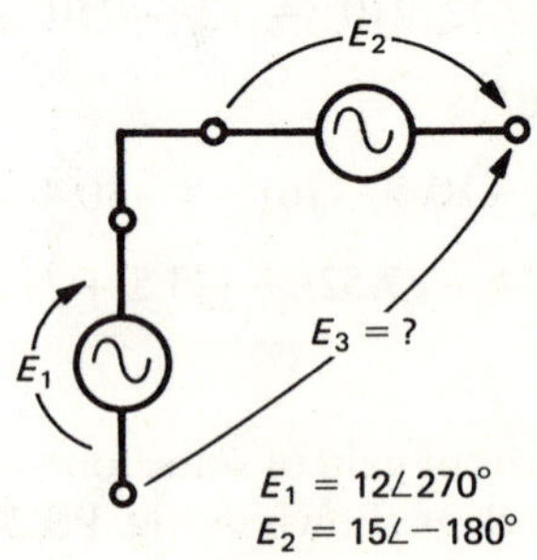

ANSWER $19.2\underline{/218.7^\circ}$ V

PA 2-3 Determine E_{ab} in Fig. PA 2-3 if $E_1 = 12\underline{/120^\circ}$ and $E_2 = 9\underline{/-200^\circ}$.

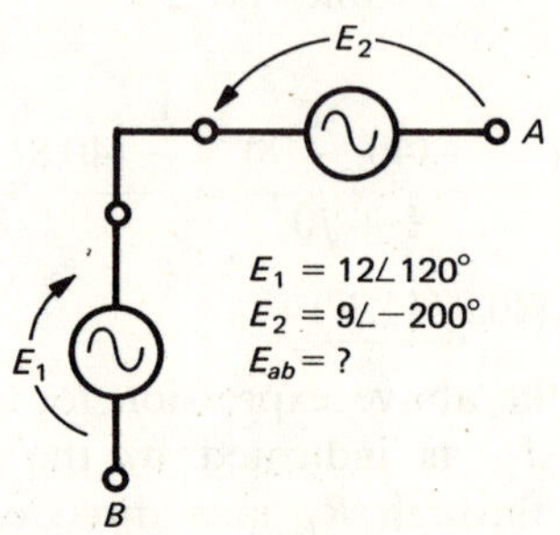

ANSWER $7.7\underline{/251^\circ}$ V

PA 2-4 Express E_x in Fig. PA 2-4 in terms of the circuit voltages.

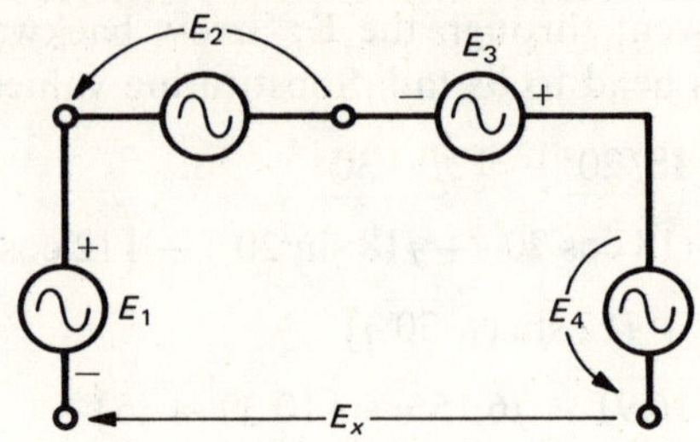

ANSWER $E_x = -E_4 - E_3 + E_2 - E_1$

PA 2-5 Determine I_1 in Fig. PA 2-5 if $I_3 = 0.8\underline{/-40^\circ}$ and $I_2 = 0.5\underline{/-20^\circ}$.

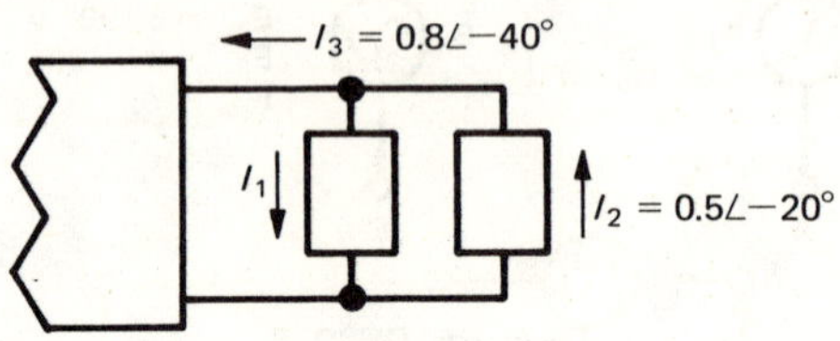

ANSWER $0.373\underline{/113^\circ}$

PA 2-6 Determine the voltage gain A_v of the four-terminal network in Fig. PA 2-6 if $e_o = -(10\underline{/12^\circ} + 8\underline{/9^\circ})$ when $e_{in} = 7\underline{/6^\circ}$. Note $A_v = \mathbf{e}_o/\mathbf{e}_{in}$.

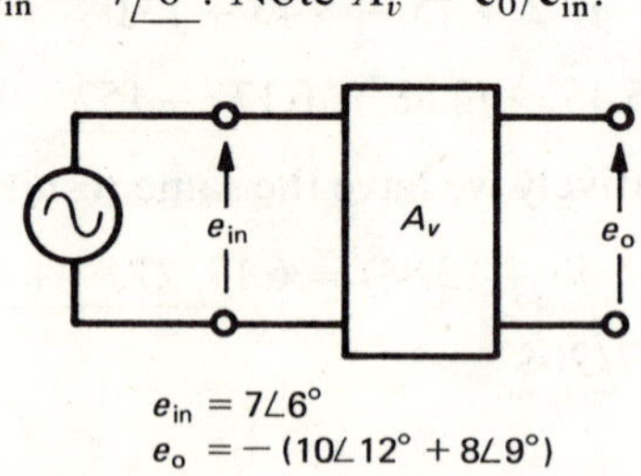

ANSWER $2.57\underline{/184.7^\circ}$

PROBLEMS WITHOUT ANSWERS

P 2-1 Two voltage sources, e_1 and e_2, are connected series-aiding. If e_1 is given by $e_1 = 68$ V and $e_2 = 85 \cos \theta$, determine, if any, the dc, positive peak, negative peak, peak-to-peak, and peak ac voltage of the resultant (e_3) waveform. Also, find the instantaneous and ac instantaneous voltage at $\theta = \pi/3$ rad.

P 2-2 Express E_{ab} in Fig. P 2-2.

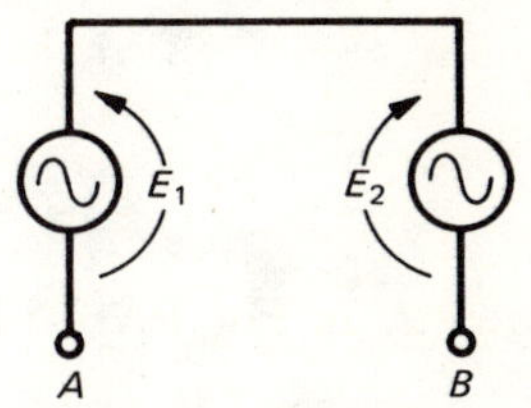

P 2-3 Express E_x in Fig. P 2-3.

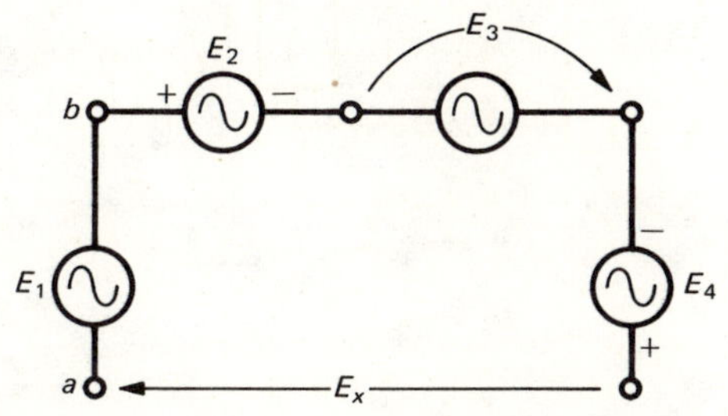

P 2-4 Determine E_{ab} in Fig. P 2-2 if $E_1 = 48\angle 65^\circ$ and $E_2 = -60\angle -180^\circ$.

P 2-5 Determine E_{ab} in Fig. P 2-3 if $E_1 = -3\angle -\dfrac{\pi}{2}$, $E_2 = -3\angle \dfrac{3\pi}{2}$, $E_3 = 4 \cos \omega t + 5 \sin \omega t$, and $E_4 = -2 + j1.5$.

P 2-6 Determine I_0 in Fig. P 2-6 if $I_1 = -2\angle -180^\circ$ and $I_2 = 1.6\angle 28^\circ$.

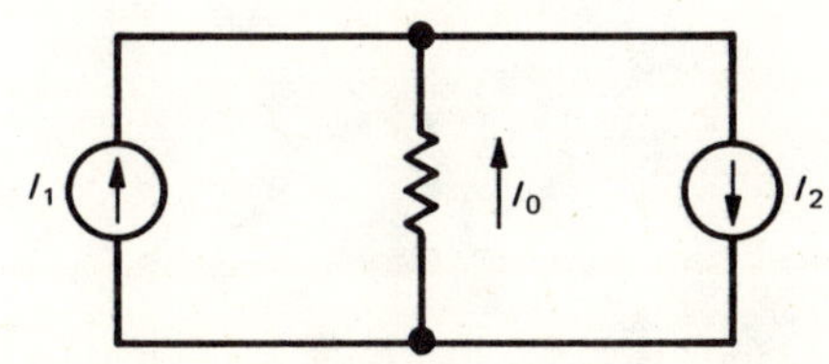

3 power in the purely resistive circuit

Determining the instantaneous and average power dissipated in some device or circuit is a most significant calculation. For example, it is necessary to know the power dissipated in a resistor, transistor, or whatever if we do not want to reduce it to ashes. We may perhaps be interested in the power some batteries may have to furnish in order to specify their size and configuration. Or we may be concerned with the power an amplifier pumps into a loudspeaker in a high fidelity sound system, or the power an amplifier can deliver to some motor which it controls as part of a control system. Consequently, let us build upon the foundation laid in Chap. 1, where the concepts of effective voltage and current and instantaneous and average power were introduced. The notation of Chap. 2 will also prove helpful in exploring the subject.

3-1 Notation in power analysis

Suppose we wish to analyze some circuit for the power dissipated (or perhaps generated) in a particular circuit element or subcircuit that is characterized by a pair of terminals as shown in Fig. 3-1*a*. To determine the in-

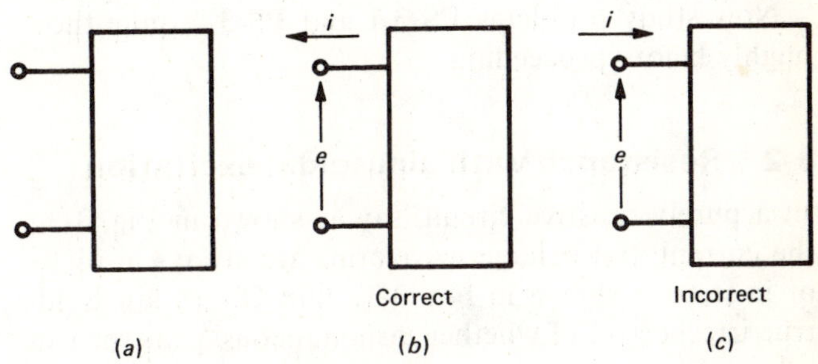

FIGURE 3-1

stantaneous power as a function of time, we might drive the input terminals with a voltage source to develop the terminal voltage *e* and then measure (or calculate) the resultant current *i*. Alternatively, we could drive the input with a current source *i* and measure (or calculate) the resultant voltage *e*. The choice generally depends upon what is in the box, but—irrespective of whether a voltage or current drive is used—the important point is that the *e* and *i* reference directions be correctly chosen.

In circuit analysis, as when writing Kirchhoff's loop equations, the *choice* of *e* and *i* reference directions is completely arbitrary. This is not true when performing power analysis. If you want to determine the power dissipated in a pair of terminals, set up the *e* and *i* reference directions shown in Fig. 3-1*b*. Note that *i* is so directed that when *e* is a positive quantity, *i* truly flows in the indicated current reference direction.[1] If upon further analysis p turns out positive with these reference directions, the box is absorbing (dissipating) power. On the other hand, if p turns out negative, it means the box is not absorbing but instead is generating power.

Figure 3-1c shows a choice of *incorrect* reference directions for power analysis. Note that i would actually have to flow in the opposite direction for e positive.

Any circuit in which current flows can in the final analysis be broken down into a power source (generator) and power sink (load), as shown in Fig. 3-2. Normally,

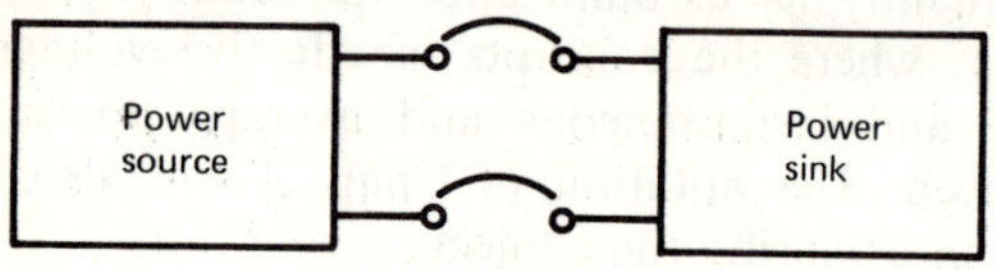

FIGURE 3-2

as with a purely resistive load, the power flow is always from the source to the sink. However, without going into details, suffice it to say that under certain conditions the role of source and sink may be exchanged. This usually involves something in the load which is capable of storing energy. We will explore this phenomenon later.

Now study problems PS 3-1 and PS 3-2 quite thoroughly before proceeding.

3-2 Resistance with sinusoidal excitation

In a purely resistive circuit, say as shown in Fig. 3-3a, the current and voltage waveforms are always in phase or in step, as shown in Fig. 3-3b, and Ohm's law holds true irrespective of whether instantaneous, peak, or rms values are used.

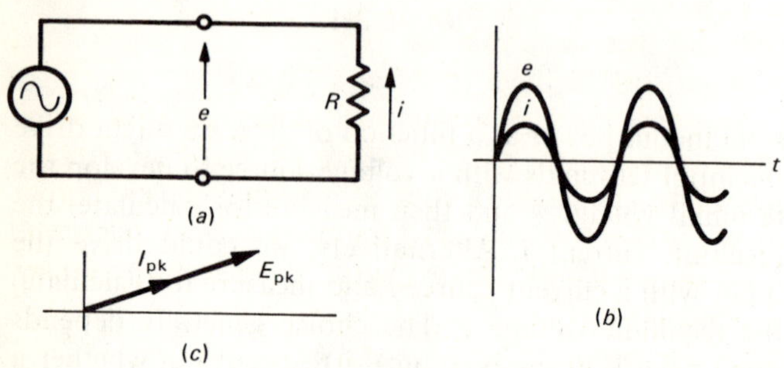

FIGURE 3-3

The e and i sinusoids of Fig. 3-3b may be considered as the vertical projections of the rotating vectors having lengths of E_{pk} and I_{pk} respectively, as shown in Fig. 3-3c.

[1] The same e and i reference directions of Fig. 3-1b must be preserved when determining input resistance or impedance. Otherwise a negative R or Z results, and if it were truly a negative R (or Z), the erroneous choice of reference directions would make it seem positive.

The relationship, in this case,

1) $$R = \frac{e}{i} = \frac{E_{pk}}{I_{pk}} \tag{3-1}$$

is always true. Since the effective value of a sinusoid is equal to the peak value divided by $\sqrt{2}$, we may also write

2) $$R = \frac{E_{pk}/\sqrt{2}}{I_{pk}/\sqrt{2}} = \frac{E_{rms}}{I_{rms}} = \frac{E}{I} \tag{3-2}$$

Thus the current and voltage symbols of Fig. 3-3a could be expressed as the effective values shown in Fig. 3-4a and by the rotating vectors of Fig. 3-4b.

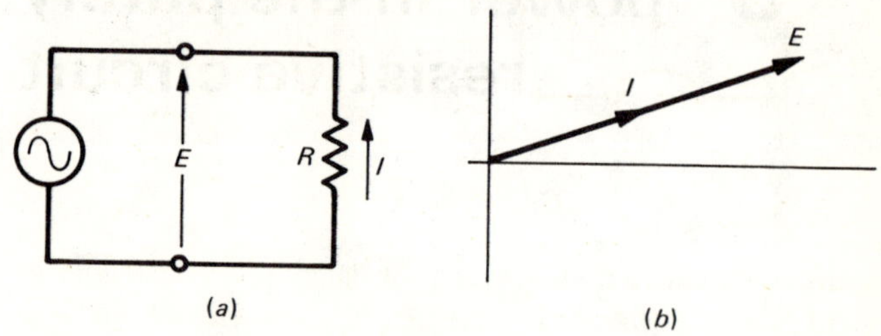

FIGURE 3-4

The average power dissipated in R is given by

3) $$P = EI \tag{3-3a}$$

4) $$P = I^2R \tag{3-3b}$$

5) $$P = E^2/R \tag{3-3c}$$

which is identical with the dc circuits case except that we now use the effective values of the sinusoids.

Now work problems PS 3-3 and PS 3-4.

3-3 Resistive circuit with identical-frequency multiple sources

Frequently we encounter situations where any of the following multiple driving sources may exist:

(a) ac sources of the same frequency

(b) ac sources of different frequencies

Case (b) can include dc sources also, since a dc source may be considered a zero-frequency ac source.

It might also be mentioned at this time that a nonsinusoidal waveform can be shown to be the superposition of two or more sinusoids having various amplitude and phase relationships between them which depend upon the waveform. A detailed study of this subject comes under the heading of Fourier analysis, the pursuit of which is not our intent at this time. The point is that since a nonsinusoidal waveform can be resolved into different frequencies (including a dc component), it falls under case (b). Of course, we would first have to know how to find these frequencies, which we do not at this time.

If we consider case (*a*) first, we might assume the general situation as shown in Fig. 3-5*a*. Clearly, the two

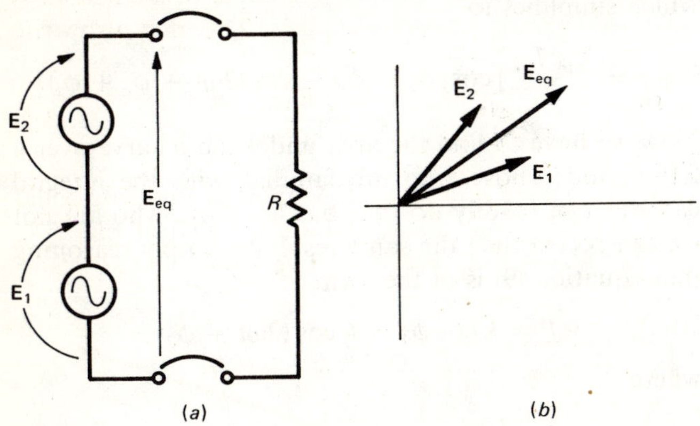

FIGURE 3-5

sources (or more if such were the case) add vectorially to yield some equivalent voltage E_{eq}, as shown in Fig. 3-5*b*. Now

1) $$\mathbf{E}_1 = E_{1x} + jE_{1y}$$

where the *x* and *y* subscripts indicate the *x* and *y* components of the vector. The magnitude of $\mathbf{E}_1$ is

2) $$|\mathbf{E}_1| = E_{1\text{rms}} = \sqrt{E_{1x}^2 + E_{1y}^2}$$

We are assuming the lengths of the vectors in Fig. 3-5*b* represent rms values. Similarly,

3) $$\mathbf{E}_2 = E_{2x} + jE_{2y}$$

4) $$|\mathbf{E}_2| = E_{2\text{rms}} = \sqrt{E_{2x}^2 + E_{2y}^2}$$

Since $\mathbf{E}_{eq}$ by vector addition is

5) $$\mathbf{E}_{eq} = \mathbf{E}_1 + \mathbf{E}_2$$

or

6) $$\mathbf{E}_{eq} = (E_{1x} + E_{2x}) + j(E_{1y} + E_{2y}) \qquad (3\text{-}4a)$$

and

7) $$|\mathbf{E}_{eq}| = E_{eq(\text{rms})} \qquad (3\text{-}4b)$$
$$= \sqrt{(E_{1x} + E_{2x})^2 + (E_{1y} + E_{2y})^2}$$

We had plenty of practice in the vector addition of sinusoids in Chap. 2.

The power dissipated in *R* in Fig. 3-5*a* is then

$$P = \frac{E_{eq\,(\text{rms})}^2}{R} \qquad (3\text{-}5)$$

Do not make the mistake of reasoning that we can determine the power P_1 that only E_1 develops in *R* and add to it the power P_2 that only E_2 develops in *R*. You cannot use superposition in this sense to determine power. To satisfy yourself on this point, consider Fig. 3-6. Surely

1) $$P = \frac{E_{eq}^2}{R} = \frac{(E_1 + E_2)^2}{R}$$

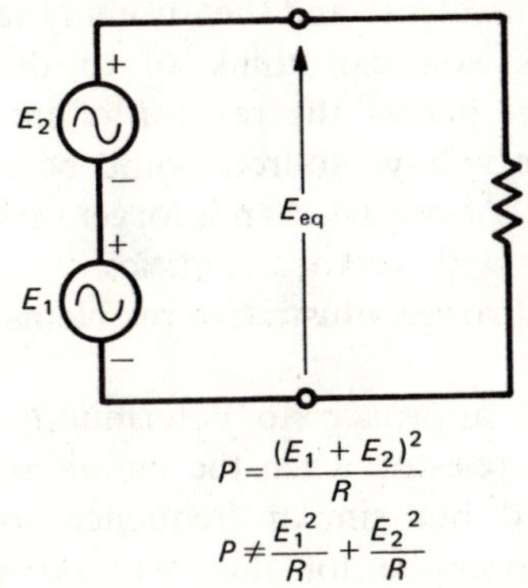

FIGURE 3-6

But if we consider the individual power contributions by superposition,

2) $$P = P_1 + P_2 = \frac{E_1^2}{R} + \frac{E_2^2}{R} = \frac{E_1^2 + E_2^2}{R}$$

Clearly, the final form of (2) is different from (1), which is correct.

If it is current sources in parallel you are working with, just follow a similar procedure to determine $I_{eq\,(\text{rms})}^2$ by vector addition and then use the relation

$$P = I_{eq\,(\text{rms})}^2 R \qquad (3\text{-}6)$$

In summary then, for multiple voltage or current sources of similar frequency but differing phase, determine the effective equivalent source by vector addition and then determine the power.

Now work through problem PS 3-5.

Another problem to consider under case (*a*) is the situation shown in Fig. 3-7, where we have both voltage and current sources which in general may have different phase angles with respect to some arbitrary reference. How do we handle this situation if we seek the power in some resistor?

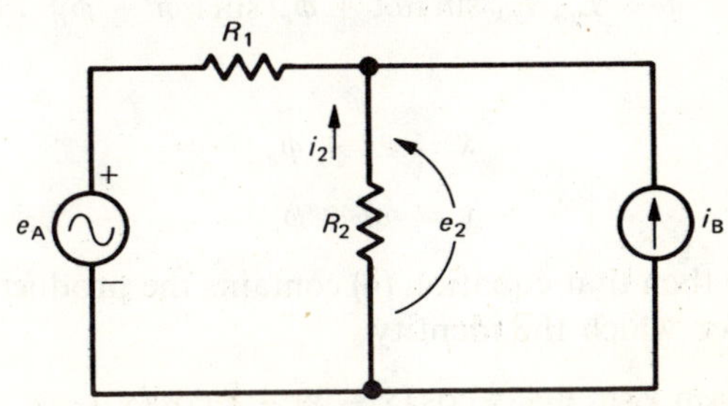

FIGURE 3-7

We could, of course, use the superposition theorem to sketch the voltage and current waveforms in the resistor of interest, multiply them together for the p waveform, and then determine the average value P. However, a much easier approach is to convert the current sources (Norton's equivalent circuits) to their voltage source (Thévenin's) equivalents and then use any techniques of circuit analysis you can think of to determine the effective voltage across the resistor of interest. Alternatively, all the voltage sources could be converted to current sources. Above all, do not forget that, in general, you are dealing with vector quantities.

Now, work through illustrative problems PS 3-6 and PS 3-7.

Still another approach to determining the power dissipated in a resistor when the circuit is driven (excited) by mixed but similar frequency sources is to develop an expression for the instantaneous voltage across and current through the resistor in question. The product of the e and i waveforms is the instantaneous power (p) waveform. The average value of the p waveform taken over a full cycle is the average power P. This is similar to some of the work you did in Chap. 1.

While this approach is not necessarily the simplest, it is most general and hence is worthy of close examination. As a starting point, consider the box of Fig. 3-8, which, incidentally, need not in general contain a resistor.

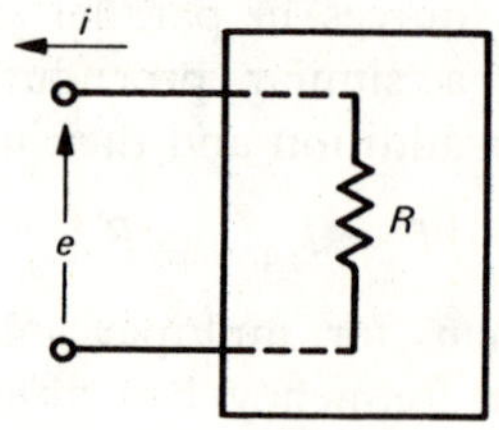

FIGURE 3-8

Let

1) $$e = E_{pk} \sin(\omega t + \phi_e)$$

2) $$i = I_{pk} \sin(\omega t + \phi_i)$$

Thus

3) $$p = ei$$

4) $$p = E_{pk} I_{pk} \sin(\omega t + \phi_e) \sin(\omega t + \phi_i)$$

Let

5) $$x = \omega t + \phi_e$$

6) $$y = \omega t + \phi_i$$

We see then that equation (4) contains the product sin x sin y, for which the identity

7) $$\sin x \sin y = \tfrac{1}{2}\cos(x - y) - \tfrac{1}{2}\cos(x + y) \quad (3\text{-}7)$$

may be written (see problem PS 3-8 for the proof).

Substituting equations (5), (6), and (7) into (4) yields

8) $$p = E_{pk} I_{pk} [\tfrac{1}{2}\cos(\omega t + \phi_e - \omega t - \phi_i) - \tfrac{1}{2}\cos(\omega t + \phi_e + \omega t + \phi_i)]$$

which simplifies to

9) $$p = \frac{E_{pk} I_{pk}}{2}[\cos(\phi_e - \phi_i) - \cos(2\omega t + \phi_e + \phi_i)]$$

Now we have to find the area under the p curve over a full period. Those who are familiar with the integral calculus can readily do this, but for those who are not we can accomplish the same result by simply reasoning that equation (9) is of the form

10) $$P = k\cos\phi_1 - k\cos(2\omega t + \phi_2)$$

where

$$k = \frac{E_{pk} I_{pk}}{2}, \phi_1 = \phi_e - \phi_i, \phi_2 = \phi_e + \phi_i$$

Now the net area under the p curve will in general be due to the area contributed by the first term in (10) plus the area under the second term. These areas may, of course, be positive or negative. Clearly, the first term is a constant since it is independent of time (t) and hence its contribution to the area under p during 2π radians is

11) $$A_1 = 2\pi k \cos\phi_1$$

The second term in equation (10) is a double-frequency sinusoid, which simply means it swings through two positive and negative cycles in the interval 2π. Since the areas under the positive and negative alternations are equal, the average area under this sinusoid is zero irrespective of the phase angle ϕ_2. Thus

12) $$A_2 = 0$$

and the average power, P, is

13) $$P = \frac{1}{2\pi}(A_1 + A_2)$$

Substituting (11) into (13), we obtain

14) $$P = k\cos\phi_1$$

Or from the substitutions in (10),

15) $$P = \frac{E_{pk} I_{pk}}{2}\cos(\phi_e - \phi_i) \quad (3\text{-}8)$$

Since $2 = (\sqrt{2})(\sqrt{2})$ and rms = peak/$\sqrt{2}$, we may rewrite (15) as

16) $$P = E_{rms} I_{rms} \cos(\phi_e - \phi_i)$$

17) $$P = EI\cos(\phi_e - \phi_i) \quad (3\text{-}9)$$

Now work through PS 3-9.

3-4 Resistive circuit with different frequency sources

The next case we will consider is that of determining the power in a resistor when the driving sources are of different frequencies. In particular, we are interested in the case where a dc or zero frequency source and an ac source are involved, because this is analogous to the actual circuit conditions under which many electronic devices and their loads operate.

For example, consider Fig. 3-9*a*, which represents some two-terminal network whose terminal voltage and current is illustrated by the waveforms of Fig. 3-9*b*.

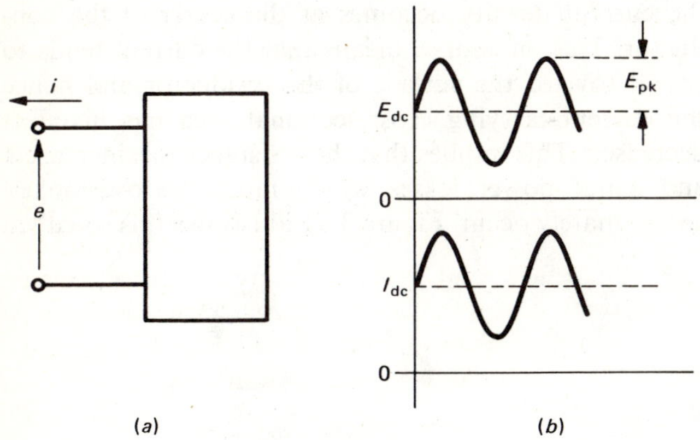

FIGURE 3-9

Let

1) $$e = E_{dc} + E_{pk} \sin(\omega t + \phi_e)$$

where E_{pk} is the peak value of the sinusoidal component only, and similarly

2) $$i = I_{dc} + I_{pk} \sin(\omega t + \phi_i)$$

Therefore, the instantaneous power is

3) $$p = ei$$

or substituting (1) and (2) into (3) and multiplying out,

4) $$p = E_{dc} I_{dc} + I_{dc} E_{pk} \sin(\omega t + \phi_e) + E_{dc} I_{pk} \sin(\omega t + \phi_i) + E_{pk} I_{pk} \sin(\omega t + \phi_e) \sin(\omega t + \phi_i)$$

To find the average power, we need to determine the area contributed under the p curve by each of the terms in (4). Clearly the first term is a constant and hence contributes an area

5) $$A_1 = 2\pi E_{dc} I_{dc}$$

The second and third terms are sinusoids with zero net area. Hence

6) $$A_2 = A_3 = 0$$

As it stands, the area under the fourth term is not obvious. However, if you will think back to the result of problem PS 3-8, it should be clear that the fourth term can be written as

7) $$\frac{E_{pk} I_{pk}}{2} [\cos(\phi_e - \phi_i) - \cos(2\omega t + \phi_e + \phi_i)]$$

Inspection of (7) indicates two terms in the bracket to be considered. Since the second term is a sinusoid, although of double frequency, the net area it encloses is zero. Hence, the area contributed by the fourth term is

8) $$A_4 = 2\pi \frac{E_{pk} I_{pk}}{2} \cos(\phi_e - \phi_i)$$
$$= 2\pi E_{rms} I_{rms} \cos(\phi_e - \phi_i)$$

Since

9) $$P = \frac{1}{2\pi}(A_1 + A_2 + A_3 + A_4)$$

we obtain from (5), (6), (7), and (8)

10) $$P = E_{dc} I_{dc} + E_{ac(rms)} I_{ac(rms)} \cos(\phi_e - \phi_i) \quad (3\text{-}10)$$

Now work through problem PS 3-10.

It is also useful to develop an expression for the effective value of a voltage or current waveform in which a dc and ac source are superimposed, say Fig. 3-10, for example.

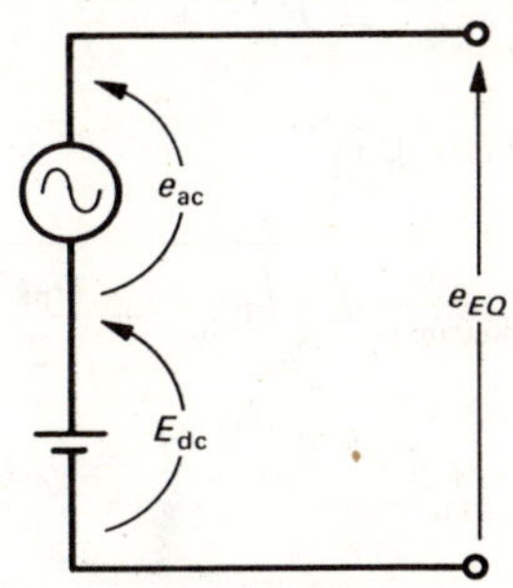

FIGURE 3-10

To analytically determine the effective or rms voltage of the e_{EQ} waveform in Fig. 3-10, we might, in a straightforward manner, write

1) $$e_{EQ} = E_{dc} + e_{ac}$$

or

2) $$e_{EQ} = E_{dc} + E_{pk} \sin(\omega t + \phi)$$

3) $$e_{EQ}^2 = E_{dc}^2 + 2E_{dc} E_{pk} \sin(\omega t + \phi) + E_{pk}^2 \sin^2(\omega t + \phi)$$

Now we have to determine the area under the e_{EQ}^2 waveform.

The first term in (3) contributes

4) $$A_1 = 2\pi E_{dc}^2$$

The second term is a sinusoid which, of course, contributes zero net area. Hence

5) $$A_2 = 0$$

The third term is not obvious, but if we recall that

6) $$\sin x \sin y = \tfrac{1}{2}\cos(x - y) - \tfrac{1}{2}\cos(x + y)$$

it should be apparent that when $x = y = \theta$, eq. (6) reduces to

7) $$\sin^2 \theta = \tfrac{1}{2} - \tfrac{1}{2}\cos 2\theta$$

Thus the third term of (3) may be written as

8) $$E_{pk}^2 [\tfrac{1}{2} - \tfrac{1}{2}\cos 2(\omega t + \phi)]$$

Since the second term in (8) is a sinusoid, it contributes zero net area, whereas the first term yields

9) $$A_3 = 2\pi \frac{E_{pk}^2}{2}$$

Therefore

10) $$A = A_1 + A_2 + A_3 = 2\pi E_{dc}^2 + 0 + 2\pi \frac{E_{pk}^2}{2}$$

Now (10) is the area under the e_{EQ}^2 curve. The average value of the e_{EQ}^2 curve over an interval 2π rad is then (the overbar implies average value)

11) $$\overline{e_{EQ}^2} = E_{dc}^2 + \frac{E_{pk}^2}{2}$$

And the effective value is

12) $$E_{eq\,(rms)} = \sqrt{E_{dc}^2 + \frac{E_{pk}^2}{2}} \qquad (3\text{-}11)$$

Since $E_{dc} = E_{dc(rms)}$ and $E_{pk}/\sqrt{2} = E_{ac(rms)}$, we may rewrite (12) as

13) $$E_{eq\,(rms)} = \sqrt{E_{dc}^2{}_{(rms)} + E_{ac}^2{}_{(rms)}} \qquad (3\text{-}12)$$

Although the rms subscript is not necessary for the dc term in (13), we will include it as it conveys the form of Eq. (3-12). Note that the equivalent rms voltage is *not* the sum of the component rms voltages. Similarly

14) $$I_{eq\,(rms)} = \sqrt{I_{dc}^2{}_{(rms)} + I_{ac}^2{}_{(rms)}} \qquad (3\text{-}13)$$

Now work through problem PS 3-11.

It may be shown that when several *different* frequency sources are involved, the equivalent rms value is

$$E_{eq\,(rms)} = \sqrt{E_{dc}^2{}_{(rms)} + E_{1ac}^2{}_{(rms} + E_{2ac}^2{}_{(rms)} + \ldots} \qquad (3\text{-}14)$$

The frequencies of E_1, E_2, etc., need not necessarily be harmonically related—that is, in integral ratios such as $\omega_3 = 3\omega_1$, $\omega_2 = 2\omega_1$, etc., It is permissible to have, say, $\omega_3 = 1.8\omega_1$, $\omega_2 = 1.1\omega_1$, or whatever. However, the frequency components must all be different.

Now work problem PS 3-12.

Although we have not explored the subject, it turns out that a nonsinusoidal waveform may be resolved into dc and harmonically related frequency components. If all the amplitudes of these frequency components were known, Eq. (3-14) could be applied.

3-5 Skin effect

It turns out that the distribution of current through the cross-sectional area of a conductor is a function of frequency, and that the higher the frequency, the less the current density becomes at the center of the conductor. This, of course, means that the current tends to crowd toward the surface of the conductor and hence the current-carrying cross-sectional area has in effect decreased. This implies that the resistance has increased, and hence power losses which might be overlooked unfortunately occur. Figure 3-11 illustrates this so-called

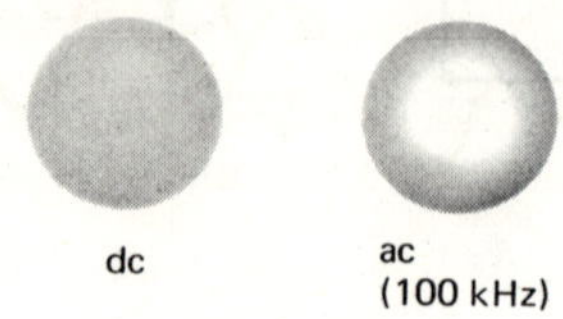

FIGURE 3-11

skin effect. This is the reason we frequently find some components or conductors fashioned from copper tubing for use at high frequencies. A copper tube would exhibit more surface area for the current to crowd toward at high frequencies. At the lower radio frequencies we often use Litz wire, which consists of a conductor made from fine twisted wire strands which are insulated from each other.

In general, the ratio of ac resistance (R_{ac}) to dc resistance (R_{dc}) for a solid, round copper wire is given by

$$\frac{R_{ac}}{R_{dc}} = k\sqrt{f_{MHz}} \qquad (3\text{-}15)$$

where f is the frequency in MHz and $k = 6.9$, 10.9, and 17.6 for #22, #18, and #14 wire sizes respectively.

PROBLEMS WITH SOLUTIONS

PS 3-1 Determine the average power dissipated in the the box of Fig. PS 3-1*a* if the *e* and *i* waveforms are as shown in Fig. PS 3-1*b*. Assume the waveforms are sinusoids.

SOLUTION We could solve the problem graphically by multiplying the *e* and *i* waveforms, point by point, to generate the *p* waveform of Fig. PS 3-1*c* and then determining the average value of *p* over the interval from

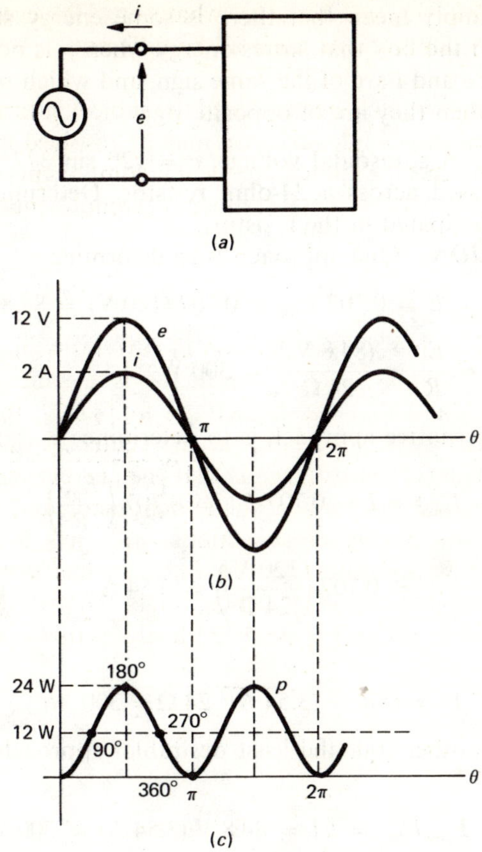

FIGURE PS 3-1

$\theta = 0$ to 2π. You may notice the p waveform of Fig. PS 3-1c is double the frequency of the e and i waveforms if we recognize that the p waveform swings symmetrically about an average value (12 W) equal to half the peak power (24 W). Thus the box should dissipate an average power $P = 12$ W.

We might just try to verify this analytically as follows. Since the waveforms are sinusoids, we may, by inspection of Fig. PS 3-1b, write

1) $$e = 12 \sin \omega t = 12 \sin \theta$$

2) $$i = 2 \sin \omega t = 2 \sin \theta$$

Note there is no phase angle between the e and i waveforms. Since

3) $$p = ei$$

we have from (1) and (2)

4) $$p = (12 \sin \theta)(2 \sin \theta) = 24 \sin^2 \theta$$

Now it was shown in Chap. 1 (PS 1-15) that

5) $$\sin^2 \theta = \tfrac{1}{2} - \tfrac{1}{2} \cos 2\theta$$

Substituting (5) into (4),

6) $$p = 24 \left[\tfrac{1}{2} - \tfrac{1}{2} \cos 2\theta\right] = 12 - 12 \cos 2\theta$$

Now we need to find

7) $$P = \frac{\text{area under the } p = 12 - 12\cos 2\theta \text{ curve between } \theta = 0 \text{ and } \theta = 2\pi}{2\pi}$$

which in the language of calculus is written as

8) $$P = \int_0^{2\pi} p \, d\theta = \int_0^{2\pi} (12 - 12 \cos 2\theta) \, d\theta$$

Once again, if the symbolism of (8) provokes undue feelings of insecurity—forget it and just consider (7). Clearly, if we were to plot (6), we should obtain Fig. PS 3-1c. This may easily be verified by trying a few values of θ and finding the corresponding p. Since the p waveform is symmetrical about 12 W, the average p is 12 W, and the area under the curve for the interval from 0 to 2π is 2π (12 W, we obtain—upon substituting this area into the numerator of (7)—

9) $$P = \frac{2\pi\,(12 \text{ W})}{2\pi} = 12 \text{ W}$$

which verifies our previous, more intuitive solution.

Actually it is not even necessary to sketch (6) to determine P since we may reason that the average value of any sinusoidal function over a full period (2π rad) or whole number multiple thereof is zero, it follows that the area contributed by the 12 cos 2θ term itself is zero. However, this term does contribute an area of 12 W × 2π rad.

It is quite useful to remember that the full period area under a strictly sinusoidal function is zero.

PS 3-2 Assume the box of Fig. PS 3-1a sees the

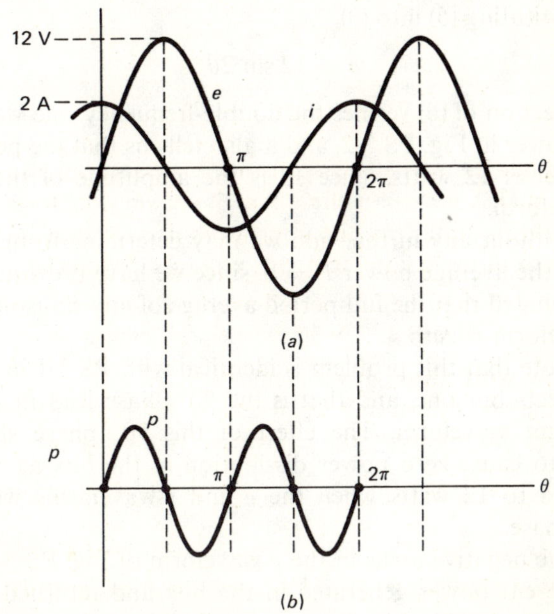

FIGURE PS 3-2

sinusoidal terminal voltage and current waveforms of Fig. PS 3-2*a*. Determine the average power the box dissipates.

SOLUTION Again we multiply the e and i waveforms to obtain the p waveform. If we roughly sketch it while paying careful attention to the signs of e and i, we obtain Fig. PS 3-2*b*. Again, we have a double-frequency power curve which this time swings about zero. Hence the average dissipation is zero, since there is just as much negative area below the p curve as above it. Thus $P = 0$.

For a more analytic solution we might examine the sinusoids of Fig. PS 3-2*a* and conclude that

1) $$e = 12 \sin \theta$$

2) $$i = 2 \sin (\theta + 90°) = 2 \cos \theta$$

Thus

3) $$p = ei = 24 \sin \theta \cos \theta$$

At this point you can turn to a handbook with a full table of identities or, with a little manipulation, you can use (3) in the table of identities in Chap. 1. We consider (3) of the table because, as shown here,

4) $$\sin (x + y) = \sin x \cos y + \cos x \sin y$$

it contains a $\sin x \cos y$ term which is what is hurting us in (3). If (4) is solved for $\sin x \cos y$ with $x = y = \theta$, we obtain

$$\sin (\theta + \theta) = \sin \theta \cos \theta + \cos \theta \sin \theta$$

$$\sin 2\theta = 2 \sin \theta \cos \theta$$

or

5) $$\sin \theta \cos \theta = \tfrac{1}{2} \sin 2\theta$$

Substituting (5) into (3),

6) $$p = 12 \sin 2\theta$$

Inspection of (6) verifies the double-frequency sine wave of power in Fig. PS 3-2, and it also tells us that the peak power is 12 watts, since 12 is the amplitude of the p waveform.

Without any further ado we may determine from (6) that the average power $P = 0$, since we have previously concluded that the full-period average of any sinusoidal waveform is zero.

Note that this problem is identical with PS 3-1 in all respects but one, and that is the 90° phase lead in the current waveform. The effect of this 90° phase shift was to cause zero power dissipation in the box as opposed to 12 watts when the e and i waveforms were in phase.

The negative areas in the p waveform of Fig. PS 3-2*b* represent power generated in the box and returned to the source. This does not necessarily mean that there is an honest-to-goodness rotary generator in the box. It could simply mean that there is some energy storage device in the box that stores energy when p is positive, as when e and i are of the same sign, and which returns power when they are of opposite sign.

PS 3-3 A sinusoidal voltage, $e = 120 \sin 377t$ volts, is impressed across a 24-ohm resistor. Determine the power dissipated in the resistor.

SOLUTION One approach is to determine

$$E_{\text{rms}} = E = 0.707\ E_{\text{pk}} = 0.707\ (120\ \text{V}) = 84.8\ \text{V}$$

$$P = \frac{E^2}{R} = \frac{(84.8\ \text{V})^2}{24\ \Omega} = 300\ \text{W}$$

An alternative approach is to determine

$$I_{\text{rms}} = I = 0.707\ I_{\text{pk}} = 0.707 \left(\frac{E_{\text{pk}}}{R}\right)$$

$$= 0.707 \left(\frac{120\ \text{V}}{24\ \Omega}\right) = 3.54\ \text{A}$$

Then

$$P = I^2 R = (3.54\ \text{A})^2\ 24\ \Omega = 300\ \text{W}$$

Still another (but the least desirable) approach is to write

$$P = E_{\text{rms}} I_{\text{rms}} = EI = 84.8\ \text{V}\ (3.54\ \text{A}) = 300\ \text{W}$$

This last approach *just happens* to work for an in-phase current and voltage sinusoid. In general, this method is not recommended because the product of averages is not the same as the average of products and rms values are a special kind of average. If this is hazy to you, review the discussion following Eq. (1-9*c*) in Chap. 1.

PS 3-4 Switch S_1 is normally closed in Fig. PS 3-4, while S_2 is normally open. Under no circumstances are

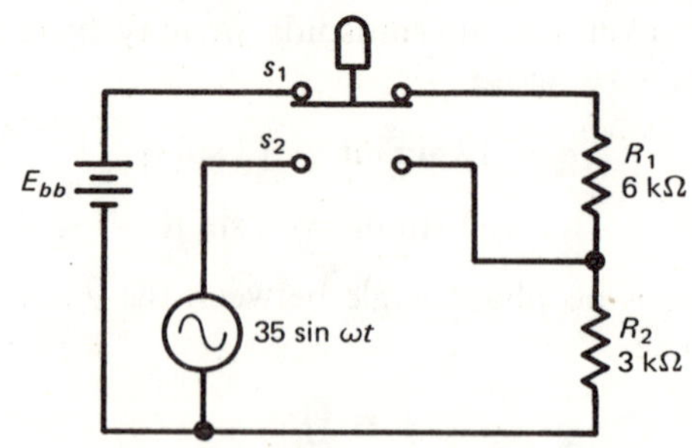

FIGURE PS 3-4

they closed simultaneously. If S_1 is open, S_2 is closed, and vice versa. Determine E_{bb} so that the power dissipated in R_2 is the same with either switch closed.

SOLUTION The effective voltage of the ac source is

$$E_{\text{rms(ac)}} = 0.707(35\ \text{V}) = 24.7\ \text{V}$$

Hence the effective current is

$$I_{\text{rms(ac)}} = \frac{24.7\ \text{V}}{3\ \text{k}\Omega} = 8.25\ \text{mA}$$

To develop the same power in the 3 kΩ resistor, the dc current I_{dc} supplied by E_{bb} must equal the effective ac current. Thus $I_{dc} = I_{\text{rms(ac)}} = 8.25$ mA. Consequently

$$E_{bb} = (R_1 + R_2)I_{dc} = 9\ \text{k}\Omega\ (8.25\ \text{mA}) = 74.2\ \text{V}$$

This problem is intended to refresh your memory that the rms value if current (or voltage) is that value which causes an equivalent power dissipation (heating effect) as some dc current (or voltage).

PS 3-5(a) Determine the power dissipated in R in the circuit of Fig. PS 3-5 if $e_1 = 70 \sin(\omega t - 30°)$ and $e_2 = 80 \sin(\omega t + 60°)$.

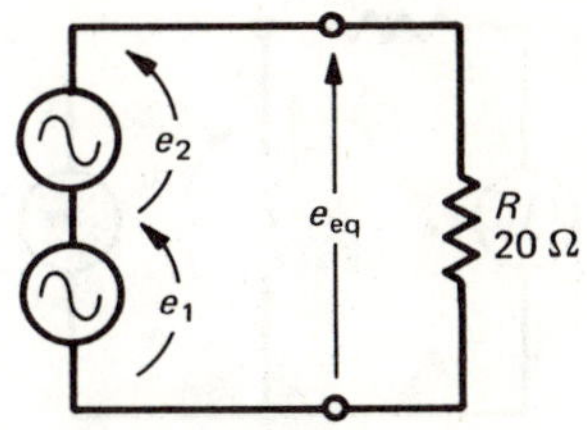

FIGURE PS 3-5

SOLUTION Vectorially, add e_1 and e_2 as follows

$$\begin{aligned} e_1 &= 70 \sin(\omega t - 30°) = 60.6 - j35.0 \\ + \quad e_2 &= 80 \sin(\omega t + 60°) = 40.0 + j69.3 \\ \hline e_{eq} &= 101 + j34.3 \end{aligned}$$

or

$$e_{eq} = 106\underline{/18.8°}\ \text{V}$$

The 106 in the above is a peak voltage. The phase angle of e_{eq} can be ignored, as it is of no concern in determining $E_{\text{eq(rms)}}$. Thus

$$E_{\text{eq(rms)}} = 0.707\,(106) = 75.2\ \text{V}$$

Hence

$$P = \frac{E_{\text{eq(rms)}}{}^2}{R} = \frac{(75.2\ \text{V})^2}{20\ \Omega} = 282\ \text{W}$$

Interestingly enough, if we try to use the incorrect superposition of powers method, we obtain

$$P = \frac{E_{1\text{rms}}{}^2}{R} + \frac{E_{2\text{rms}}{}^2}{R} = \frac{(49.5)^2}{20\ \Omega} = \frac{(56.6)^2}{20\ \Omega} = 282\ \text{W}$$

which is so similar to the correct answer of 282 watts that it is profoundly disturbing. To reassure ourselves, let us consider the problem again with different values.

PS 3-5(b) Assume $e_1 = 3\sqrt{2} \sin \omega t$ and $e_2 = 4\sqrt{2} \sin(\omega t + 45°)$ in the circuit of Fig. PS 3-5. Determine the power developed in R.

SOLUTION Converting e_1 and e_2 to rms values, we obtain

$$\begin{aligned} \mathbf{E}_{1\text{(rms)}} &= 3\underline{/0°} = 3 + j0 \\ + \quad \mathbf{E}_{2\text{(rms)}} &= 4\underline{/45°} = 2.828 + j2.828 \\ \hline E_{\text{eq(rms)}} &= 5.828 + j2.828 = 6.478\underline{/25.89°} \end{aligned}$$

$$P = \frac{(6.478\ \text{V})^2}{20\ \Omega} = 2.098\ \text{W}$$

If we erroneously use superposition of powers

$$P = P_1 + P_2$$

$$= \frac{(3\ \text{V})^2}{20\ \Omega} + \frac{(4\ \text{V})^2}{20\ \Omega} = \frac{25\ \text{V}^2}{20\ \Omega} = 1.25\ \text{W}$$

which happily does not check.

PS 3-6 Assume in Fig. PS 3-6*a* that $e_A = 24 \sin(\omega t + 30°)$ volts, $i_B = 2 \sin(\omega t - 60°)$ amps, $R_1 = 3$ ohms, and $R_2 = 6$ ohms. Determine P_2, the power in R_2.

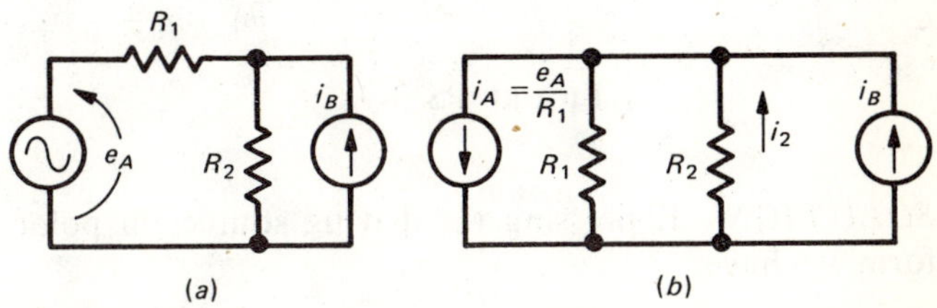

FIGURE PS 3-6

SOLUTION Convert e_A and R_1 to their Norton's equivalent as shown in Fig. PS 3-6*b*. Thus in polar form

$$i_{A\text{(pk)}} = \frac{e_A}{R_1} = \frac{24\underline{/30°}\ \text{V}}{3\ \Omega} = 8\underline{/30°}\ \text{A}$$

Converting to rms value

$$I_a = 0.707\,(8\underline{/30°}\ \text{A}) = 5.66\underline{/30°}\ \text{A}$$

Similarly

$$I_b = 0.707\,(2\underline{/-60°}\ \text{A}) = 1.414\underline{/-60°}\ \text{A}$$

Using the current divider relationship and the superposition theorem developed in the preceding text, we may write[2]

$$\mathbf{I}_2 = \frac{R_1}{R_1 + R_2}\mathbf{I}_a - \frac{R_1}{R_1 + R_2}\mathbf{I}_b = \frac{R_1}{R_1 + R_2}(\mathbf{I}_a - \mathbf{I}_b)$$

$$= \frac{3\ \Omega}{3\ \Omega + 6\ \Omega}(5.66\underline{/30°}\ \text{A} - 1.414\underline{/-60°}\ \text{A})$$

[2] Phillip Cutler, "DC Circuit Analysis with Illustrative Problems," McGraw-Hill Book Company, New York, 1968.

$$= \frac{1}{3}[(4.9 + j2.83) - (0.707 - j1.225)]$$

$$= \frac{1}{3}(4.193 + j4.055)$$

or

$$\mathbf{I}_2 = (1.398 + j1.352)\ \text{A} = 1.94\underline{/44.03^\circ}\ \text{A}$$

The magnitude of this rms current is 1.94 A, and hence

$$P_2 = I_2{}^2R_2 = (1.94\ \text{A})^2(6\ \Omega) = 22.7\ \text{W}$$

PS 3-7 Use Thévenin's theorem to determine the power, P_2, dissipated in R_2. This problem is identical to PS 3-6 except that we are using another approach.

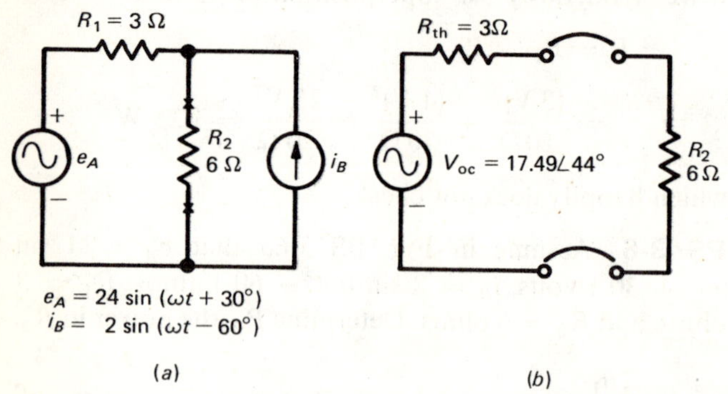

FIGURE PS 3-7

SOLUTION Expressing the driving sources in polar form, we have

$$E_{A(\text{pk})} = 24\underline{/30^\circ} \qquad I_{B(\text{pk})} = 2\underline{/-60^\circ}$$

and converting to rms values

$$E_a = 16.97\underline{/30^\circ} \qquad I_b = 1.414\underline{/-60^\circ}$$

The open-circuit voltage V_{oc} that R_2 looks back into is

$$V_{oc} = 16.97\underline{/30^\circ} - (3\ \Omega)(1.414\underline{/-60^\circ})$$

$$= 16.97\underline{/30^\circ} - 4.243\underline{/-60^\circ}$$

$$= (14.69 + j8.485) - (2.123 - j3.674)$$

$$V_{oc} = 12.58 + j12.16 = 17.49\underline{/44^\circ}\ \text{V}$$

By Ohm's law,

$$I = \frac{17.49\underline{/44^\circ}\ \text{V}}{3\ \Omega + 6\ \Omega} = 1.94\underline{/44^\circ}\ \text{A}$$

The power in P_2 is determined by the magnitude of the rms current as follows:

$$P_2 = I_{\text{rms}}{}^2R = I^2R$$

$$P_2 = (1.94)^2 6 = 22.7\ \text{W}$$

which checks the result obtained in PS 3-6.

PS 3-8 Using the short table of identities in Chap. 1, prove that $\sin x \sin y = \frac{1}{2}\cos(x - y) - \frac{1}{2}\cos(x + y)$.

SOLUTION From Eq. (4) in the table of identities, repeated here as (1) and (2),

1) $\cos(x - y) = \cos x \cos y + \sin x \sin y$

2) $\cos(x + y) = \cos x \cos y - \sin x \sin y$

Subtracting (2) from (1) yields, after some simplification,

3) $\sin x \sin y = \frac{1}{2}\cos(x - y) - \frac{1}{2}\cos(x + y)$

which is the desired result.

PS 3-9 Solve the circuit of Fig. PS 3-9 for P_2, the power dissipated in R_2. Use Eq. (3-9), which is repeated here for convenience, and compare this result with that obtained in identical problem PS 3-7.

$$P = EI\cos(\phi_e - \phi_i) \tag{3-9}$$

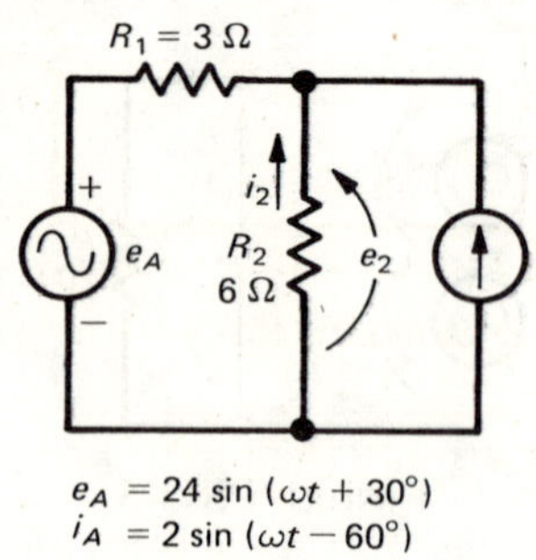

FIGURE PS 3-9

SOLUTION Let us first express e_A and i_B in rms polar form. Thus

1) $$\mathbf{E}_a = (0.7071)\ 24\underline{/30^\circ} = 16.97\underline{/30^\circ}\ \text{V}$$

2) $$\mathbf{I}_b = (0.7071)\ 2\underline{/-60^\circ} = 1.414\underline{/-60^\circ}\ \text{A}$$

Applying the superposition theorem to Fig. PS 3-9, we may express the voltage, e_2, due to $\mathbf{E}_a$ and $\mathbf{I}_b$ as

3) $$\mathbf{E}_2 = \frac{R_2}{R_1 + R_2}\mathbf{E}_a - \frac{R_1R_2}{R_1 + R_2}\mathbf{I}_b$$

$$= \frac{R_2}{R_1 + R_2}(\mathbf{E}_a - R_1\mathbf{I}_b)$$

And the current $\mathbf{I}_2$ as

4) $$\mathbf{I}_2 = \frac{1}{R_1 + R_2}\mathbf{E}_a - \frac{R_1}{R_1 + R_2}\mathbf{I}_b$$

$$= \frac{1}{R_1 + R_2}(\mathbf{E}_a - R_1\mathbf{I}_b)$$

Evaluating (3), we have

$$\mathbf{E}_2 = \frac{6}{9}[16.97\underline{/30^\circ} - 3(1.414\underline{/-60^\circ})]$$

$$= \frac{6}{9}(12.58 + j12.16)$$

5) $$\mathbf{E}_2 = \frac{6}{9}(17.49\underline{/44.03^\circ}) = 11.66\underline{/44^\circ}\ \text{V}$$

Evaluating (4), we have

6) $\mathbf{I}_2 = \frac{1}{9}(17.49\angle 44.03°) = 1.943\angle 44°$ A

The magnitudes of $\mathbf{E}_2$ and $\mathbf{I}_2$ in Eq. (6) are rms values, and both ϕ_e and ϕ_i, the voltage and current phase angles, are 44°. Hence, substituting into Eq. (3-9), we obtain

7) $P_2 = (11.66)(1.943)\cos(44° - 44°) = 22.7$ W

which verifies the solution previously obtained in problem PS 3-7.

PS 3-10 Determine the power dissipated in resistor R in Fig. PS 3-10.

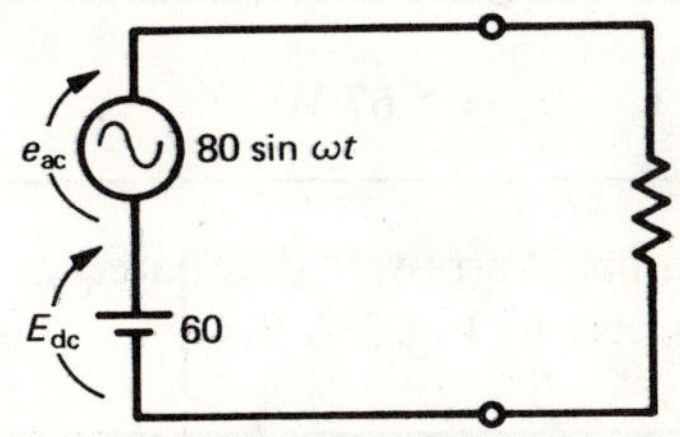

FIGURE PS 3-10

SOLUTION To apply Eq. (3-10), we first note that

$$E_{dc} = 60 \text{ V} \qquad E_{ac(rms)} = 0.7071\,(80 \text{ V}) = 56.57 \text{ V}$$

$$I_{dc} = \frac{60 \text{ V}}{12\,\Omega} = 5 \text{ A} \qquad I_{ac(rms)} \frac{56.57 \text{ V}}{12\,\Omega} = 4.714 \text{ A}$$

$$\phi_e = 0 \qquad \phi_i = 0$$

Equation (3-10) indicates that

$$P = E_{dc}I_{dc} + E_{ac(rms)}I_{ac(rms)}\cos(\phi_e - \phi_i)$$
$$= (60)(5) + (56.57)(4.714)\cos 0° = 300 + 266.7$$
$$P = 566.7 \text{ W}$$

PS 3-11 Determine with the aid of Eq. (3-12) the power dissipated in R in the circuit of Fig. PS 3-10. Check your solution against that previously obtained in problem PS 3-10.

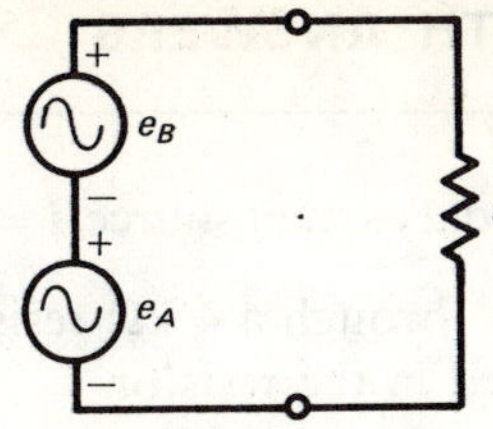

FIGURE PS 3-12

SOLUTION The effective equivalent voltage impressed across R is

$$E_{eq(rms)} = \sqrt{E_{dc}{}^2{}_{(rms)} + E_{ac(rms)}{}^2}$$
$$= \sqrt{60^2 + [0.7071\,(80)]^2} = 82.46 \text{ V}$$

The power in R is then

$$P = \frac{E_{eq}{}^2{}_{(rms)}}{R}$$
$$= \frac{(82.46 \text{ V})^2}{12\,\Omega} = 566.7 \text{ W}$$

which verifies the previous result.

PS 3-12 If, in Fig. PS 3-12, $e_A = 10 \sin \omega t$, $e_B = 10 \sin(\omega t + \pi)$, and $R = 5$ ohms, determine the power in R.

SOLUTION If you blindly plunge into the problem and use Eq. (3-14), there results

$$E_{eq(rms)} = \sqrt{E_a{}^2{}_{(rms)} + E_b{}^2{}_{(rms)}}$$
$$= \sqrt{\left(\frac{10}{\sqrt{2}}\right)^2 + \left(\frac{10}{\sqrt{2}}\right)^2} = 10 \text{ V}$$

and

$$P = \frac{(10 \text{ V})^2}{5\,\Omega} = 20 \text{ W}$$

This is an erroneous solution because Eq. (3-14) must *not* be used for similar frequency sources. A little thought will reveal that e_A and e_B are equal and opposite at all instants of time and hence that the actual equivalent voltage is always zero.

PROBLEMS WITH ANSWERS

PA 3-1 A sinusoidal current source, $i = 8 \cos\left(3t - \frac{\pi}{6}\right)$ amp drives current through a 4 kΩ resistor. Determine the power dissipated in the resistor.

ANSWER $P = 128$ kW

PA 3-2 Determine the power dissipated in the box of Fig. PA 3-2 if the terminal voltage and current are respectively given by $e = 5 \sin(\omega t + 65°)$ V, $i = 3 \sin(\omega t + 40°)$ A.

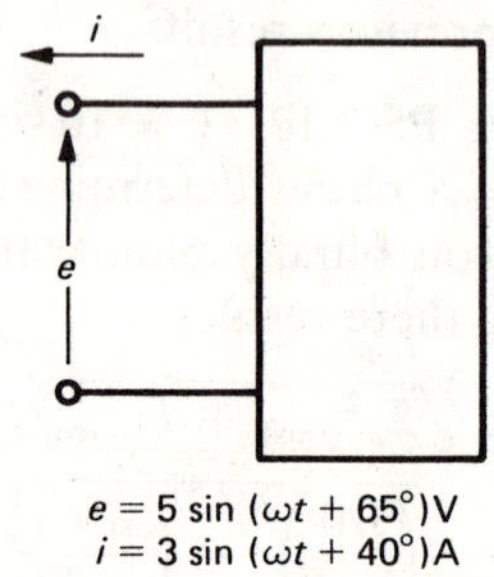

ANSWER $P = 6.797$ W

PA 3-3 Determine the power dissipated in the 4-ohm resistor in the circuit of Fig. PA 3-3.

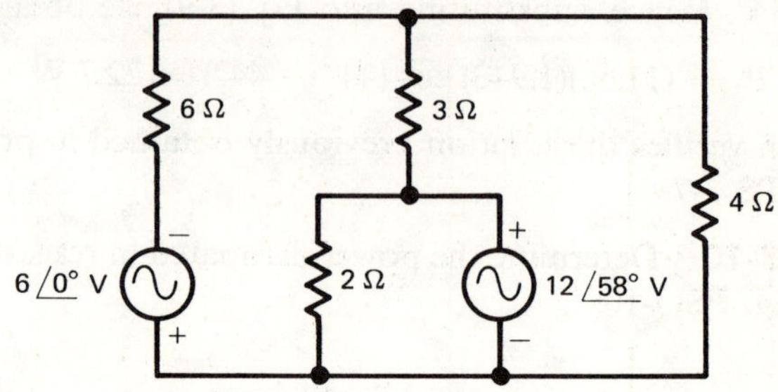

ANSWER $P = 5.67$ W

PA 3-4 Determine the power dissipated in the 1-ohm resistor in the circuit of Fig. PA 3-4.

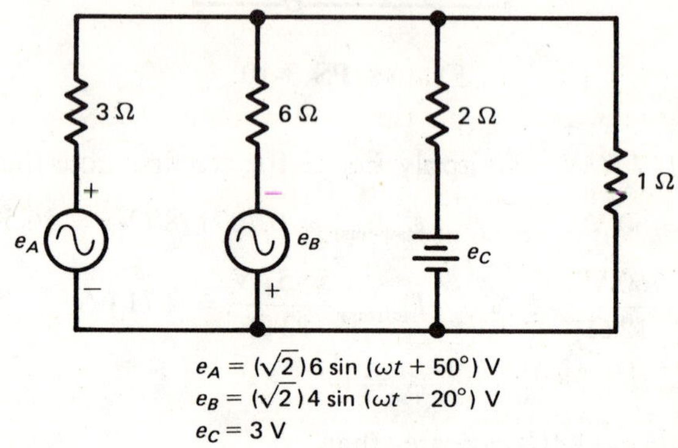

ANSWER $P = 1.45$ W

P 3-1 Insofar as the load R_L and V_{CC} supply are concerned, the transistor amplifier of Fig. P 3-1*a* is equivalent to the circuit of Fig. P 3-1*b*. The current source i_C will in general consist of a dc and ac component so that i_C is of the form

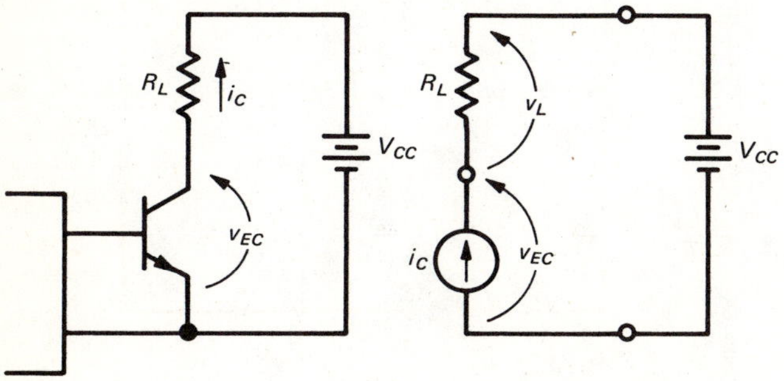

(*a*) (*b*)

$$i_C = I_{CQ} + I_{c(\mathrm{pk})} \sin \omega t \qquad \text{(P 3-1)}$$

where I_{CQ} (the quiescent collector current) is the dc component and $I_{c(\mathrm{pk})}$ is the peak value of the ac (signal) component.

Assuming $V_{CC} = 45$ volts, determine R_L so that the current source dissipates 15 watts (P_{ECQ}) when the ac or signal component of $i_C = 0$ and $v_{EC} = V_{ECQ} = V_{CC}/2$.

P 3-2 Using the information in P 3-1 determine the power (P_{CC}) put out by the V_{CC} supply in Fig. P 3-1*b* under the quiescent conditions of $V_{ECQ} = V_{CC}/2 = 22.5$ V.

P 3-3 What is the quiescent power dissipation (P_{LQ}) in R_L?

P 3-4 If a signal (ac) component of collector current i_C manifests itself such that

$$i_C = I_{CQ} + I_{c(\mathrm{pk})} \sin \omega t$$

where $I_{c(\mathrm{pk})} = I_{CQ}$, determine the power put out by the V_{CC} supply, the power P_L in the load, and the power dissipated in the current source. Also determine the conversion efficiency $\eta = P_{L(\mathrm{ac})}/P_{CC}$, which compares the ac or signal power in the load to the power put out by the V_{CC} supply.

P 3-5 Determine the power dissipated in the 6-ohm resistor in Fig. P 3-5 if $e_A = 12\angle 160^\circ$ volts, and $i_B = 6\angle 70^\circ$ amps.

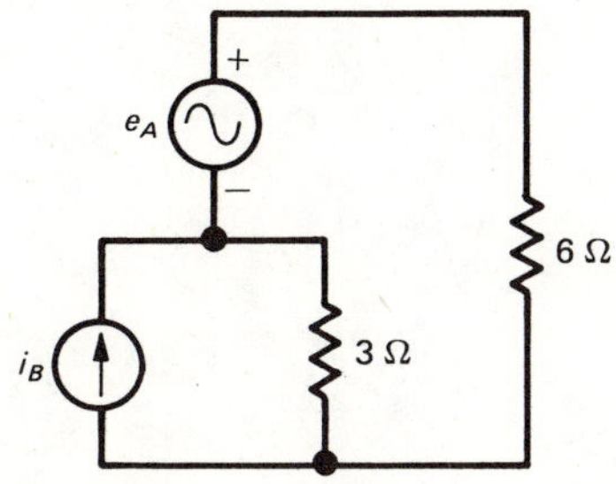

P 3-6 Repeat problem P 3-5 with $e_A = 12\sqrt{2} \sin \omega t$ volts and $i_B = I_B = 6\sqrt{2}$ amps (a dc current source).

P 3-7 Repeat problem P 3-5 with $e_A = 12\sqrt{2} \sin \omega t$ volts and $i_B = 6\sqrt{2} \sin 2\omega t$ amps.

P 3-8 One thousand feet of #22 wire has a dc resistance of 16.1 ohms. What is the apparent resistance of 20 ft of this wire at 12 MHz?

4 capacitance

Resistance, you will recall, is that property of an electric circuit which opposes the flow of current. We identify this circuit parameter with the symbol R (or its reciprocal which is conductance G) and measure it in units of ohms.

4-1 Capacitance

Now we shall investigate another circuit parameter called capacitance, which is given the symbol C and measured in units called farads.[1] Capacitance is that property of a circuit which causes it to oppose a *change* in voltage. To illustrate the effect of capacitance, let us first consider Fig. 4-1*a*. Assume the switch SW is initially

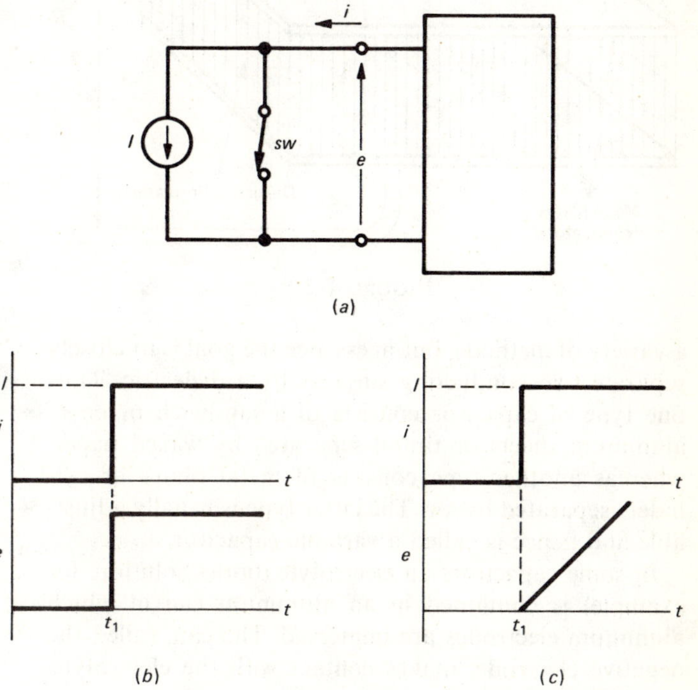

FIGURE 4-1

closed. Then, assuming there are no voltage or current sources in the box, we have $i = 0$ and $e = 0$. Now, let the switch be opened at time t, as shown in Fig. 4-1*b*. Opening the switch forces the terminal current i to equal the current I due to the current source. If the circuitry in the box is or appears to be a pure resistance, we should find that the step (jump) in the current i is instantaneously followed by a step in the voltage e as shown in Fig. 4-1*b*. Should the box, however, exhibit the property of pure capacitance, we should then find that the step in the current does *not* cause a corresponding step in voltage but instead produces a ramp of voltage as in Fig. 4-1*c*. Note that capacitance prevented

[1] The farad is a rather large amount of capacitance and most practical circuits involve microfarads (μF) or picofarads ($\mu\mu$F or pF) where 1 $\mu F = 10^{-6}$ F and 1 $\mu\mu F = 10^{-12}$ F = 1 pF.

or opposed a corresponding *change* of voltage. To reiterate—capacitance is that property of a circuit which causes it to oppose a change in voltage.

Physically, a capacitor consists of two conductors separated by some insulating medium called the dielectric, as shown in Fig. 4-2. Capacitors may be constructed by

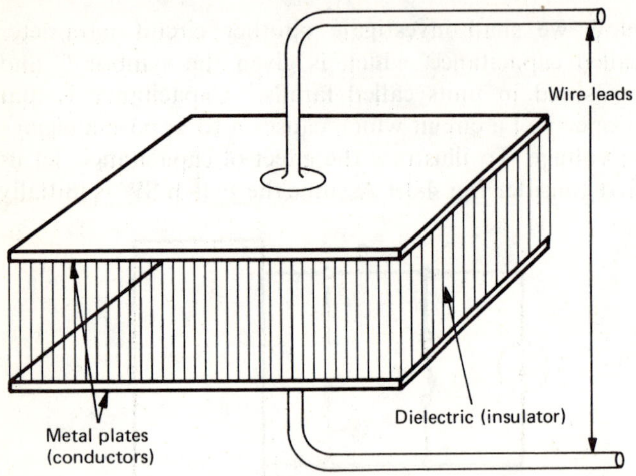

FIGURE 4-2

a variety of methods, but in essence the goal is to closely separate two conducting surfaces by a dielectric. Thus one type of capacitor consists of a sandwich made of aluminum sheets or tinfoil separated by waxed paper, whereas another type consists of metal plates or cylinders separated by air. The latter type is usually adjustable and hence is called a variable capacitor.

In some capacitors an electrolyte (borax solution, for example) is contained in an aluminum can in which aluminum electrodes are immersed. The can, called the negative electrode, makes contact with the electrolyte, which is actually one of the conductors. The immersed electrodes are connected together, and they form the other conductor. As such there is no dielectric. However, if a dc voltage is impressed between the can (negative electrode) and the immersed electrodes (positive electrode), a molecule-thin dielectric film of aluminum oxide is formed at the surface of the positive electrode. Because this dielectric film is polarity-sensitive, it is necessary to observe the polarity of the voltage applied to these electrolytic-type capacitors. Electrolytic capacitors provide huge capacitance in a relatively small volume.

Some schematic representations of capacitors are shown in Fig. 4-3.

Because of the narrow dielectric region, it is necessary to limit the voltage impressed across capacitors to less than that which will cause rupture (voltage breakdown) of the dielectric. For this reason capacitors carry a working voltage rating which should not be exceeded under worst-case conditions of temperature, humidity, etc.

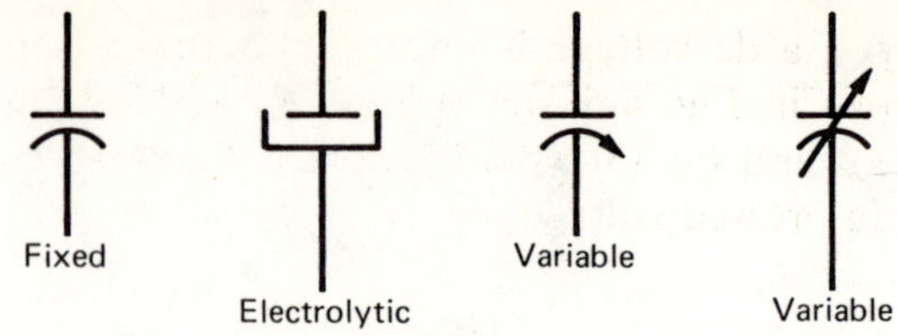

FIGURE 4-3

4-2 Capacitor charging and discharging

To gain some insight into the properties of capacitance, consider Fig. 4-4. If we open the switch at time zero (t_0), the current source starts forcing electrons into the lower

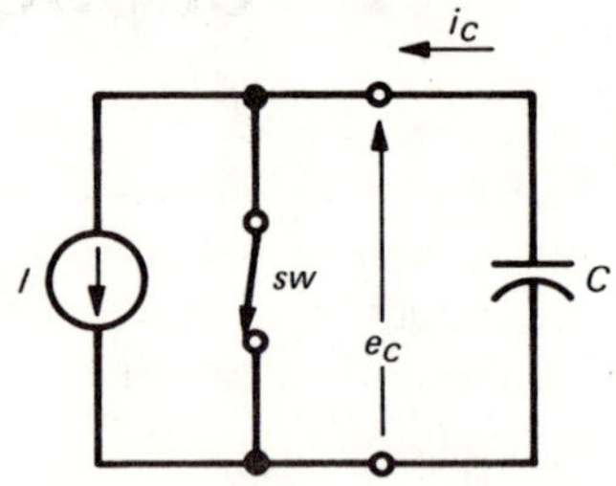

FIGURE 4-4

plate of the capacitor. Each electron forced into the lower plate forces a corresponding electron out of the upper plate (like charges repel) and back into the current source. Hence, electrons accumulate on the lower plate relative to the upper plate and an apparent current actually flows in the circuit despite the insulating medium between the plates. The capacitor under these conditions is said to be charging.

The charge Q is measured in coulombs and is given by

1) $$Q = i_C\,(\Delta t) = I\,(\Delta t) \qquad (4\text{-}1)$$

where Δt is the duration of the charging interval under consideration. Now, whenever a surplus of charge carriers exists at one point relative to some other point, there also exists a potential difference or voltage between these same two points. The magnitude of this voltage is related to the charge by

2) $$E_C = \frac{Q}{C} \qquad (4\text{-}2)$$

where Q = charge in coulombs
C = capacitance in farads
E_C = voltage in volts

Alternatively we may write

3) $$C = \frac{Q}{E_C} \qquad (4\text{-}3)$$

which indicates that capacitance is a measure of how much charge you can store for each volt developed across the capacitor in the circuit of Fig. 4-4*a*. Conversely, if a dc voltage is impressed across a capacitor as shown in Fig. 4-5, the voltage E_C is determined by $E_C = E_{bb}$ and Eq. (3) would yield the charge stored due to the impressed voltage.

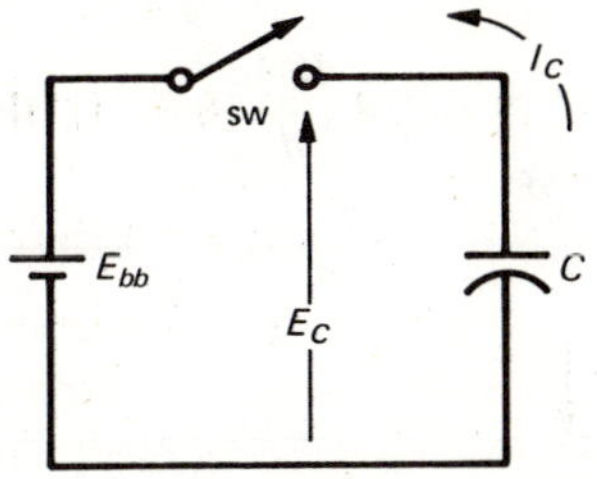

FIGURE 4-5

Equation (4-3) in effect defines the unit of capacitance. Specifically, if 1 coulomb of charge stored in a capacitor produces 1 volt across· its terminals, $C = 1$ farad. Alternatively, we might say that a capacitor has a capacitance of 1 farad if impressing 1 volt across its terminals forces 1 coulomb of charge into it.

A more practical way of regarding capacitance is derived as follows. Recall from Eq. (4-2) that

1) $$Q = CE_C \tag{4-4}$$

Now consider that a voltage change ΔE_C across the capacitor in a time interval Δt causes a corresponding change in charge ΔQ. Thus the average change in charge due to the average change in voltage is

2) $$\frac{\Delta Q}{\Delta t} = C\frac{\Delta E_C}{\Delta t}$$

But you may also recall that

3) $$I = \frac{\Delta Q}{\Delta t}$$

Substituting (3) into (2) yields

4) $$I_C = C\frac{\Delta E_C}{\Delta t} \tag{4-5a}$$

or in terms of instantaneous rather than average values,

5) $$i_C = C\frac{de_C}{dt} \tag{4-5b}$$

Equation (4-5) may be interpreted to mean that if a voltage variation of 1 volt/sec across a capacitor causes 1 amp of current to flow, that capacitor exhibits 1 farad of capacitance.

Now work through problems PS 4-1, PS 4-2, and PS 4-3.

Let us again examine Fig. 4-5 in more detail. If we assume the capacitor is initially uncharged, the voltage across it is zero. Now, since a capacitor opposes a change of voltage across its terminals, we may in effect state that *the voltage across a capacitor cannot change instantaneously*.[2] Yet in Fig. 4-5, closing the switch surely would seem to force E_C to instantaneously assume the value of E_{bb}. Actually it does not in a practical case due to the fact that there is some internal resistance in the voltage source as well as resistance in the connecting wires, so that the circuit is better represented by Fig. 4-6*a*. This resistance limits the initial charging current

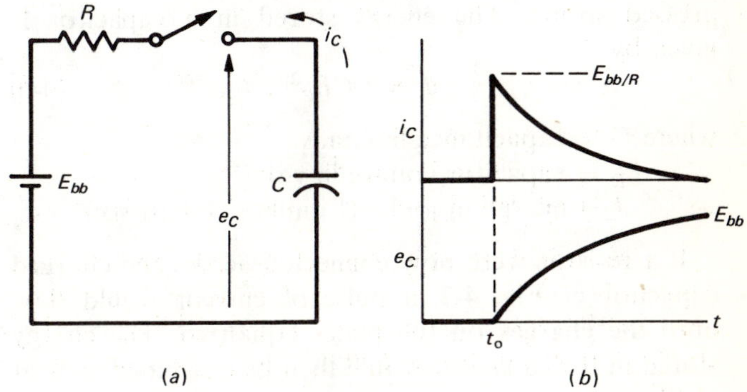

FIGURE 4-6

into the capacitor. Thus at time zero (switch closes) the capacitor voltage is zero and

$$i_C = \frac{E_{bb} - e_C}{R} = \frac{E_{bb} - 0}{R} = \frac{E_{bb}}{R}$$

As time goes by, charge accumulates and e_C builds up, which bucks E_{bb} and thereby reduces i_C. After a sufficient time interval, e_C for all practical purposes equals E_{bb}, so that the capacitor is fully charged and $i_C = 0$, as shown in Fig. 4-6*b*. Note that once this transient interval is ended, no further current flows. Hence dc current cannot flow through a capacitor.

If the capacitor is now disconnected from E_{bb} as shown in Fig. 4-7, it will still hold the accumulated charge and hence maintain E_{bb} volts across it unless some conducting or leakage path is provided for the charge on the plates to equalize. In that sense a charged capacitor may be considered as a battery.

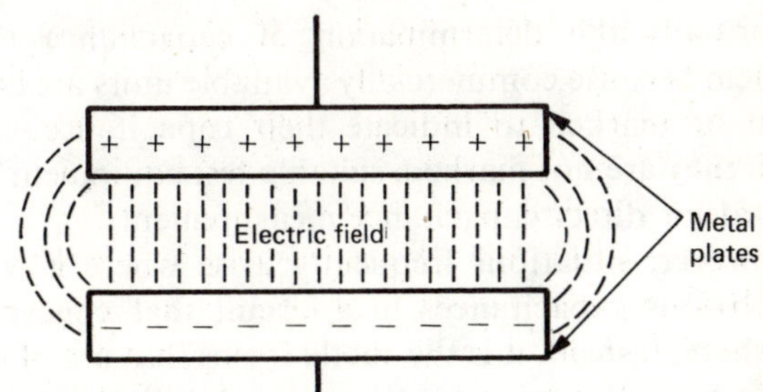

FIGURE 4-7

[2]An exception to this is the theoretical case in which an ideal (zero resistance) voltage source is impressed across a capacitor or the case in which one charged capacitor is switched across another.

4-3 Capacitor-stored energy

We may, in fact, reason that since E_{bb} did move some charge into C in Fig. 4-6*a*, some work was done and hence energy was expended. This energy, however, is not dissipated as heat loss since there is no resistance. Instead, it is stored as stress in the electric field between the charged plates in a manner analogous to a compressed spring. The energy stored in a capacitor is given by

$$J = \tfrac{1}{2} C E_C^2 \qquad (4\text{-}6)$$

where C = capacitance in farads
E_C = capacitor voltage in volts
J = energy in joules (1 joule = 1 watt-sec)

If a resistor were now connected across the charged capacitor of Fig. 4-7, a pulse of current would flow until the charges on the plates equalized. The energy stored in the capacitor would then be dissipated as heat in the resistor.

Now work through problem PS 4-4.

4-4 Capacitor geometries

As previously mentioned, capacitance manifests itself whenever two conductors are separated by an insulating medium. The exact value of capacitance depends upon factors such as the geometry of the conductors, the spacing between them, and the dielectric constant (K) of the insulating medium the conductors are immersed in. Typical values of K are listed in the following table.

Table 4-1 Typical values of dielectric constant (K)

Material	K
Vacuum	1
Air	1
Glass	4–7
Mica	7
Plastic (soft)	2–3
Plastic (hard)	4–12

Normally the determination of capacitance is no problem because commercially available units are color-coded or marked to indicate their capacitance value. And if they are not marked, suitable test equipment may be used for direct capacitance measurement.

However, situations frequently arise where it is not the obvious capacitances in a circuit that concern us but where, instead, it is the subtle forms that are of most significance. For example, a patient hospitalized in an intensive care unit may be killed by stray capacitances between components used to directly measure heart activity. Or it may be the stray capacitance between the input and output circuits of a high-gain amplifier that causes distorted frequency response or outright oscillations.

Although these stray capacitances can be measured with suitable test equipment, it is worthwhile, when designing or analyzing circuits, to acquire some familiarity with formulas for predicting them.[3]

For uniform parallel plates, as shown in Fig. 4-8,

$$C \cong 0.224 \frac{KA}{d}(n - 1)\ \text{pF} \qquad (4\text{-}7)$$

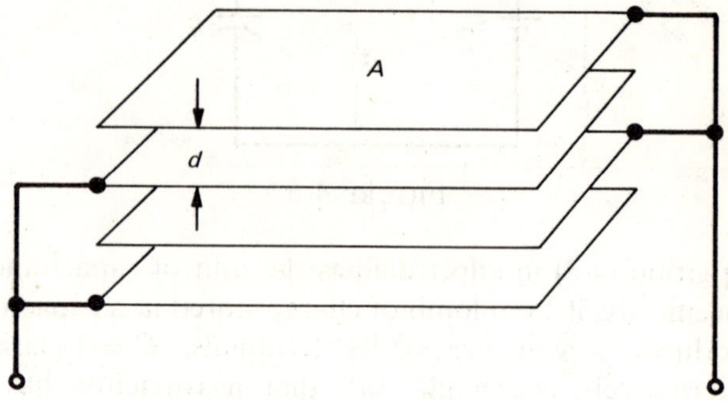

FIGURE 4-8

where C = capacitance in pF
A = area of one plate
n = number of plates
d = spacing between plates

Now work through problem PS 4-5.

For an isolated wire conductor (see Fig. 4-9) running

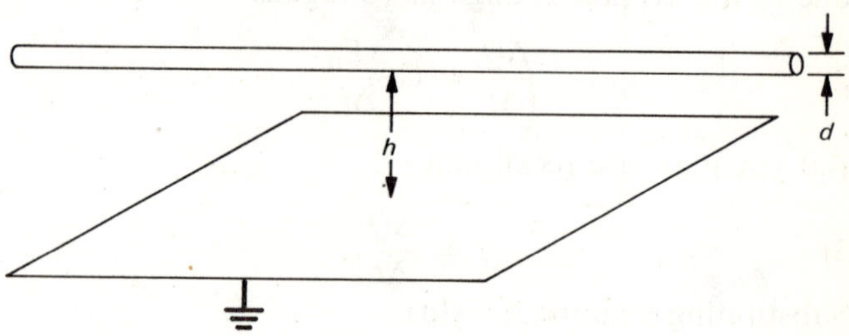

FIGURE 4-9

over a metal chassis (or other ground plane) whose dimensions are large relative to the wire conductor, the capacitance per foot is given by

$$C = \frac{7.35}{\log_{10}\left[\frac{2h}{d}\left(1 + \sqrt{1 - \frac{1}{(2h/d)^2}}\right)\right]}\ \frac{\text{pF}}{\text{ft}} \qquad (4\text{-}8)$$

where C = capacitance in pF/ft
h = height to center of wire above ground plane
d = diameter of wire

Now work through PS 4-6.

[3]For an excellent discussion of capacitors, their geometries, and a table of dielectric constants, see F. E. Terman, "Radio Engineer's Handbook," McGraw-Hill Book Company, New York, 1943.

The capacitance per foot between two parallel wire conductors above a ground plane and whose diameter is small compared to their length and spacing, as shown in Fig. 4-10, is given by Eq. (4-9).

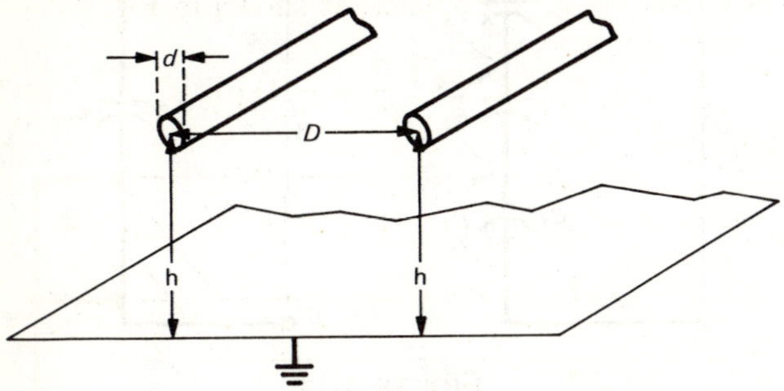

FIGURE 4-10

$$C = \frac{3.68}{\log_{10}\dfrac{2D}{d}\dfrac{1}{\sqrt{1+\left(\dfrac{D}{2h}\right)^2}}}\ \frac{\text{pF}}{\text{ft}} \qquad (4\text{-}9)$$

where C = capacitance in pF/ft
D = distance between wires
d = diameter of wire
h = height of wires above ground plane

For a coaxial (concentric) line, as shown in Fig. 4-11,

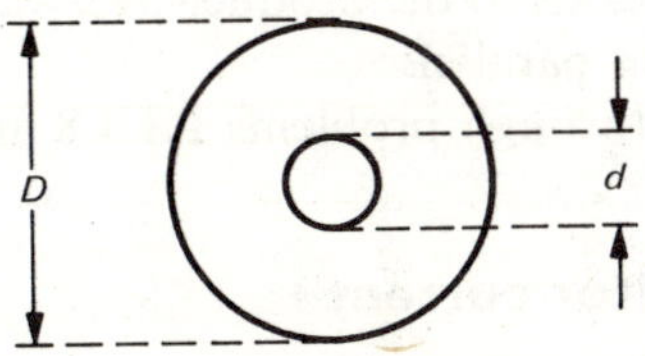

FIGURE 4-11

the capacitance per foot is given by

$$C = \frac{7.35K}{\log_{10}\left(\dfrac{D}{d}\right)}\ \frac{\text{pF}}{\text{ft}} \qquad (4\text{-}10)$$

where C = capacitance in pF/ft
K = dielectric constant of medium between the inner and outer conductors
D = inside diameter of outer conductor (shield)
d = diameter of inner conductor

4-5 Capacitor working voltage

If the voltage applied to a capacitor is continuously increased, a point is reached where the electric field intensity between the plates or conductors is sufficient to rupture the dielectric. When this occurs, a spark or miniature lightning bolt jumps between the conductors. Consequently capacitors carry a working voltage rating which should not be exceeded. This working voltage usually decreases with frequency when sinusoidal or other periodic waveforms are applied.

The voltage at which breakdown occurs also depends upon temperature, humidity, and dielectric used as well as the thickness of the dielectric. Unfortunately, trying to increase the working voltage by using a thicker dielectric results in a smaller capacitance, since the plates are spread further apart. The working voltage rating should equal or exceed the *peak* value of the waveform applied.

Electrolytic capacitors also have working voltage ratings, but these must be equal to or greater than the peak dc voltage applied. Electrolytics cannot normally be used with an ac voltage, since the polarity reversal depolarizes the dielectric film. If one must utilize the large capacitance inherent in electrolytic capacitors with an ac voltage, it is possible to place two of them back to back, as shown in Fig. 4-12. In this manner, as the dielectric film in one depolarizes, the other polarizes.

FIGURE 4-12

4-6 Capacitance temperature coefficient

Capacitance, like most circuit parameters, is temperature-sensitive, and it depends upon the materials used and manner in which the capacitor is constructed. For specific information, manufacturers' data sheets should be consulted.

A particularly stable and low-temperature coefficient of capacitance is obtained by using mica as a dielectric, with a metal film deposited on each side for the plates.

In some cases the use of capacitors with positive temperature coefficients and others with negative temperature coefficients results in equivalent capacitors with some desired or almost negligible temperature coefficient.[4]

4-7 Capacitors in parallel

Frequently capacitors are connected in parallel, as shown in the left-hand side of Fig. 4-13, and it is desired

[4]Capacitance increases (decreases) with temperature when the temperature coefficient is positive (negative).

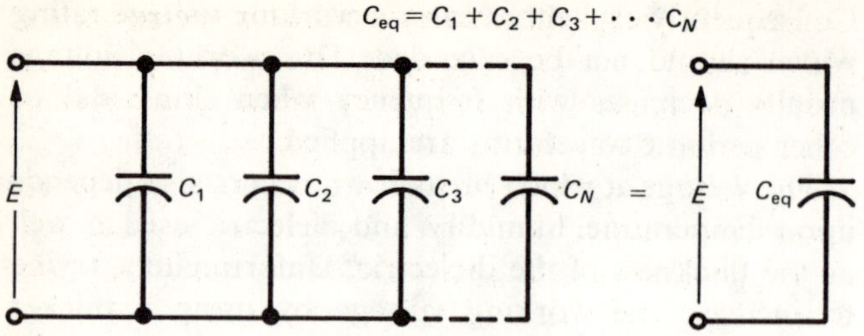

FIGURE 4-13

to determine the equivalent capacity of the combination as shown on the right. Intuitively we might reason that connecting capacitors in parallel is simply forming a capacitor whose plate area is the sum of the individual plate areas and that hence the equivalent capacity C_{eq} would be the sum of the individual capacities.

For a more formal proof, we might write for Fig. 4-13,

1) $$Q_{eq} = Q_1 + Q_2 + Q_3 + \cdots + Q_N$$

which simply states that the total charge in C_{eq} must equal the sum of the charges stored on the individual capacitors. Since in general

2) $$Q = CE$$

we may rewrite (1) as

3) $$C_{eq}E = C_1E + C_2E + C_3E + \cdots + C_NE$$

Dividing both sides by E yields

4) $$C_{eq} = C_1 + C_2 + C_3 + \cdots + C_N \qquad (4\text{-}11)$$

Note that capacitors in parallel combine in a manner analogous to resistors in series.

Now work through PS 4-7.

4-8 Capacitors in series

It is also important to know what the equivalent capacity C_{eq} of capacitors connected in series as shown in Fig. 4-14 is equal to. To determine C_{eq}, we might begin with

1) $$E = E_1 + E_2 + E_3 + \cdots + E_N$$

Hence

2) $$\frac{Q_{eq}}{C_{eq}} = \frac{Q_1}{C_1} + \frac{Q_2}{C_2} + \frac{Q_3}{C_3} + \cdots + \frac{Q_N}{C_N}$$

Now the current I which flows through each capacitor is the same in this series circuit. Hence, over any time interval t, the charge on any capacitor is the same and equal to $I \times t$. Thus we may rewrite (2) as

3) $$\frac{It}{C_{eq}} = \frac{It}{C_1} + \frac{It}{C_2} + \frac{It}{C_3} + \cdots + \frac{It}{C_N}$$

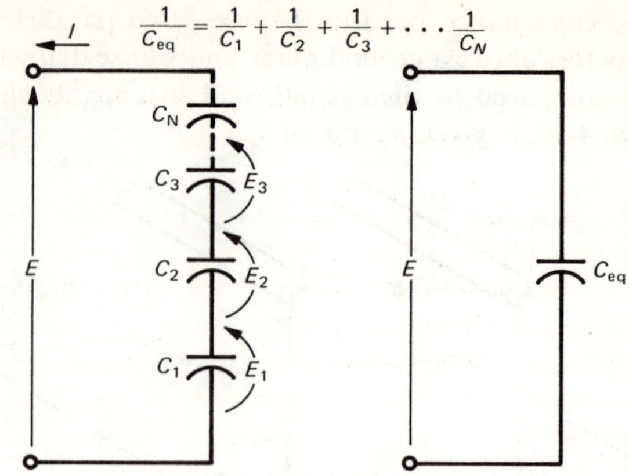

FIGURE 4-14

Dividing both sides by It yields

4) $$\frac{1}{C_{eq}} = \frac{1}{C_1} + \frac{1}{C_2} + \frac{1}{C_3} + \cdots + \frac{1}{C_N} \qquad (4\text{-}12)$$

Note that capacitors in series combine in a manner analogous to resistors in parallel.

As an exercise the reader may prove that, taking any two capacitors in series, their equivalent capacitance is

5) $$C_{eq} = \frac{C_1C_2}{C_1 + C_2} \qquad (4\text{-}13)$$

which is analogous to the product-over-sum formula for two resistors in parallel.

Now work through problems PS 4-8 and PS 4-9.

4-9 Capacitor current

You will recall that the charge on a capacitor is given by

1) $$Q = CE$$

Now if the voltage changes by some amount ΔE in an interval Δt, the corresponding change in charge is ΔQ. Thus

2) $$\Delta Q = C(\Delta E)$$

The average change in charge due to the average voltage change ΔE in this interval Δt is expressed by

3) $$\frac{\Delta Q}{\Delta t} = C\frac{\Delta E}{\Delta t}$$

But $\Delta Q/\Delta t$ is the average current that flowed through the capacitor in order to permit the charge to change. Thus we may write

4) $$I = \frac{\Delta Q}{\Delta t} = C\frac{\Delta E}{\Delta t} \qquad (4\text{-}14)$$

Now if we consider instantaneous rather than average values, we use the following notation

5) $$i = C\frac{de}{dt} \tag{4-15}$$

to indicate capacitor current. Equation (4-15) is most significant in determining the effects of capacitors in circuits.

Equation (5) relates capacitor current to a *change* in capacitor voltage. Clearly, if a dc voltage is impressed across a capacitor, there is no voltage change and hence i equals zero. Thus a capacitor does not permit dc current flow. Capacitors used to block a current path to dc are called blocking capacitors, whereas capacitors used to permit current flow due to fast voltage changes are called coupling or bypass capacitors, depending on their function.

4-10 The capacitor with sinusoidal excitation

Suppose we have a resistor driven by a sinusoidal voltage source, as shown in Fig. 4-15*a*. The current is given by

$$i = \frac{e}{R}$$

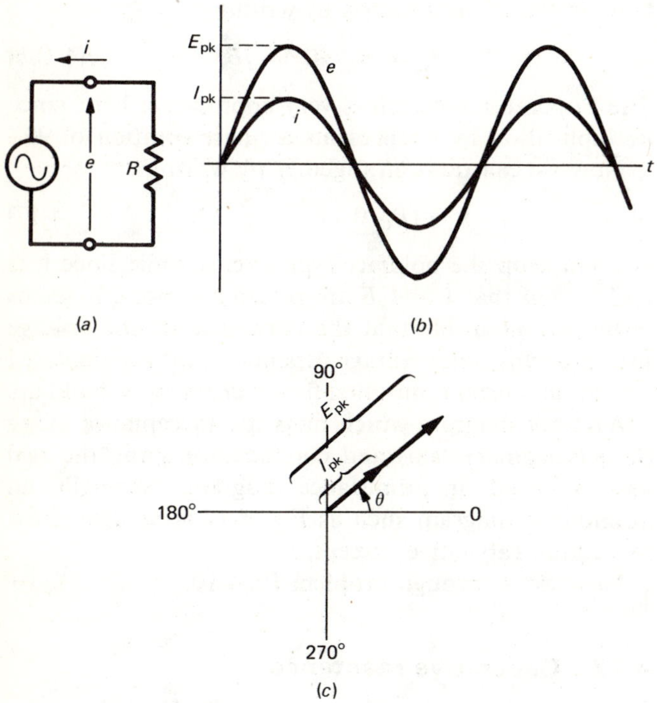

FIGURE 4-15

and the resultant e and i waveforms would appear as shown in Fig. 4-15*b*. Note that i in (1) depends only upon the value of e and not its rate of change and that the e and i waveforms of Fig. 4-15*b* are in phase. The corresponding vector diagram is shown in Fig. 4-15*c*.

Now, let us substitute a capacitor for the resistor so that we have, in effect, Fig. 4-16*a*. Here we are trying to determine the relation between current and voltage in a purely capacitive circuit. The voltage waveform across C must surely be the same as that of the driving source (the sinusoidal generator), and hence it will appear as the e waveform in Fig. 4-16*b*. What, however, does the i

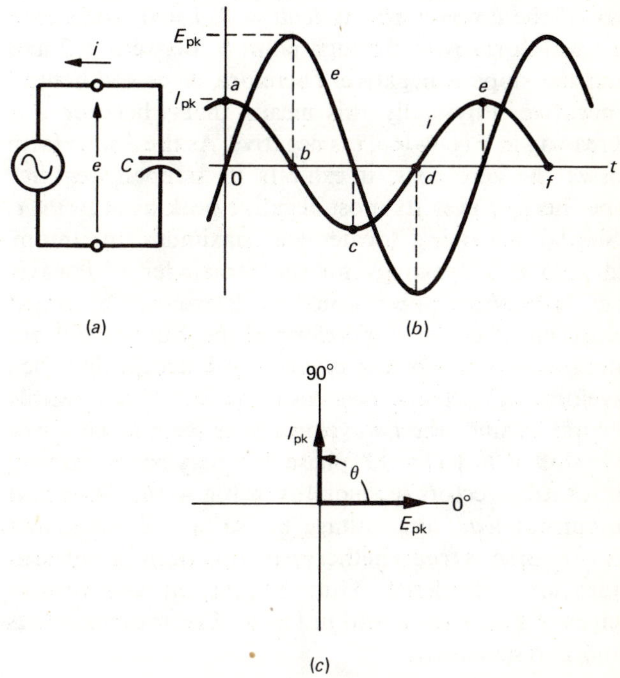

FIGURE 4-16

waveform look like? To determine this, let us go back to fundamentals [Eq. (4-15)] and recall that the instantaneous current through a capacitor is given by

1) $$i = C\frac{de}{dt}$$

where de/dt is the instantaneous rate of change of voltage with respect to time. In other words, the faster e varies with respect to time (the steeper the e-versus-t curve is), the larger the instantaneous current. Note that it is the magnitude of the voltage *change* and not the magnitude of the voltage itself that determines i. Since a rapid voltage change yields a larger current, we would suspect that the opposition that a capacitor presents to an ac voltage would decrease as the frequency increases. The frequency of the applied voltage is a measure of how fast e varies with time, and hence the instantaneous rate of change of voltage with respect to time is proportional to the angular velocity ω. We also see from (1) that i is proportional to C.

To specifically analyze the shape of the i curve in the circuit of Fig. 4-16*a*, we will take the e waveform of Fig.

4-16b and apply (1) to it. Let us start reckoning time when the e waveform goes through zero. At $t = 0$, the slope de/dt of the e curve is at its maximum value, and hence i will be a maximum, as shown at point a. Note that although $e = 0$, we still have some i, in fact the largest value of i.[5] As the e waveform builds up in magnitude, its slope (steepness) decreases until, at a time corresponding to $\pi/2$ rad, $de/dt = 0$. This is at the peak of the e waveform. If $de/dt = 0$, $i = 0$, and hence the i waveform goes through point b. Between $\pi/2$ and π rad the slope is negative. Therefore, de/dt and hence i is negative. Physically this means that whenever e is decreasing in Fig. 4-16a, i is negative. As the e waveform crosses the zero axis, it exhibits its steepest negative slope; hence i is at its most negative peak as at point c.

Similar reasoning locates the maximum, minimum, and zero axis crossings for the remainder of i waveform. In-between points could be determined by actual measurement of the e waveform slope, but we will not bother with that. For the moment, just accept that the i waveform will actually be a cosine wave.[6] This is readily acceptable, since the i waveform does seem to be a sine wave shifted 90°. This 90° phase shift may be shown with our rotating vectors (phasors) as in Fig. 4-16c. Note that the current *leads* the voltage by 90° at all times, and this, of course, is true whether peak, rms, or instantaneous values are considered. Thus the current and voltage vectors of Fig. 4-16c could just as well be the rms values I and E respectively.

4-11 Capacitive susceptance

Now Ohm's law holds for circuits with ac as well as dc or any other waveform of excitation. However, it must be modified when the circuit is not purely resistive. For a purely resistive circuit with ac voltage applied,

1) $$I = \frac{E}{R} = \left(\frac{1}{R}\right) E = GE$$

where G, you will recall, is the conductance (reciprocal of resistance). The parameter G in Eq. (1) has the dimensions of mhos and it relates how much I results for a given E in a circuit such as Fig. 4-15a. Conductance is a measure of how readily a resistor permits current flow.

The analogous situation exists in Fig. 4-16a except that this time the circuit is purely capacitive. How do we relate I to E? Evidently, we need some parameter analogous to conductance having the dimension of mhos which will be a measure of how readily a capacitor permits current flow.[7] Let us call this parameter the capacitive susceptance and give it the symbol B_C. What factors determine B_C? Well, since in a capacitor

2) $$i = C\frac{de}{dt}$$

we see that as C increases, i increases in direct proportion. Furthermore, the greater de/dt becomes, the greater i becomes, again in direct proportion. Now de/dt is rate of change of the voltage waveform, and intuitively we may suspect that de/dt is dependent upon the angular velocity ω.

It turns out that

3) $$|B_C| = \omega C \tag{4-16a}$$

We use the magnitude symbol $|\ |$ about B_C because B_C is a directed (vector) quantity. Remember that I leads E by 90° in a capacitor. Hence, the susceptance (B_C) must have the property of causing a 90° phase rotation when it multiplies the voltage. We can mathematically build in this phase rotation by writing

4) $$\mathbf{B}_C = B_C\underline{/90^\circ} = jB_C \tag{4-16b}$$

The j operator notation is very appropriate here, since multiplication by j represents a phase rotation of 90°.

Now we can tie it all together by writing

5) $$\mathbf{I} = (B_C\underline{/0^\circ})\,\mathbf{E} = (jB_C)\,\mathbf{E} \tag{4-17}$$

We will drop the boldface type after a while since it is understood that I and E are rotating vectors. It seems intuitively plausible that the current leads the voltage in a capacitor, since voltage depends on the accumulated charge but current must first flow if charge is to build up.

A vector diagram which plots the susceptance along the j (imaginary) axis and conductance along the real axis is called an admittance diagram. Normally an admittance diagram such as Fig. 4-17 is used to show more than capacitive susceptance.

Now work through problem PS 4-10.

4-12 Capacitive reactance

Just as resistance is the reciprocal of conductance, reactance is the reciprocal of susceptance. Capacitive reactance, which is given the symbol X_C, has the units

[5] This, of course, is not necessarily true at the very instant the circuit is first turned on. However, once any initial turn-on transient has ended, the waveforms assume the shapes of Fig. 4-16b in the steady state. Thus what we choose to call time zero is a convenient moment of time in the steady state.

[6] That the i waveform is a cosine wave is rigorously shown as follows:

$$i = C\frac{de}{dt} = C\frac{d(E_{pk}\sin\omega t)}{dt} = \omega C E_{pk}\cos\omega t$$

[7] It should be clear that when we speak of current flowing through a capacitor it does not mean actual movement of electrons through the insulating dielectric. Rather, it means the equivalent flow due to electrons moving into one plate and being repelled from the other. Hence when the voltage varies and charge is being moved into and out of the plate, the capacitor appears to permit current flow through it.

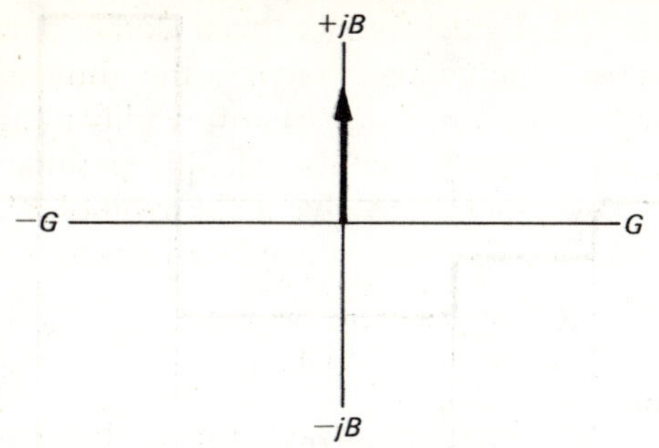

FIGURE 4-17

of ohms and is a measure of the opposition a capacitor presents to the flow of current. Mathematically,

1) $$\mathbf{X}_C = \frac{1}{\mathbf{B}_C} \tag{4-18}$$

Since

2) $$\mathbf{B}_C = j\omega C$$

3) $$\mathbf{X}_C = \frac{1}{j\omega C} = -j\frac{1}{\omega C} = -j\,|X_C| \tag{4-19}$$

From (3) we see that

4) $$|X_C| = \frac{1}{\omega C} \tag{4-20}$$

A vector diagram in which reactance is plotted along the axis of imaginaries, (j) axis, and resistance along the axis of real numbers is called an impedance diagram. From Eq. (4-19) we see that $\mathbf{X}_C$ would plot as shown in Fig. 4-18.

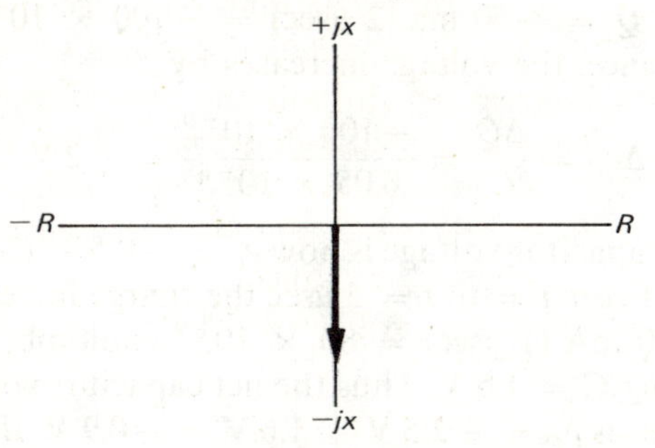

FIGURE 4-18

Now work through problem PS 4-11.

4-13 Power in the purely capacitive circuit

A pure capacitor cannot dissipate any power. To see this more clearly, consider Fig. 4-19, where the voltage and current waveforms of a capacitor are sketched in

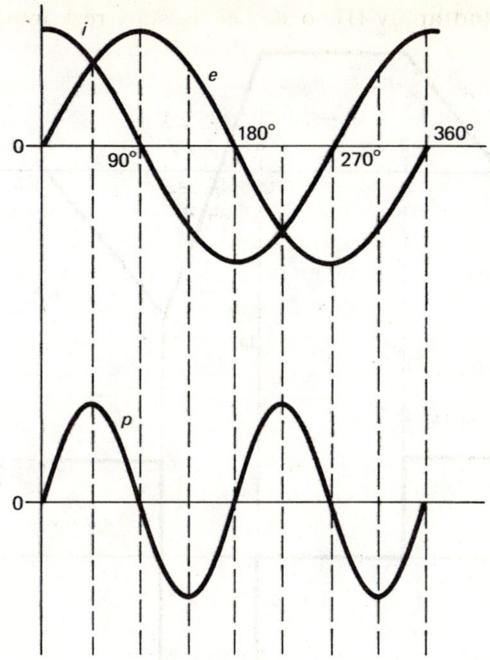

FIGURE 4-19

the upper part of the figure and the instantaneous power waveform in the lower part. The p waveform, as previously, is generated by point-by-point multiplication of the e and i waveforms.

Clearly at $t = 0$, $p = 0$ since $e = 0$, and this is also true at 90, 180, 270, and 360°. Also, the ei product is positive from 0 to 90° (both e and i are positive) and from 180 to 270° (both e and i are negative). In the interval from 90 to 180° and 270 to 360°, either e or i is negative, and hence p is negative. It turns out that the p waveform is once again a double-frequency sinusoid except that in the purely capacitive circuit it is symmetrical about the zero axis, hence the average power dissipated is also zero. Physically the capacitor is storing energy when p is positive and returning it to the source when p is negative.

Now work through problem PS 4-12.

PROBLEMS WITH SOLUTIONS

PS 4-1 A 0.5-mA constant current source drives a 0.2-μF capacitor for 0.03 sec. What is the charge stored in the capacitor and the voltage at that time?

SOLUTION

$$Q = I(\Delta t) = 0.5\text{ mA}(0.03\text{ sec}) = 0.015 \times 10^{-3}\text{ coul}$$

$$E_C = \frac{Q}{\text{coul}} = \frac{0.015 \times 10^{-3}\text{ coul}}{0.2 \times 10^{-6}\text{ F}} = 75\text{ V}$$

PS 4-2 The voltage across a 0.05-μF capacitor varies, as shown in Fig. PS 4-2*a*. Sketch the corresponding current waveform.

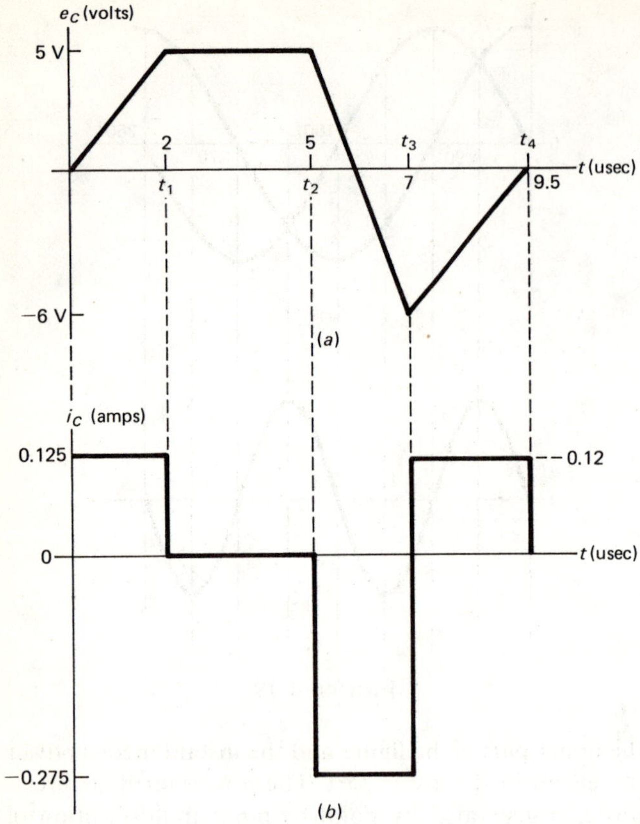

FIGURE PS 4-2

SOLUTION

$$i_C = C\frac{de_C}{dt} \cong C\frac{\Delta e_C}{\Delta t}$$

From $t = 0$ to $t = \mu$sec

$$i_C = 0.05 \times 10^{-6}\,\text{F}\left(\frac{5\text{ V} - 0\text{ V}}{2 \times 10^{-6}\text{ sec} - 0\text{ sec}}\right) = 0.125\text{ A}$$

From $t = 2$ μsec to $t = 5$ μsec

$$i_C = 0.05 \times 10^{-6}\,\text{F}\left(\frac{5\text{ V} - 5\text{ V}}{5 \times 10^{-6}\text{ sec} - 2 \times 10^{-6}\text{ sec}}\right) = 0\text{ A}$$

From $t = 5$ to $t = 7$ μsec

$$i_C = 0.05 \times 10^{-6}\,\text{F}\left(\frac{-6\text{ V} - 5\text{ V}}{2 \times 10^{-6}\text{ sec}}\right) = -0.275\text{ A}$$

From $t = 7$ to $t = 9.5$ μsec

$$i_C = 0.05\,\mu\text{F}\left(\frac{6\text{ V}}{2.5\,\mu\text{sec}}\right) = 0.12\text{ A}$$

The resultant waveform is shown in Fig. PS 4-2*b*.

PS 4-3 The current through a 0.05 μF capacitor is given by the waveform of Fig. PS 4-3*a*. Sketch the corresponding voltage waveform.

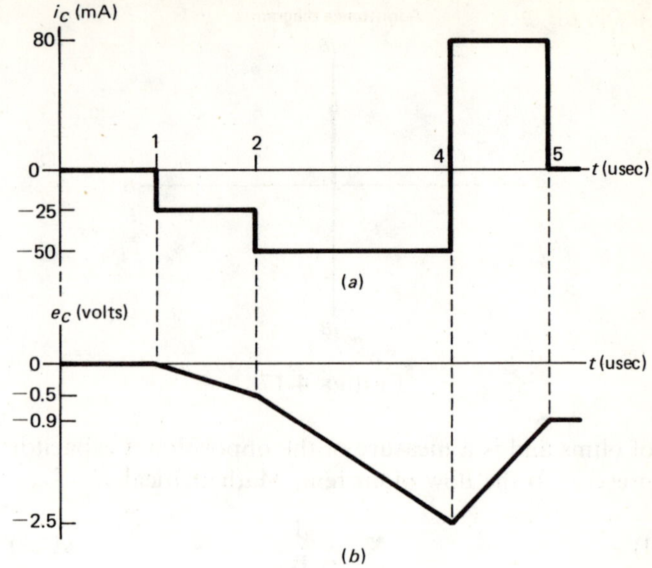

FIGURE PS 4-3

SOLUTION

1) $$e_C = \frac{Q}{C} = \frac{i_C(\Delta t)}{C}$$

From $t = 0$ to $t = 1$ μsec $i_C = 0$ and hence $e_C = 0$. From $t = 1$ to $t = 2$ μsec or $\Delta t = 1$ μsec we have $i_C = -25$ mA. Thus $Q = -25$ mA (1 μsec) $= -25 \times 10^{-9}$ coulomb. Hence, at $t = 2$ μsec.

$$e_C = \frac{-25 \times 10^{-9}}{0.05 \times 10^{-6}} = -0.5\text{ V}$$

This constant charging current results in a ramp of voltage. This is mathematically indicated in Eq. (1).

From $t = 2$ to $t = 4$ μsec the capacitor gains a charge given by $Q = -50$ ma (2 μsec) $= -100 \times 10^{-9}$ coulomb. Hence, the voltage increases by

$$\Delta e_C = \frac{\Delta Q}{C} = \frac{-100 \times 10^{-9}}{0.05 \times 10^{-6}} = -2\text{ V}$$

The net capacitor voltage is now $e_C = -0.5 + (-2.0) = -2.5$ V. From $t =$ to $t = 5$ μsec the charge increment is $\Delta Q = 80$ mA (1 μsec) $= 80 \times 10^{-9}$ coulomb. Hence, $\Delta e_C = \Delta Q/C = 1.6$ V. Thus the net capacitor voltage at $t = 5$ μsec is $e_C = -2.5\text{ V} + 1.6\text{ V} = -0.9$ V. If i_C then remains at 0 mA, the capacitor will maintain this voltage (see Fig. PS 4-3*b*).

PS 4-4 A small and precise spot welder uses the stored energy of a capacitor to form a weld by discharging the capacitor through the joint to be welded. If it takes 2.6 joules to form the weld, determine the dc voltage required to charge a 4-μF oil-filled capacitor to the required level.

SOLUTION $J = \frac{1}{2} C E_C^2$

$$E_C = \sqrt{\frac{2J}{C}} \quad \sqrt{\frac{2\,(2.6)}{4 \times 10^{-6}}} = 1{,}140 \text{ V}$$

PS 4-5 Two identical square metal plates are separated by 0.125 in. in air. If each plate is 1.2 in. on a side, determine the capacitance between them.
SOLUTION By Eq. (4-7),

$$C \cong 0.224 \frac{KA}{d} (n - 1) \text{ pF}$$

$$= 0.224 \frac{(1)(1.2)^2}{0.125} (2 - 1) = 2.58 \text{ pF}$$

PS 4-6 Six inches of #18 wire having a diameter of approximately 0.04 in. is supported at a height of approximately 0.1 in. Estimate the capacitance between the wire and ground plane.
SOLUTION This configuration corresponds to Fig. 4-9 and calls for the application of Eq. (4-8). Thus

$$C = \frac{7.35}{\log_{10}\left[\frac{2h}{d}\left(1 + \sqrt{1 - \frac{1}{(2h/d)^2}}\right)\right]} \frac{\text{pF}}{\text{ft}}$$

$$= \frac{7.35}{\log_{10}\left[\frac{2(0.1)}{0.04}\left(1 + \sqrt{1 - \frac{1}{[2(0.1)/0.04]^2}}\right)\right]} \frac{\text{pF}}{\text{ft}}$$

$$= \frac{7.35}{\log_{10}[9.899]} = \frac{7.35}{0.9956}$$

$$= 7.38 \frac{\text{pF}}{\text{ft}} \text{ or } 3.69 \text{ pF for 6 in.}$$

PS 4-7 Three capacitors, 1 μF, 0.7 μF, and 1.2 μF, are connected in parallel. Determine the equivalent capacitance.
SOLUTION

$$C_{eq} = C_1 + C_2 + C_3$$

$$= 1\ \mu\text{F} + 0.7\ \mu\text{F} + 1.2\ \mu\text{F} = 2.9\ \mu\text{F}$$

PS 4-8 Determine the equivalent capacitance if the capacitors of the previous problem are connected in series.

SOLUTION $\frac{1}{C_{eq}} = \frac{1}{C_1} \quad \frac{1}{C_2} \quad \frac{1}{C_3}$

$$= \frac{1}{1\ \mu\text{F}} + \frac{1}{0.7\ \mu\text{F}} + \frac{1}{1.2\ \mu\text{F}}$$

Solving for C_{eq} yields

$$C_{eq} = 0.307\ \mu F$$

PS 4-9 Determine C_{eq} in Fig. PS 4-9.

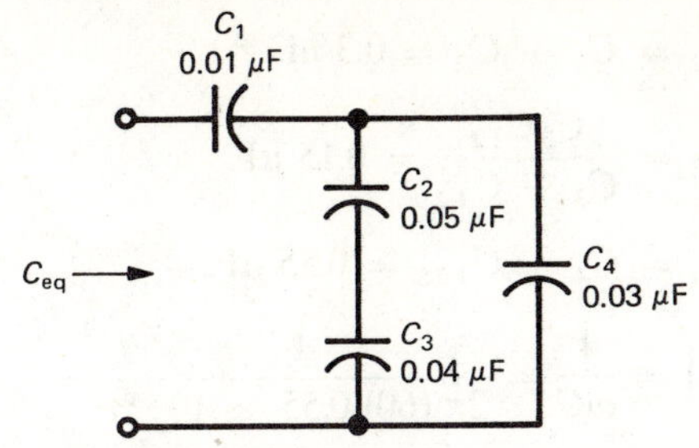

FIGURE PS 4-9

SOLUTION First determine the equivalent capacity of the series combination of C_2 and C_3. Thus

$$C_{23} = \frac{C_2 C_3}{C_2 + C_3} = \frac{(0.05)(0.04)}{0.05 + 0.04} = 0.0222\ \mu\text{F}$$

Taking the parallel combination of C_{23} and C_4 yields

$$C_{234} = C_{23} + C_4 = 0.0522\ \mu\text{F}$$

Taking the series combination of C_{234} and C_1 yields

$$C_{eq} = \frac{C_1 C_{234}}{C_1 + C_{234}} = \frac{(0.01)(0.0522)}{0.01 + 0.0522} = 0.00839\ \mu\text{F}$$

PS 4-10 What is the rms current through a 5-pF capacitor if a sinusoidal voltage source having a peak amplitude of 15 volts and a frequency of 28 MHz is applied across it?
SOLUTION

$$|B_C| = \omega C = 2\pi f C = 2\pi\,(28 \text{ MHz})\ 5 \text{ pF}$$

$$= 2\pi\,(28 \times 10^6 \text{ Hz})(5 \times 10^{-12} \text{ F})$$

$$= 8.796 \times 10^{-4} \text{ mhos}$$

$$\mathbf{I} = (jB_C)E$$

$$= (8.796 \times 10^{-4} \angle 90^\circ \text{ mhos})\ 15 \angle 0^\circ \text{ V} = 13.2 \angle 90^\circ \text{ mA}$$

PS 4-11 Determine the current I in the circuit of Fig. PS 4-11 if E = 120 V at 60 Hz, C_1 = 0.1 μF, C_2 = 0.2 μF, C_3 = 0.3 μF, and C_4 = 0.4 μF.

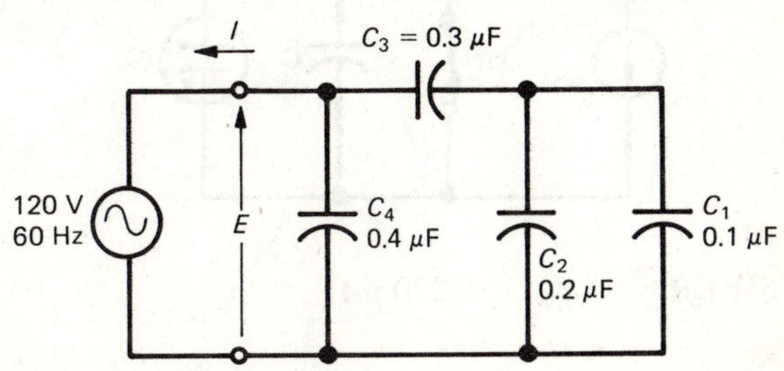

FIGURE PS 4-11

SOLUTION First we determine C_{eq} that the voltage source looks into. Thus

$$C_{12} = C_1 + C_2 = 0.3\ \mu\text{F}$$

$$C_{123} = \frac{C_3 C_{12}}{C_3 + C_{12}} = 0.15\ \mu\text{F}$$

$$C_{eq} = C_4 + C_{123} = 0.55\ \mu\text{F}$$

$$|X_C| = \frac{1}{\omega C} = \frac{1}{2\pi\,(60)(0.55 \times 10^{-6})} = 4.82\ \text{k}\Omega$$

Consequently, taking E at a reference angle of zero degrees,

$$\mathbf{I} = \frac{E/0^\circ}{-jX_C} = \frac{E/0^\circ}{X_C/-90^\circ} = \frac{120\ \text{V}}{4.82\ \text{k}\Omega/-90^\circ}$$

$$= 24.9\ \text{mA}/90^\circ$$

PS 4-12 Prove that the average power dissipated in a pure capacitor is zero.

SOLUTION

1) $p = ei$

2) $p = (E_{pk} \sin \omega t)(I_{pk} \cos \omega t)$

3) $p = E_{pk} I_{pk} \sin \omega t \cos \omega t$

Recalling that

4) $\sin(x + y) = \sin x \cos y + \cos x \sin y$

and letting

5) $x = y = \theta$

enables us to rewrite (4) as

6) $\sin(2\theta) = 2 \sin \theta \cos \theta$

Therefore, from (6),

7) $\sin \theta \cos \theta = \frac{1}{2} \sin 2\theta$

Letting $\theta = \omega t$ we have, by analogy with (3),

8) $p = \dfrac{E_{pk} I_{pk}}{2} \sin 2\,\omega t$

Clearly if equation (8) were sketched, it would plot as a sine curve with a frequency of twice the e or i curves and peak value equal to $E_{pk} I_{pk}/2$. Since the average value of any sinusoid is zero, we need not bother plotting (8) but can immediately conclude that the average power $P = 0$.

PROBLEMS WITH ANSWERS

PA 4-1 The neon bulb or glow tube GT in Fig. PA 4-1 is a device which looks like an open circuit until the voltage across its terminals equals a critical value E_i called the ignition potential. Once E_i is reached, the gas in the bulb ionizes and glows. If $E_i = 108$ V and $C = 0.5\ \mu$F, determine the current that the current source I must deliver if the glow tube is to ignite 0.2 sec after switch SW is opened. Do not worry about what happens once the bulb ignites.

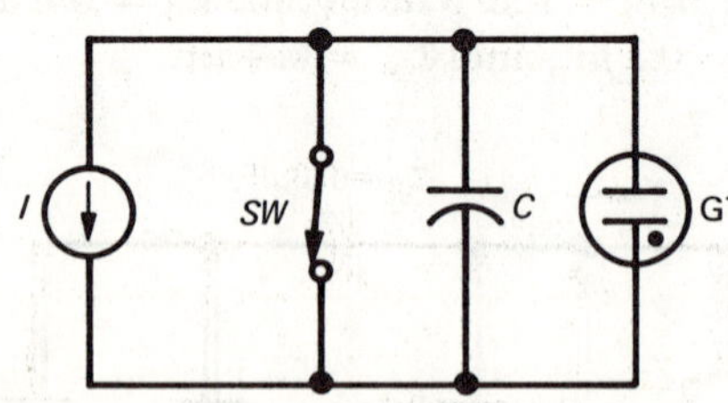

ANSWER 270 μA

PA 4-2 A 0.05 μF capacitor must store 1.2 joules of energy in a particular application. What is the minimum permissible working voltage?

ANSWER 6.93 kV

PA 4-3 Determine the capacitance of a 3-ft section of coaxial line filled with a plastic of dielectric constant K = 3 and whose dimensions are $D = 0.2$ in. (inside diameter of shield) and $d = 0.04$ in. (diameter of inner conductor).

ANSWER 94.6 pF

PA 4-4 Determine the equivalent capacitance of the network shown in Fig. PA 4-4.

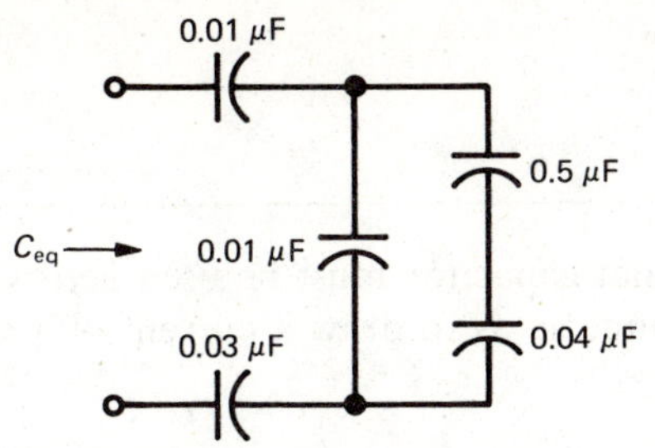

ANSWER 0.00647 μF

PA 4-5 What voltage is developed across a 22 pF capacitor when it is driven by a 0.1 mA, 15 kHz current source? Draw a vector diagram with the current at a reference angle of zero degrees.

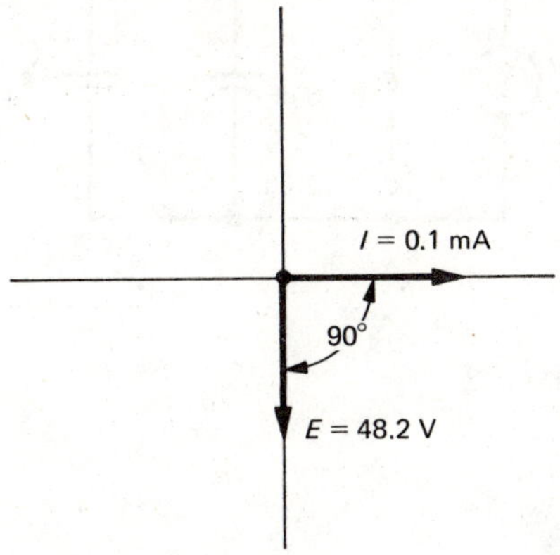

ANSWER See Fig. PA 4-5.

PA 4-6 Determine the voltage across C_2 in Fig. PA 4-6.

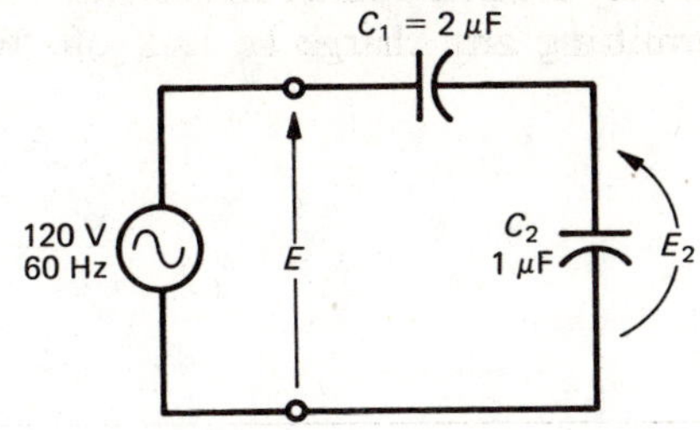

ANSWER $E_2 = 80\text{ V}\angle 0^\circ$

PA 4-7 Determine the current through C_5 in Fig. PA 4-7.

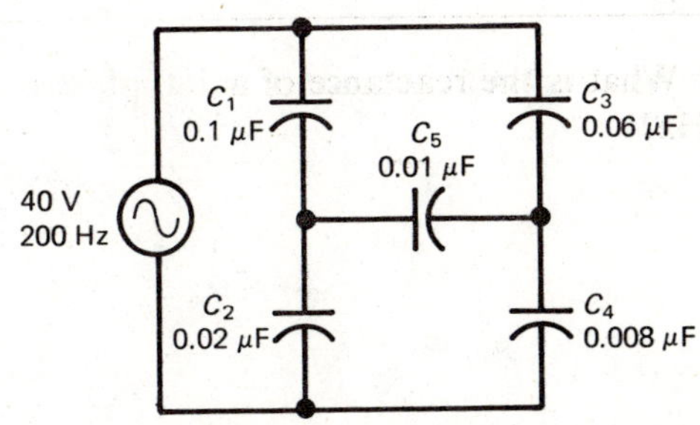

ANSWER $I_5 = 39.6\angle 90^\circ$ μA

PROBLEMS WITHOUT ANSWERS

P 4-1 The current into a 0.5-μF capacitor is given by the waveform of Fig. P 4-1. Determine the capacitor voltage in 1.5 and 2 sec.

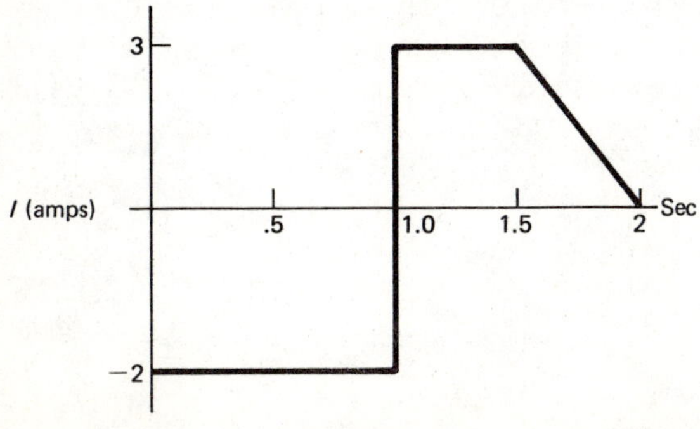

P 4-2 A 2-μF capacitor must be charged so as to store 3 joules in 5 sec. Determine the required average charging current and minimum working voltage.

P 4-3 Two circular metal disks having a diameter of 2 in. are separated in air by 0.05 in. What is the charge in this capacitor if it is connected across a 50-volt dc source? What is the voltage across it?

P 4-4 Suppose we disconnect the 50-volt source from the capacitor in the previous problem and then separate the plates so they are 2 in. apart. Assume that this is done without permitting any charge to leak off. What will happen?

P 4-5 What capacitance must be connected in series with a 4-μF capacitor to yield an equivalent capacitance of 2.5 μF?

P 4-6 What is the reactance of a 180-pF capacitor at 10.7 MHz?

P 4-7 What capacitance is required at 20 kHz for a reactance of 500 Ω?

P 4-8 What capacitor must be used across a 120-volt 120-Hz source so as to draw a current of 3 amp?

P 4-9 Determine the current I_3 in C_3 in Fig. P 4-9.

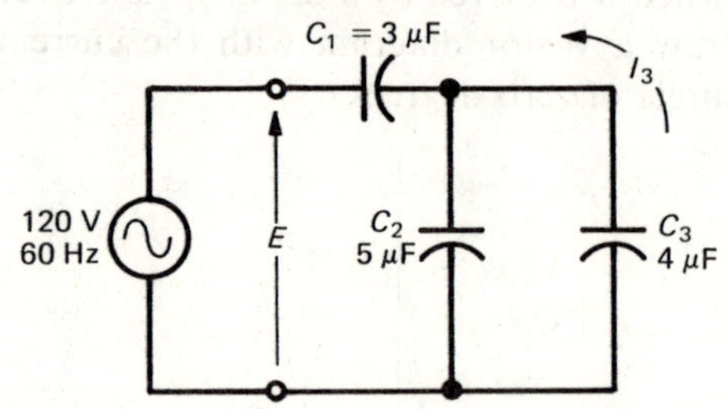

5 inductance

We previously learned that resistance is that property of a circuit which causes it to oppose the flow of current, whereas capacitance is a circuit property which causes it to oppose a change in voltage.

5-1 Inductance

Now we shall investigate a third circuit parameter called inductance. Inductance is given the symbol L, is measured in a unit called the henry, and is that property of a circuit which causes it to oppose a *change* in current. Or, in other words, the current through an inductor cannot change instantaneously.

To illustrate the effect of circuit inductance, first consider Fig. 5-1*a*. If the switch is closed at t_0, $e = E$, and if the box contained a pure resistance, the current would immediately jump in step with the voltage to some maximum value determined by Ohm's law. This is shown in Fig. 5-1*b*. However, if the box instead contained a pure inductor, the current would not immediately jump up to some maximum value as with a resistor but instead would gradually rise as shown in Fig. 5-1*c*.

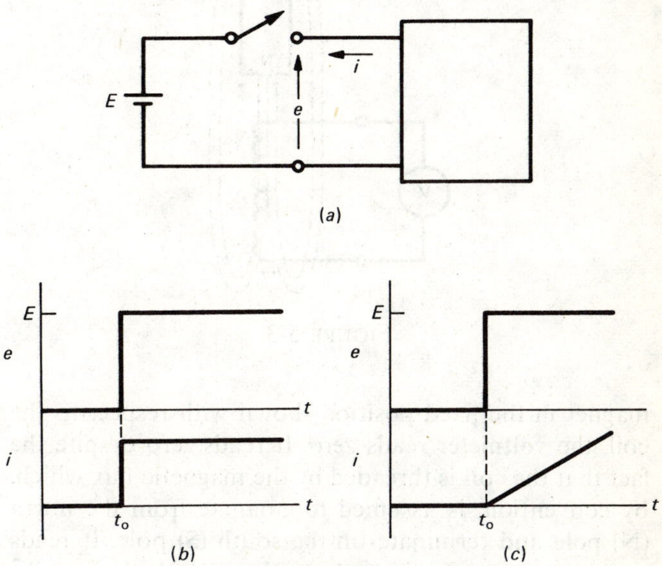

FIGURE 5-1

In theory, it would continue to rise without limit and with the same slope. However, in practice the inevitable internal resistance of the voltage source and connecting wires cause the current to approach some limiting value.

Physically any length of conductor has some inductance associated with it. If the conductor is wound into the form of a coil, the inductance is greatly increased. This is due to the fact that inductance is related to the magnetic field about a conductor, and since wrapping the conductor into the shape of a coil concentrates the magnetic flux within it, it likewise increases the inductance.[1] Inductance is further increased by placing an iron core within the coil since this further concentrates the magnetic flux.

The symbol for inductance is L, and some schematic representations are shown in Fig. 5-2.

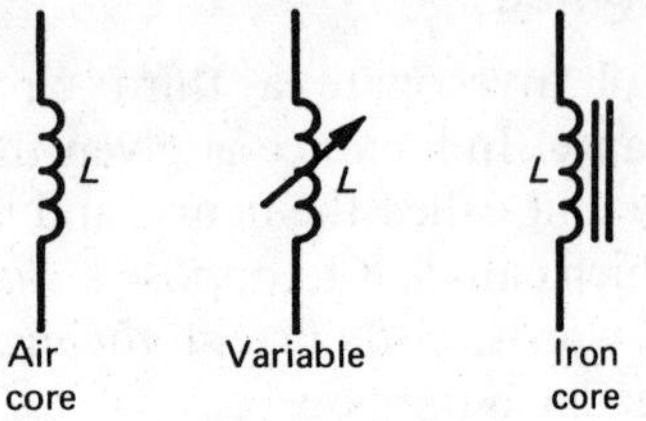

FIGURE 5-2

5-2 Induced voltage

The voltage which is developed across an inductor depends upon a *change* in the magnetic flux which threads it. For example, in Fig. 5-3, if we hold the

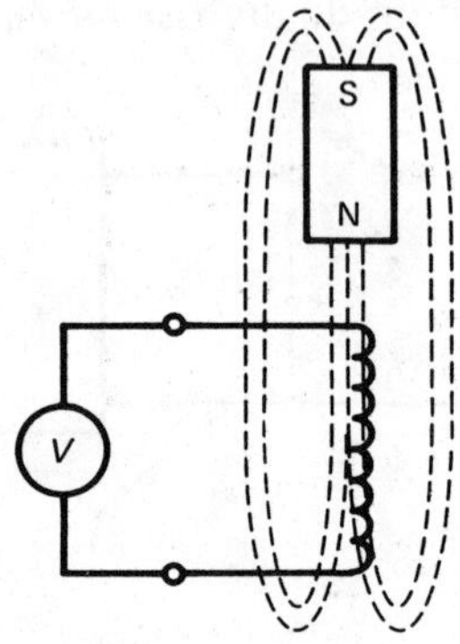

FIGURE 5-3

magnet in the fixed position shown with respect to the coil, the voltmeter reads zero. It reads zero despite the fact that the coil is threaded by the magnetic flux which, by convention, is assumed to emanate from the north (N) pole and terminate on the south (S) pole. It reads zero because the flux is not in motion relative to the coil.[2]

Now if we were to suddenly push the magnet into the coil (or the coil onto the magnet), there would result a momentary voltmeter deflection which ceases when we stop moving the magnet. If we now yank the magnet back out of the coil, the voltmeter would again deflect but in the opposite direction. The most significant point of this experiment is that only when there was a changing magnetic field did we find a voltage developed, or induced, across the coil.

[1]The magnetic field is portrayed by considering lines of magnetic flux which emanate from the north pole and terminate on the south pole.

[2]This is grossly oversimplified, but we wish to avoid the details of Lenz's law, flux linkage concepts, etc., in this discussion.

The magnetic field need not have been due to the permanent magnet. Instead, we could have used the principle that any current-carrying coil has a magnetic field about it. Thus, in Fig. 5-4 we have L_1 and L_2 closely

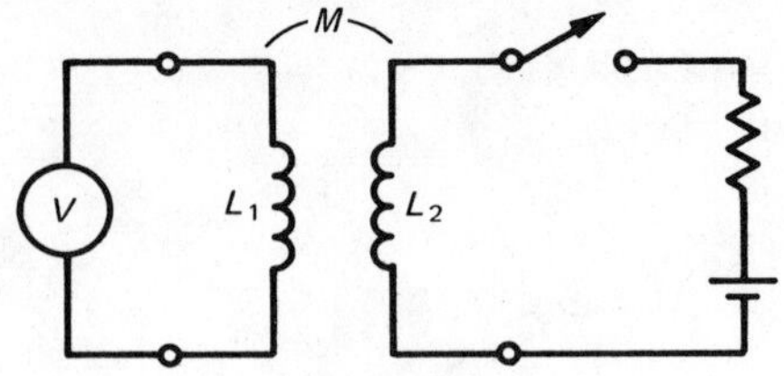

FIGURE 5-4

coupled, so that when we close the switch and current starts to flow in L_2, the growing magnetic field about L_2 threads L_1. Since the field is growing (in motion), a voltage is induced in L_1. When the current in L_2 finally reaches its maximum value, the field becomes fixed (unvarying) and no further voltage is induced in L_1, so the voltmeter reading drops to zero. If the switch is reopened, the field about L_2 collapses and in doing so it again cuts through L_1. This again induces a voltage in L_1, although of opposite polarity.

When the voltage across a coil is due to the flux change caused by a current variation within that coil only, we often call that voltage a self-induced voltage (or back emf), and the inductance is called the self-inductance. When the coil voltage in question is due to a current change in a different coil, we speak of a mutual inductance. This will be explored further in the chapter on coupled circuits.

It turns out that the voltage across an inductor is given by

1) $$e = L\frac{di}{dt} \tag{5-1a}$$

where e, the instantaneous self-induced voltage, is in volts; L is the inductance in henries; and di/dt is the instantaneous time rate of change of current in amperes/second.

If, instead of instantaneous values, we choose to consider an average current change with respect to time ($\Delta i/\Delta t$), voltage e becomes the average voltage developed during the interval Δt, and Eq. (1) may be rewritten as

2) $$e = L\frac{\Delta i}{\Delta t} \tag{5-1b}$$

The significance of Eq. 5-1b (or 5-1a) is twofold. First, it clearly indicates that the inductor voltage exists only when the inductor current is changing. A steady current, no matter how large, yields no voltage across an ideal (zero resistance) inductor. Second, though not as obvious, there is the implication that the current in an inductor cannot change instantaneously. This becomes more evident if we solve (2) for Δi to obtain[3]

$$\text{3)} \qquad \Delta i = \frac{e}{L}(\Delta t) \qquad \text{(5-2)}$$

Equation (3) tells us that if $\Delta t = 0$, $\Delta i = 0$, which means no current change can occur unless some time has passed.

From a physical viewpoint it boils down to the fact that an inductor opposes a change of current through it, and it does so by developing a self-induced voltage which is of a polarity to oppose the current change. To appreciate the mechanics of this, consider Fig. 5-5*a*, where we have assigned some voltage and current reference directions to the inductor terminals. If I is constant, then $i = I$ is also constant and $e = 0$.

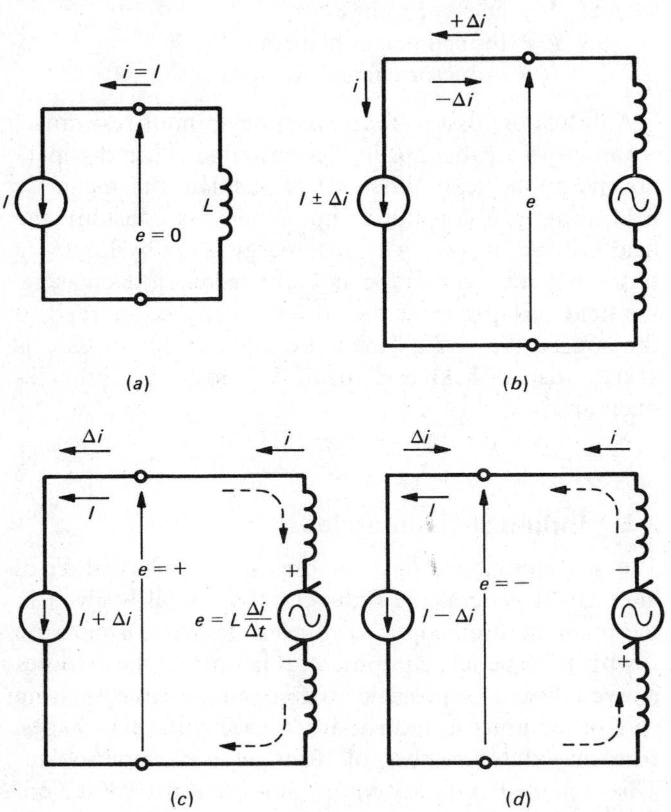

FIGURE 5-5

Now let us consider Fig. 5-5*b*, where the possibility of a change in i due to a component $\pm\Delta i$ in the current source is considered. A positive Δi means an increase of i, which is indicated by the $+\Delta i$ current arrow, while a decrease of current pumped by the current source is indicated by a negative Δi and a $-\Delta i$ current arrow opposite to the direction of I. Hence, $\pm\Delta i$ simply represents some variation about the average value of I. Any time we have some $\pm\Delta i$, we will find a voltage induced in the coil, and this voltage can be represented by a voltage source buried in the coil. Hence, in Fig. 5-5*b* we have separated the inductor into two parts in order to insert our fictitious self-induced voltage source.

Now let us see how we determine the polarity of the self-induced voltage. Consider Fig. 5-5*c*, where a positive Δi is assumed to take place. A positive Δi means that in addition to I another current component, Δi, is trying to be established in the coil. The coil will try to oppose this Δi by developing a voltage of such a polarity that it tries to force a current in a direction opposite to that of Δi. This current direction is indicated by the dashed line in Fig. 5-5*c*. For the self-induced voltage source to force current in this direction, it must develop the polarity shown, and hence e swings positive by $L(\Delta i/\Delta t)$ volts.

Next, consider a decrease in the current forced by the current source. This is indicated in Fig. 5-5*d* by the Δi arrow going opposite to the I arrow. Since this means the inductor current i is decreasing, the self-induced voltage will assume a polarity so as to try to pump a current in the dashed arrow direction. The dashed arrow current bucks the Δi current component and hence tries to prevent i from changing. To force a current in the direction of the dashed arrow, the inductor must develop the indicated polarity across the self-induced voltage source. Note that this time e swings negative. You will note in these examples that it is only the direction of Δi that determines e_L. The direction of the steady-state component (I) is of no consequence.

Now work through problems PS 5-1, PS 5-2, and PS 5-3.

5-3 Inductor current

Upon occasion we may be interested in the current through an inductor. If the voltage across the inductor is held constant, as in Fig. 5-6*a*, the resultant current

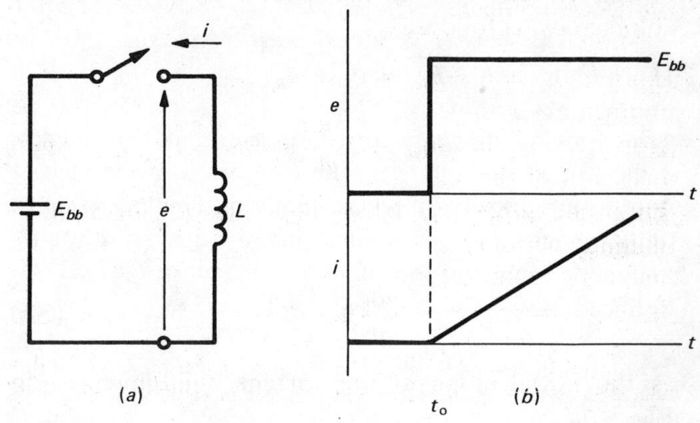

FIGURE 5-6

[3]For the reader familiar with calculus,

$$i = \frac{1}{L}\int_0^t (e)dt$$

waveform will be a ramp whose slope is readily derived from Eq. (5-1*b*) by solving for $\Delta i/\Delta t$. Thus

1) $$\frac{\Delta i}{\Delta t} = \frac{e}{L} = \frac{E_{bb}}{L} \qquad (5\text{-}3)$$

and the current at any time is

2) $$i = \left(\frac{e}{L}\right)\Delta t + I_0 \qquad (5\text{-}4)$$

You may recognize the above as being in the form of a straight-line equation, $y = mx + b$, where $b = I_0$, which allows for any initial current prior to initiating Δt. In Fig. 5-6*a*, $I_0 = 0$.

Suppose the voltage applied to the inductor is not constant. What then? This situation may arise if a resistor is connected in series with the inductor, as in Fig. 5-7*a*. Even though E_{bb} is a constant, the voltage across L need not be since it equals E_{bb} minus the voltage across R. To analyze this circuit, let us begin at time $= t_0$ when the switch closes as shown in Fig. 5-7*b*.

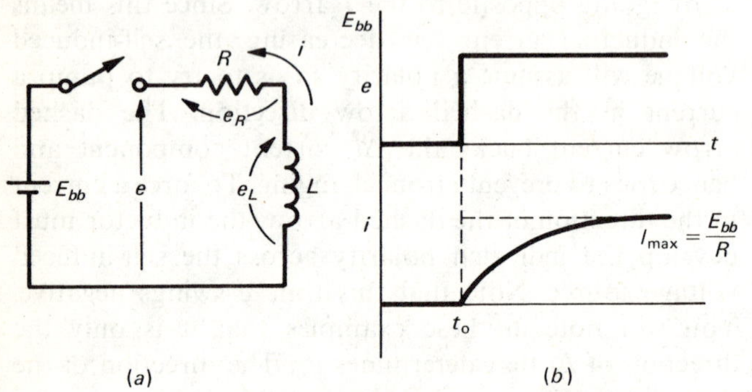

FIGURE 5-7

Since the inductor current was zero prior to closing the switch and since the inductor current cannot change instantaneously, it follows that $i = 0$ the moment after the switch closes (t_0^+). With $i = 0$, $e_R = 0$, and $e_L = E_{bb} - e_R = E_{bb}$. We can even determine the initial slope of the current rise in this circuit from Kirchhoff's voltage law as follows:

1) $$E_{bb} = e_L + e_R$$

2) $$E_{bb} = L\frac{di}{dt} + Ri \qquad (5\text{-}5)$$

Since at $t = t_0^+$, $i = 0$, we may solve (2) for di/dt to obtain

3) $$\frac{di}{dt} = \frac{E_{bb}}{L} \qquad (5\text{-}6)$$

as the initial slope of the current buildup curve in Fig 5-7*b*.

As the current builds up, a voltage drop appears across R and hence less voltage is left for the coil. The exact nature of this situation need not concern us here.[4] Suffice it to say that in time the current will approach a limiting value of E_{bb}/R.

A broader study of circuits involving arbitrary waveforms and inductors, inductors and resistors, and/or capacitors is beyond the scope of this text, which is primarily concerned with the response of these circuit parameters to sinusoidal excitation.

Now work problem PS 5-4.

5-4 Inductor energy storage

An inductor stores energy in its magnetic field as given by the relation

$$J = \tfrac{1}{2}LI^2 \qquad (5\text{-}7)$$

where J = energy in joules
L = inductance in henries
I = inductor current in amps

An ideal or lossless inductor (one without resistance) is similar to a capacitor in the sense that it can dissipate no energy as heat. When Δi is positive, the magnetic field about the coil builds up. When i is constant, the field is constant with a stored energy given by Eq. (5-7). If Δi is negative so that the inductor current is decreasing, the field collapses and the stored energy is returned to the source that established the current. In no case is energy lost as heat due to an I^2R loss in a pure inductance.[5]

Now work through problem PS 5-5.

5-5 Inductor geometries

The self-inductance of a conductor or coil is often of interest. In general the inductance of a coil is given in the manufacturer's data or, if need be, we can measure L with suitable test equipment. If neither of these routes is available, it is possible to find information relating coil or conductor inductance to various coil shapes, core materials, number of turns, etc., in handbooks.[6] This type of handbook information is most valuable if one has to design a coil or estimate stray inductance in the wiring of some electronic circuits. Normally the inductance of short wires does not particularly worry us, but many modern electronic circuits operate at tremendous speeds and current swings of, say, 15 mA

[4]See any text devoted to transient analysis or Chap. 3 of Cutler, "Electronic Circuit Analysis—Passive Networks," McGraw-Hill Book Company, New York, 1960.

[5]We are assuming negligible loss due to electromagnetic radiation when the field varies.

[6]F. E. Terman, "Radio Engineer's Handbook," Mc-Graw-Hill Book Company, New York, 1943.

in 0.5 μsec are quite common. Now 15 mA/0.5 μsec is a $\Delta i/\Delta t$ of 30,000 amps/sec! Under these circumstances relatively large voltages may appear along the connecting wires due to their inductance. This may in some cases completely disrupt the proper operation of a high-speed circuit. Hence, it is of some interest to examine the inductance of a few configurations if only to get some feeling for the magnitudes involved.

All the equations in the following discussion approximate the low-frequency inductance of the configuration in question. The high-frequency inductance tends to be somewhat less due to skin effect.

Figure 5-8 illustrates an isolated length of round wire. The inductance of this wire in μH is given by

$$L = 0.00508l\left[\left(2.303 \log_{10} \frac{4l}{d}\right) - 0.75\right] \mu\text{H} \quad (5\text{-}8)$$

with l and d in inches

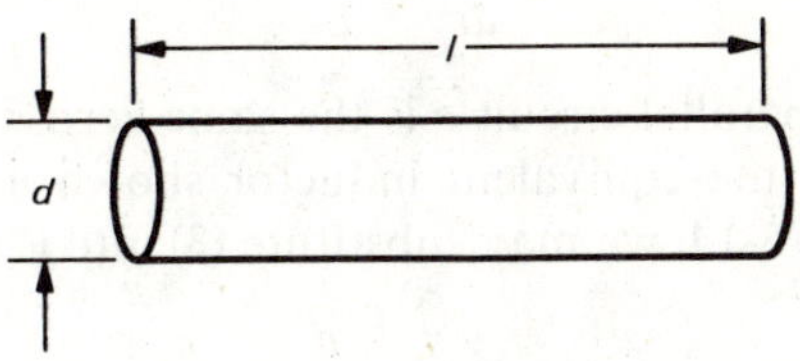

FIGURE 5-8

For two identical conductors carrying current in opposite directions (shown in Fig. 5-9) and separated by a distance D which is large compared to their diameter, the equivalent inductance of both is given by

$$L \cong 0.01016l\left[\left(2.303 \log_{10} \frac{2D}{d}\right) - \frac{D}{l}\right] \mu\text{H} \quad (5\text{-}9)$$

with l and d in inches

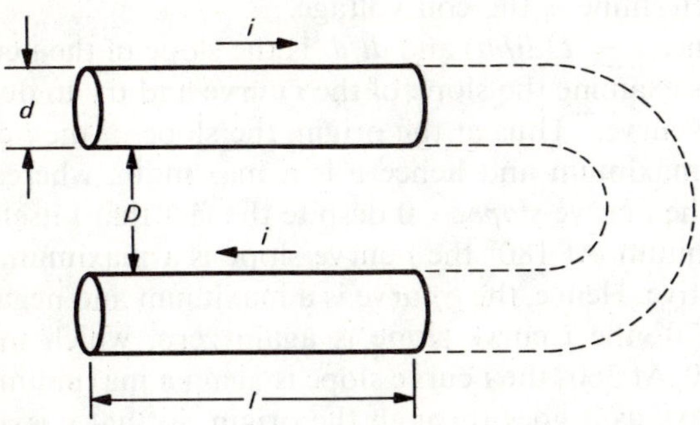

FIGURE 5-9

Equation (5-9) does not allow for any connecting link at their ends. If the link is run essentially perpendicular to these two conductors, the link inductance may be estimated by Eq. (5-8) and that inductance added to that of Eq. (5-9).

Figure 5-10 illustrates a single-layer coil (solenoid).

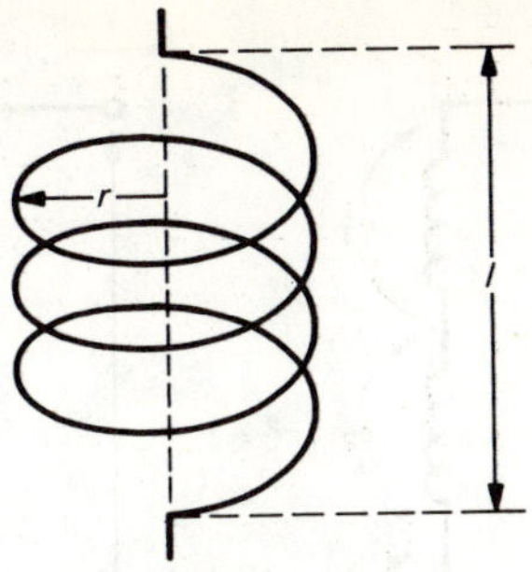

FIGURE 5-10

The inductance (in μH) is given by the following equation to within 1 percent accuracy if $l > 0.8r$.

$$L \cong \frac{r^2N^2}{9r + 10l} \mu\text{H} \quad (5\text{-}10)$$

where N = number of turns
r = radius in inches
l = length of winding

This is a fairly common configuration and we frequently desire to know how many turns (N) must be used to obtain a given L with an assumed r and l (usually determined by any coil form available). In that case, we may solve Eq. (5-10) for N to obtain

$$N = \sqrt{\frac{L(9r + 10l)}{r^2}} \quad (5\text{-}11)$$

Now work through problems PS 5-6 and PS 5-7.

5-6 Inductors in series

If inductors are connected in series as shown in Fig. 5-11 and if no coupling exists between their magnetic

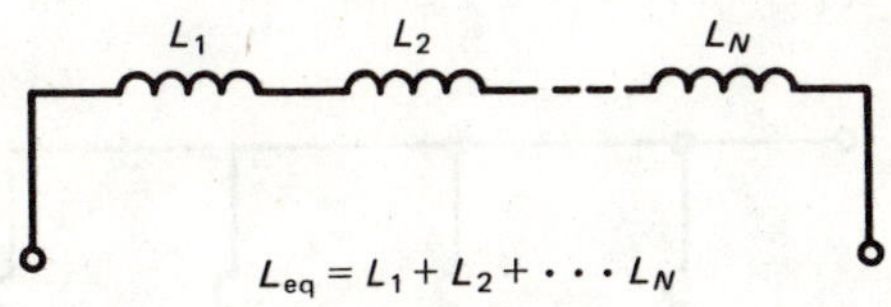

FIGURE 5-11

fields (mutual inductances = 0), the equivalent inductance is given by

1) $$L_{eq} = L_1 + L_2 + L_3 + \ldots + L_N \quad (5\text{-}12)$$

Note that inductors in series are analogous to resistors in series.

We can prove Eq. (5-12) with the aid of Fig. 5-12. Clearly, for the left-hand figure,

2) $e = e_1 + e_2 + e_3$

3) $$e = L_1 \frac{di}{dt} + L_2 \frac{di}{dt} + L_3 \frac{di}{dt} = (L_1 + L_2 + L_3) \frac{di}{dt}$$

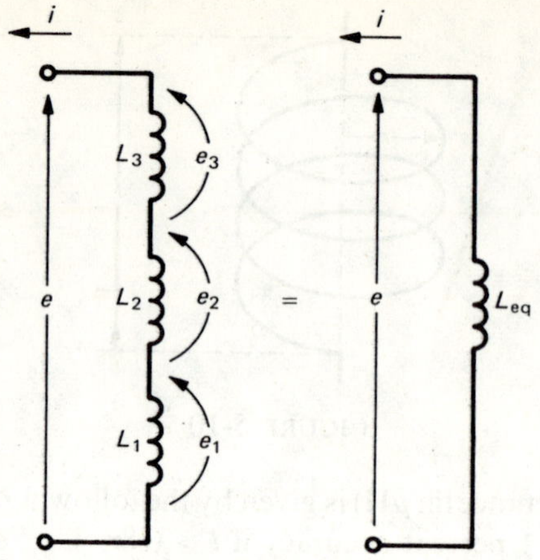

FIGURE 5-12

And for the right-hand figure,

4) $$e = L_{eq}\frac{di}{dt}$$

If the networks are to be equivalent insofar as their terminal conditions are concerned, it follows that

5) $$L_{eq}\frac{di}{dt} = L_1\frac{di}{dt} + L_2\frac{di}{dt} + L_3\frac{di}{dt}$$

Dividing through by di/dt yields Eq. (5-12).

5-7 Inductors in parallel

When inductors are connected in parallel, as in Fig. 5-13, and with no magnetic coupling between them, the equivalent inductance is given by

$$\frac{1}{L_{eq}} = \frac{1}{L_1} + \frac{1}{L_2} + \frac{1}{L_3} + \cdots + \frac{1}{L_N} \tag{5-13}$$

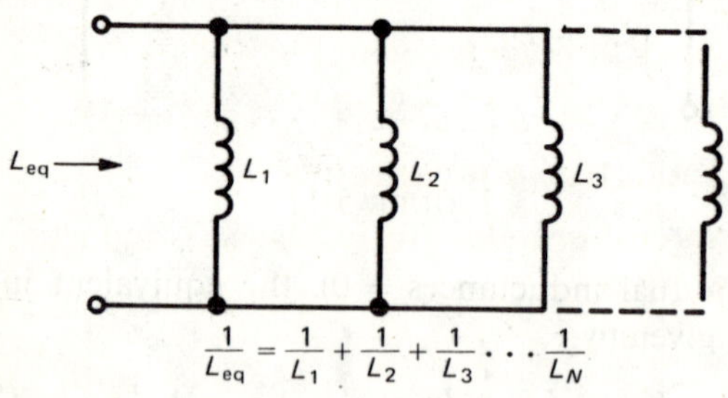

FIGURE 5-13

Note the similarity between Eq. (5-13) and that for resistors in parallel.

To verify Eq. (5-13), we may write for the left side of Fig. 5-14

1) $$i = i_1 + i_2 + i_3$$

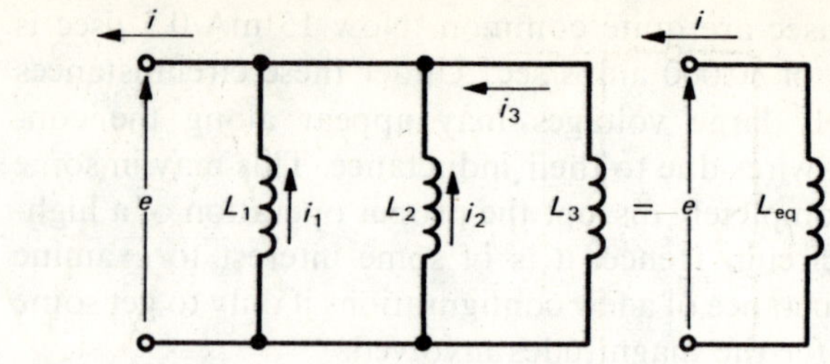

FIGURE 5-14

Consequently, the rate at which i varies with time is the sum of the individual rates. Thus

2) $$\frac{di}{dt} = \frac{di_1}{dt} + \frac{di_2}{dt} + \frac{di_3}{dt}$$

Now in general, $e = L\,(di/dt)$, from which

3) $$\frac{di}{dt} = \frac{e}{L}$$

Since in a parallel circuit e is the same across all the inductors, or the equivalent inductor shown in the right side of Fig. 5-14, we may substitute (3) into (2) to obtain

4) $$\frac{e}{L_{eq}} = \frac{e}{L_1} + \frac{e}{L_2} + \frac{e}{L_3}$$

Dividing equation (4) by e then yields Eq. (5-13).

Now work through problems PS 5-10 and PS 5-11.

5-8 Inductors with sinusoidal excitation

To study the response of an inductor to sinusoidal excitation, let us assume that the inductor in Fig. 5-15*a* is being driven by a sinusoidal current $i = I_{pk} \sin \omega t$. The i waveform is plotted in Fig. 5-15*b*. The problem is to determine e, the coil voltage.

Since $e = L(di/dt)$ and di/dt is the slope of the i curve, let us examine the slope of the i curve and try to deduce the e curve.[7] Thus at the origin, the slope of the i curve is a maximum and hence e is a maximum, whereas at 90° the i curve *slope* = 0 despite the fact that i itself is a maximum. At 180° the i curve slope is a maximum and negative. Hence, the e curve is a maximum and negative. At 270° the i curve slope is again zero, which means $e = 0$. At 360° the i curve slope is again a maximum and positive as it goes through the origin, so that e is again positive and a maximum.

It may be rigorously shown by methods of calculus that the e curve will actually be a cosine wave—which is a sine wave shifted by 90°— and the points on the e

[7]The student familiar with calculus may immediately proceed to differentiate $i = I_{pk} \sin \omega t$ to obtain $di/dt = \omega I_{pk} \cos \omega t$. Therefore, $e = L\omega I_{pk} \cos \omega t = \omega L I_{pk} \sin(\omega t + 90°)$, where ωL is the magnitude of the inductive reactance.

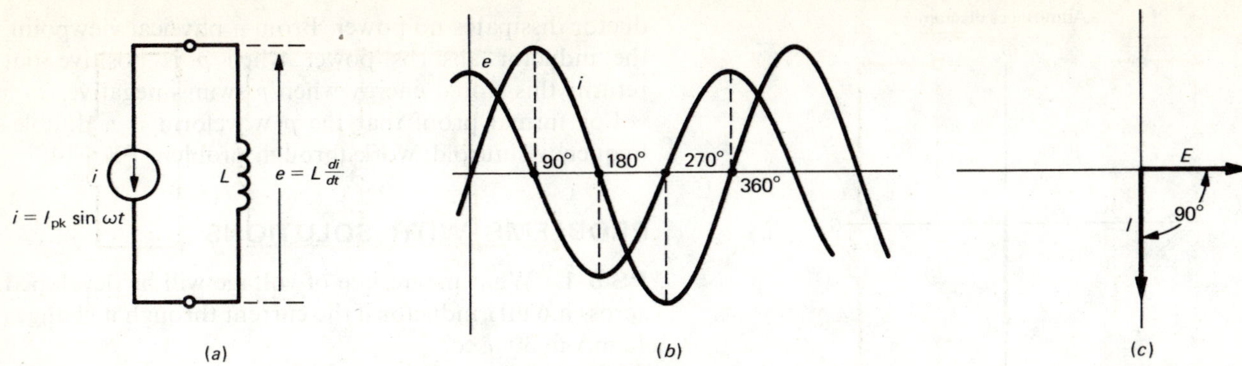

FIGURE 5-15

curve previously determined are certainly in keeping with this.

Note that the e waveform leads the i waveform by 90°, which is the same as saying that the i waveform lags the e waveform by 90°. This is reflected in the vector diagram of Fig. 5-15c, where E and I are the rms values.

5-9 Inductive reactance

By Ohm's law we know that the inductor voltage in the vector diagram of 5-15c is given by

1) $$\mathbf{E} = (X_L\underline{/90^\circ})\,\mathbf{I}$$

where the quantity $(X_L\underline{/90^\circ})$ must have the dimensions of ohms if it is to yield volts when multiplied by the current. This quantity, which represents the opposition of an inductor to a sinusoidal current, is called *inductive reactance*, X_L. Intuitively we might suspect that since inductor voltage is proportional to L and to the rate of change of current, and since rate of change of current depends on ω, X_L should be proportional to ω and L, and indeed it is. Specifically,

2) $$|X_L| = \omega L = 2\pi f L \qquad (5\text{-}14)$$

However, there is more to it. Note from Fig. 5-15c that E leads I by 90°. That means that multiplying I by the inductive reactance must also have the effect of giving the right-hand side of Eq. (1) a +90° phase shift in order to yield E at 0°. Since multiplication by the j operator effects a 90° phase rotation, the full description of inductive reactance is

3) $$\mathbf{X}_L = jX_L = X_L\underline{/90^\circ} \qquad (5\text{-}15)$$

Thus

4) $$\mathbf{E} = (jX_L)\mathbf{I} = (X_L\underline{/90^\circ})\mathbf{I} \qquad (5\text{-}16)$$

and for Fig. 5-15c,

5) $$\mathbf{E} = E\underline{/0^\circ} = (X_L\underline{/90^\circ})(I\underline{/-90^\circ}) = X_L I\underline{/0^\circ}$$

A vector diagram which plots inductive reactance along the $+j$ (imaginary) axis and resistance along the real axis is called an impedance diagram. This is shown in Fig. 5-16a. Since capacitive reactance was plotted along the $-j$ axis, we can relabel the imaginary axis so that reactances appear as shown in Fig. 5-16b.

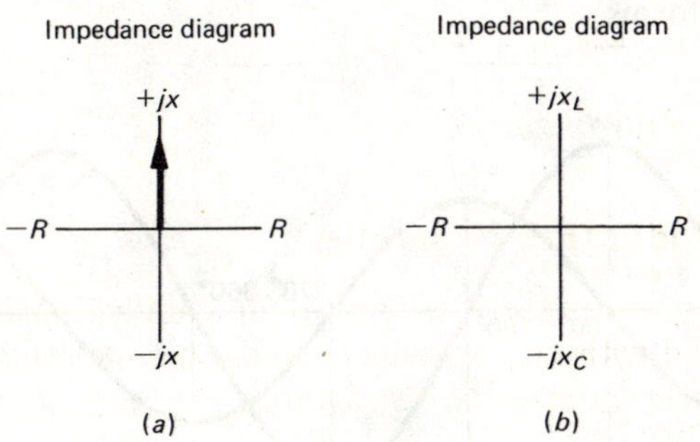

FIGURE 5-16

Now work through problem PS 5-12.

5-10 Inductive susceptance

The reciprocal of inductive reactance is inductive susceptance. Thus

$$\mathbf{B}_L = \frac{1}{jX_L} = \frac{1}{X_L\underline{/90^\circ}} = B_L\underline{/-90^\circ} = -jB_L \qquad (5\text{-}17)$$

where B_L is in mhos, since X_L is in ohms.

Inductive susceptance may be plotted on an admittance diagram, as shown in Fig. 5-17. Since capacitive susceptance was written as $+jB_C$ and inductive susceptance is $-jB_L$, it is convenient to label the imaginary axis $+jB_C$ and $-jB_L$ as shown. The horizontal, or axis of real numbers, is again conductance $G = 1/R$.

Now work through problem PS 5-13.

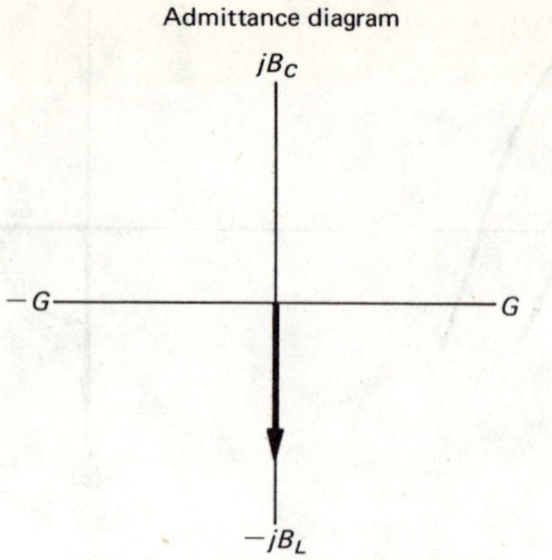

FIGURE 5-17

5-11 Power in the purely inductive circuit

A pure inductor cannot dissipate any power. To see this more clearly, consider Fig. 5-18, where the voltage and current waveforms of an inductor are sketched in the upper part of the figure and the instantaneous power waveform in the lower part. Note that i lags e by 90°, as is characteristic of an inductor. The p waveform is again generated by point-by-point multiplication of the e and i waveforms.

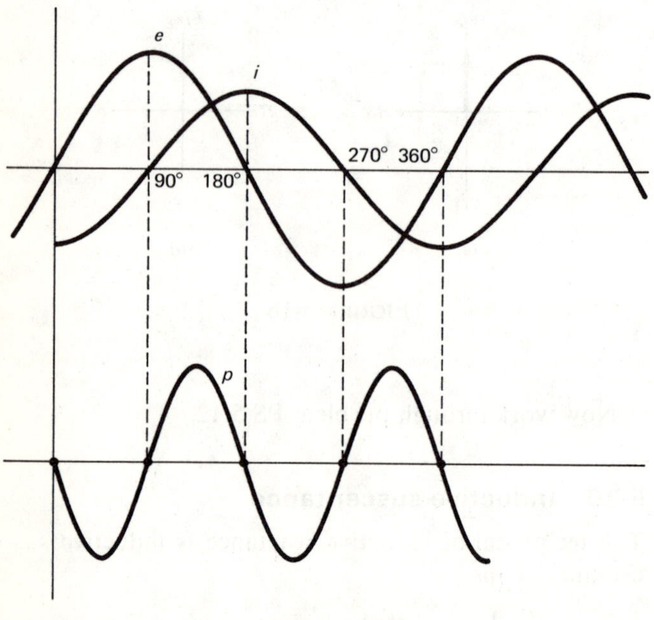

FIGURE 5-18

Clearly at $t = 0$, $p = 0$ since $e = 0$, and this is also true at 180 and 360°. Also the ei product must be zero at 90 and 270°, since the i waveform crosses the zero axis there. Thus the p waveform has zero crossings at 0, 90, 180, 270, and 360°, which is indicated by the double-frequency p waveform of Fig. 5-18. Note that the p waveform is symmetrical about the horizontal axis, and hence its average value is zero. This means that a pure inductor dissipates no power. From a physical viewpoint, the inductor absorbs power when p is positive but returns this stored energy when p swings negative.

For formal proof that the p waveform is a double-frequency sinusoid, work through problem PS 5-14.

PROBLEMS WITH SOLUTIONS

PS 5-1 What magnitude of voltage will be developed across a 6 mH inductor if the current through it changes 45 mA in 30 μsec?

SOLUTION

$$e = L\frac{\Delta i}{\Delta t}$$

$$= (6 \times 10^{-3}\text{H})\frac{45 \times 10^{-3}\text{ A}}{30 \times 10^{-6}\text{ sec}} = 9\text{ V}$$

PS 5-2 What is the inductance of the coil if closing the switch in Fig. PS 5-2*a* causes the current waveform of Fig. PS 5-2*b*?

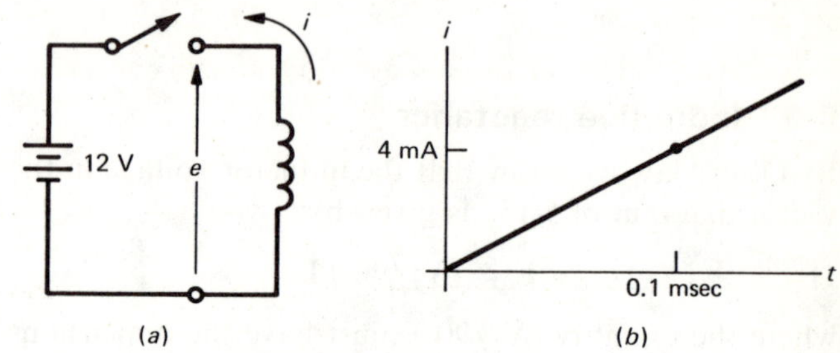

FIGURE PS 5-2

SOLUTION

1) $$e = L\frac{di}{dt}$$

Solving for L,

2) $$L = \frac{e}{di/dt} = \frac{12\text{ V}}{4 \times 10^{-3}\text{A}/0.1 \times 10^{-3}\text{ sec}} = 0.3\text{ H}$$

PS 5-3 Determine the inductor voltage in Fig. PS 5-3*a* if the inductor current follows the i waveform in Fig. PS 5-3*b*.

SOLUTION From $t = 0$ to $t = 1$ sec, $di/dt = 0$ and hence $e = 0$. From $t = 1$ sec to $t = 2$ sec, $di/dt = -4$ A/1 sec. Hence

$$e = L\frac{di}{dt} = 0.2\text{ H}\left(\frac{-4\text{ A}}{1\text{ sec}}\right) = -0.8\text{ V}$$

From $t = 2$ to $t = 3$ sec, $di/dt = 0$ and hence $e = 0$. At $t = 3$ sec, $di/dt = +\infty$, since the slope is vertical. Hence, e would jump, at least in theory, to plus infinity. In actual practice the inevitable circuit capacitance pre-

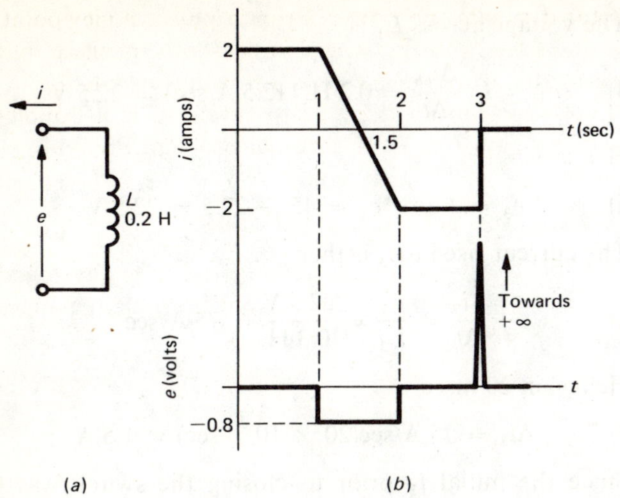

FIGURE PS 5-3

vents an instantaneous voltage change and instead we would see a large positive spike. The resultant e waveform is plotted below the i waveform in Fig. PS 5-3.

PS 5-4 Determine the current in a 0.8-henry inductor 3 sec after a 24-volt dc source is switched across it. Assume zero initial inductor current.
SOLUTION

$$e = L\frac{\Delta i}{\Delta t}$$

Therefore

$$\Delta i = \frac{e}{L}\Delta t$$

$$= \frac{24\ \text{V}}{0.8\ \text{H}}(3\ \text{sec}) = 90\ \text{A}$$

Since there is zero initial current, $i = \Delta i = 90$ amps.

PS 5-5 A 12-volt battery is connected across a lossless 10-henry inductor. How much energy is stored in the inductor at the end of 4 sec?
SOLUTION

$$i = \Delta i = \frac{e}{L}(\Delta t)$$

$$i = \frac{12\ \text{V}}{10\ \text{H}}(4\ \text{sec}) = 4.8\ \text{A}$$

$$J = \tfrac{1}{2}Li^2$$

$$= \tfrac{1}{2}(10)(4.8^2) = 115\ \text{J}$$

PS 5-6 Determine the inductance of a straight 4-in. length of #18 wire (diameter = 0.040 in.).
SOLUTION From Eq. (5-8),

$$L = 0.00508l\left(2.303\log_{10}\frac{4l}{d} - 0.75\right)\mu\text{H}$$

$$= 0.107\ \mu\text{H}$$

PS 5-7 What is the voltage drop across the 4-in.-long wire of the previous problem if the current through it swings 15 mA in 0.5 μsec?
SOLUTION

$$\frac{\Delta i}{\Delta t} = \frac{15\ \text{mA}}{0.5\ \mu\text{sec}} = 30{,}000\ \text{A/sec}$$

$$e = L\frac{\Delta i}{\Delta t}$$

$$= (0.107 \times 10^{-6}\text{H})(3 \times 10^4\ \text{A/sec})$$

$$= 3.21\ \text{mV}$$

PS 5-8 What is the equivalent inductance of 1.8 H, 2 H, 3 H, and 5 H inductors connected in series?
SOLUTION

$$L_{\text{eq}} = L_1 + L_2 + L_3 + L_4$$

$$= 1.8 + 2 + 3 + 5 = 11.8\ \text{H}$$

PS 5-9 Determine the voltage across L_2 in Fig. PS 5-9 at $t = 0.01$, 0.1, 1, and 10 sec.

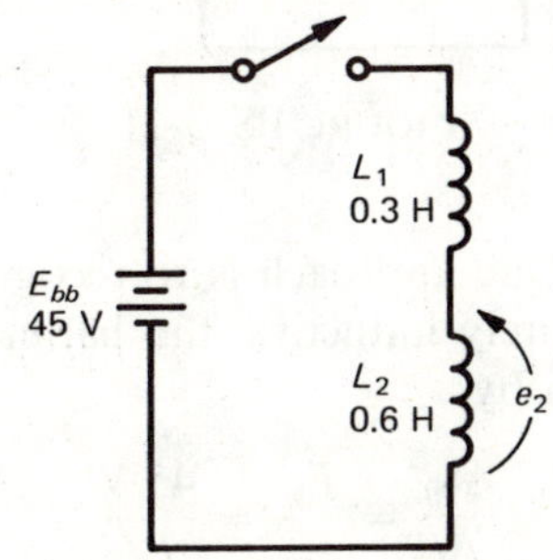

FIGURE PS 5-9

SOLUTION First we find

1) $$L_{\text{eq}} = L_1 + L_2 = 0.9\ \text{H}$$

Then from

2) $$e = L\frac{\Delta i}{\Delta t}$$

we obtain

3) $$\frac{\Delta i}{\Delta t} = \frac{e}{L} = \frac{E_{bb}}{L_{\text{eq}}} = \frac{45\ \text{V}}{0.9\ \text{H}} = 50\ \text{A/sec}$$

Note that $\Delta i/\Delta t$ is a constant, and since

4) $$e_2 = L_2\frac{\Delta i}{\Delta t}$$

$$= 0.6\ \text{H}\ (50\ \text{A/sec}) = 30\ \text{V}$$

and since $\Delta i/\Delta t$ is constant, $e_2 = 30$ V for all values of t.

PS 5-10 Inductors of 2, 3, and 6 mH are connected in parallel. Determine the equivalent inductance.

SOLUTION

$$\frac{1}{L_{eq}} = \frac{1}{L_1} + \frac{1}{L_2} + \frac{1}{L_3}$$

$$= \frac{1}{2\text{ mH}} + \frac{1}{3\text{ mH}} + \frac{1}{6\text{ mH}}$$

or

$$L_{eq} = 1\text{ mH}$$

PS 5-11 In L_3, determine the current 20 msec after the switch in Fig. PS 5-11 closes.

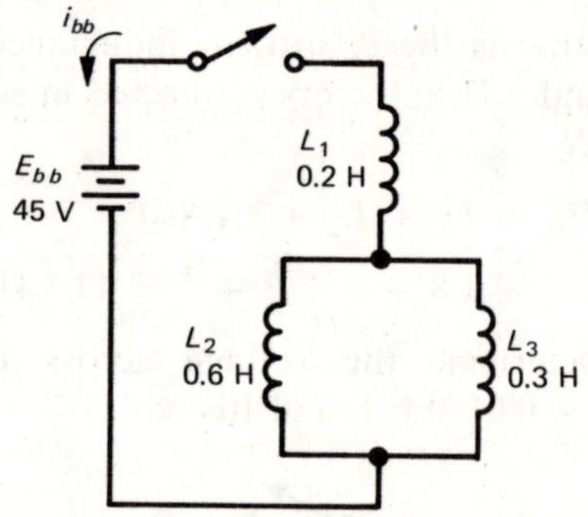

FIGURE PS 5-11

SOLUTION One approach is to recognize that since the circuit is purely inductive, the battery current will be a ramp given by

1) $$\frac{\Delta i_{bb}}{\Delta t} = \frac{E_{bb}}{L_{eq}} = \frac{45\text{ V}}{L_{eq}}$$

where

2) $$L_{eq} = L_1 + \frac{L_2L_3}{L_2 + L_3} = 0.2\text{ H} + 0.2\text{ H} = 0.4\text{ H}$$

Substituting (2) into (1) yields

3) $$\frac{\Delta i_{bb}}{\Delta t} = \frac{45\text{ V}}{0.4\text{ H}} = 112.5\text{ A/sec}$$

The voltage across L_1 is

4) $$e_1 = L_1\frac{\Delta i_{bb}}{\Delta t} = 0.2\text{ H}(112.5\text{ A/sec}) = 22.5\text{ V}$$

Hence

5) $$e_3 = E_{bb} - e_1 = 45 - 22.5 = 22.5\text{ V}$$

The current rise in L_3 is then

6) $$\frac{\Delta i_3}{\Delta t} = \frac{e_3}{L_3} = \frac{22.5\text{ V}}{0.3\text{ H}} = 75\text{ A/sec}$$

Hence, in 20 msec,

7) $$\Delta i_3 = 75\text{ A/sec}(20 \times 10^{-3}\text{ sec}) = 1.5\text{ A}$$

Since the initial i_3 prior to closing the switch was 0, $i_3 = \Delta i_3 = 1.5$ amp in 20 msec.

PS 5-12 Determine the current $\mathbf{I}_5$ in Fig. PS 5-12*a*.
SOLUTION One approach is to write some loop equations. However, an easier method is to apply Thévenin's theorem by temporarily removing L_5 and then thevenizing to the right and to the left of it. But first let us determine ω, as we usually need it several times.

1) $$\omega = 2\pi f = 2\pi(2 \times 10^3) = 12.57 \times 10^3\text{ rad/sec}$$

2) $$V_{OCA} = \frac{jX_{L2}}{jX_{L1} + jX_{L2}}\mathbf{E}_{in} = \frac{j\omega L_2}{j\omega L_1 + j\omega L_2}\mathbf{E}_{in}$$

$$= \frac{L_2}{L_1 + L_2}\mathbf{E}_{in} = \frac{40\text{ mH}}{60\text{ mH} + 40\text{ mH}}(36\angle 0^\circ)$$

$$= 14.4\angle 0^\circ\text{ V}$$

3) $$X_{thA} = \frac{(j\omega L_1)(j\omega L_2)}{j\omega L_1 + j\omega L_2} = j\omega\left(\frac{L_1L_2}{L_1 + L_2}\right)$$

$$= j12.57 \times 10^3\left(\frac{60\text{ mH} \times 40\text{ mH}}{60\text{ mH} + 40\text{ mH}}\right) = j301.7\ \Omega$$

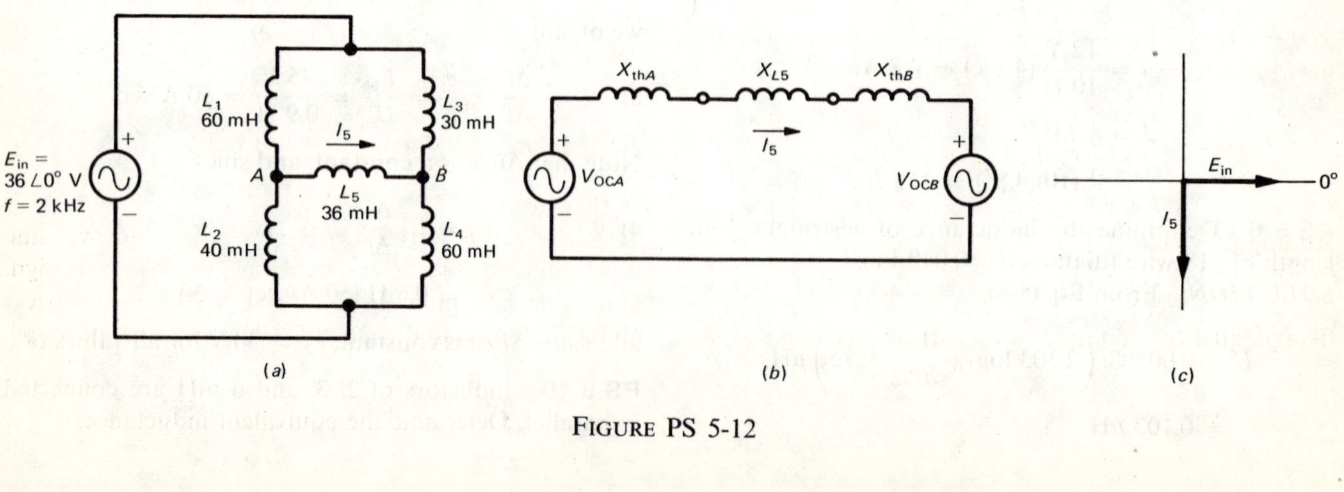

FIGURE PS 5-12

4) $V_{OCB} = \dfrac{jX_{L4}}{jX_{L3} + jX_4}\mathbf{E}_{in} = \dfrac{L_4}{L_3 + L_4}\mathbf{E}_{in}$

$$= \frac{60\text{ mH}}{60\text{ mH} + 30\text{ mH}}(36\underline{/0°}\text{ V}) = 24\underline{/0°}\text{ V}$$

5) $X_{thB} = jX_{LeqB} = j\omega\dfrac{L_3L_4}{L_3 + L_4}$

$$= j12.57 \times 10^3(20\text{ mH}) = j251.4\ \Omega$$

6) $X_{L5} = j\omega L_5 = j(12.57 \times 10^3\text{ rad/sec})(36 \times 10^{-3}\text{ H})$

$$= j\,452.5\ \Omega$$

Reinserting L_5 as shown in Fig. PS 5-12*b* yields

7) $\mathbf{I}_5 = \dfrac{V_{OCB} - V_{OCA}}{jX_{thA} + jX_{L5} + jX_{thB}}$

$$= \frac{14.4\underline{/0°} - 24\underline{/0°}}{j301.7 + j452.5 + j251.4} = \frac{-9.6\underline{/0°}}{j1006}$$

$$= \frac{9.6\underline{/180°}\text{ V}}{1006\underline{/90°}\ \Omega} = 9.55\underline{/-90°}\text{ mA}$$

A vector diagram which illustrates the phase relationships between E_{in} and I_5 is shown in Fig. PS 5-12*c*.

PS 5-13 Determine I in Fig. PS 5-13 if $E = 120\text{ V}\underline{/0°}$ at $f = 60$ Hz, $L = 1$ H, $L_2 = 2$ H, $L_3 = 3$ H, $L_4 = 4$ H, and $L_5 = 5$ H. Use the susceptance concept whenever it seems appropriate.

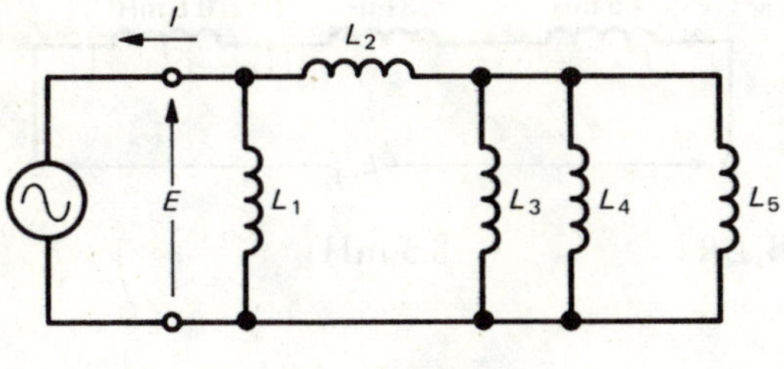

FIGURE PS 5-13

SOLUTION First find the equivalent inductance the network presents to the voltage source. Thus

$$L_{eq345} = \frac{1}{\dfrac{1}{L_3} + \dfrac{1}{L_4} + \dfrac{1}{L_5}} = \frac{1}{0.7833} = 1.277\text{ H}$$

Adding L_2 yields

$$L_{eq2345} = 2\text{ H} + 1.277\text{ H} = 3.277\text{ H}$$

The parallel combination of L_1 and the above is

$$L_{eq} = \frac{1\text{ H}(3.277\text{ H})}{1\text{ H} + 3.277\text{ H}} = 0.7662\text{ H}$$

$$\mathbf{B}_{L_{eq}} = \frac{1}{j\omega L_{eq}} = \frac{1\underline{/-90°}}{2\pi(60)(0.7662)}$$

$$= 3.462 \times 10^{-3}\underline{/-90°}\text{ mhos}$$

Therefore

$$\mathbf{I} = \mathbf{B}_{L_{eq}}\mathbf{E} = (3.462 \times 10^{-3}\underline{/-90°}\text{ mhos})(120\underline{/0°}\text{ V})$$

$$= 0.4155\underline{/-90°}\text{ A}$$

As might be expected, the current lags the voltage 90° in this purely inductive circuit.

PS 5-14 Prove that a pure inductance dissipates zero power.

SOLUTION If the voltage across the inductor is given by

1) $e = E_{pk}\sin\omega t$

the current, which lags the voltage by 90°, must then be

2) $i = I_{pk}\sin(\omega t - 90°) = -I_{pk}\cos\omega t$

Remember, a cosine wave leads a sine wave by 90°, and hence a cosine wave retarded 180° is equivalent to a sine wave with a (−90°) phase lag. Since

3) $p = ei$

we have from (1) and (2)

4) $p = -E_{pk}I_{pk}\sin\omega t\cos\omega t$

We find a sin x cos y (where $x = y$) form in (4). Just to refresh our memories, let us derive a more meaningful form of this from the basic identities in Table 1-1 of Chapter 1. Let us see how we can obtain this form. Recalling that

5) $\sin(x + y) = \sin x\cos y + \cos x\sin y$

6) $\sin(x - y) = \sin x\cos y - \cos x\sin y$

Adding (5) and (6) and letting $x = y = \theta$ yields

7) $\sin 2\theta = 2\sin\theta\cos\theta$

from which

8) $\sin\theta\cos\theta = \frac{1}{2}\sin 2\theta$

Setting $\theta = \omega t$ and using (8) in (4) yields

9) $p = -\frac{1}{2}E_{pk}I_{pk}\sin 2\omega t$

Thus the power waveform is a double-frequency sine wave shifted 180°, as indicated by the negative sign. Since this is a pure sinusoid, it must be symmetrical about the horizontal axis and have zero average value.

PROBLEMS WITH ANSWERS

PA 5-1 A 24-volt dc source is switched across a 0.08 H inductor. If the maximum current to be drawn from the dc source is 3 amps, what is the maximum time the switch can be closed?

ANSWER $t = 0.01$ sec

PA 5-2 Sketch the voltage-versus-time curve if a 1 H inductor is driven by a current source with the waveform of Fig. PA 5-2*a*.

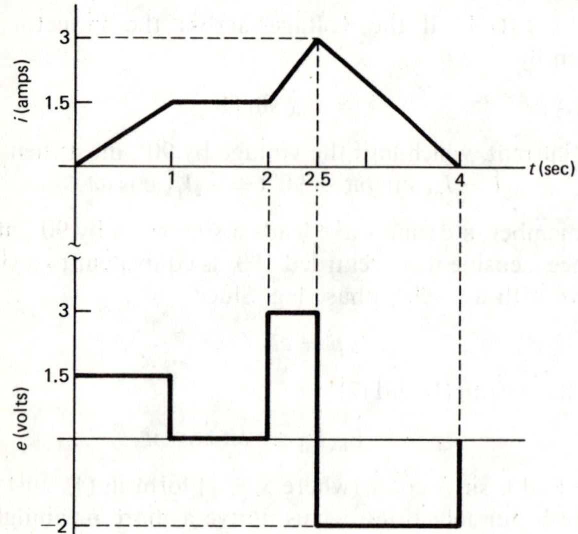

ANSWER See Fig. PA 5-2*b*.

PA 5-3 The switch in Fig. PA 5-3 is closed for 45 msec and then opened instantaneously. What is the peak voltage developed across the resistor?

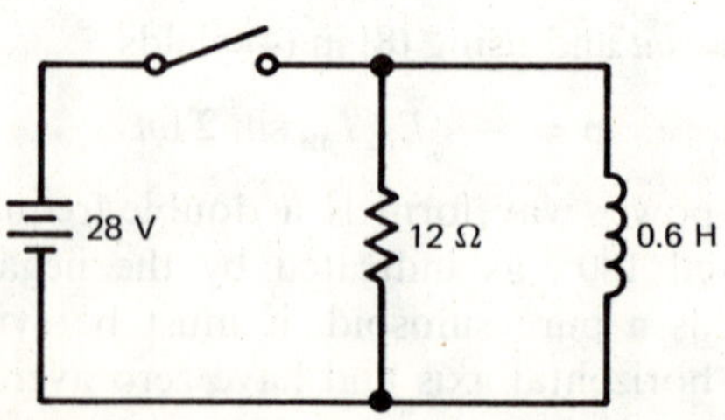

ANSWER $|E_{pk}| = 25.2$ V

PA 5-4 What is the approximate inductance of the configuration shown in Fig. PA 5-4?

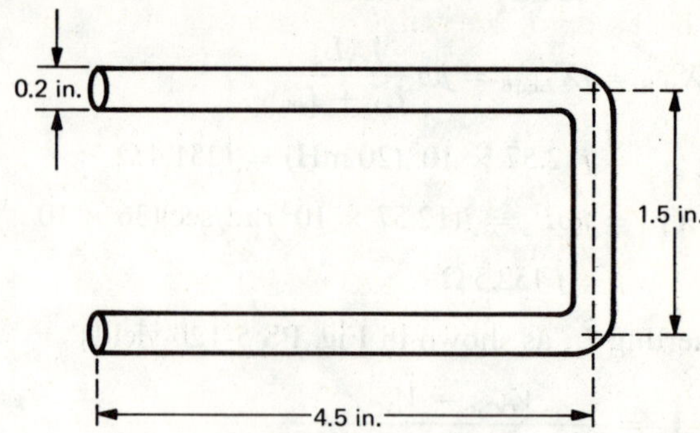

ANSWER 0.129 μH

PA 5-5 A 1/2-in. wide, 1-in. long, 50-turn winding exhibits what inductance?

ANSWER 12.8 μH

PA 5-6 Determine L_{eq} in Fig. PA 5-6.

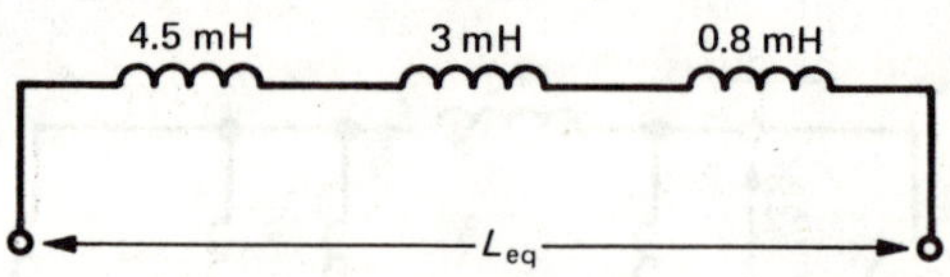

ANSWER 8.3 mH

PA 5-7 Determine L_x in Fig. PA 5-7 so that $L_{eq} =$ 2.5 mH

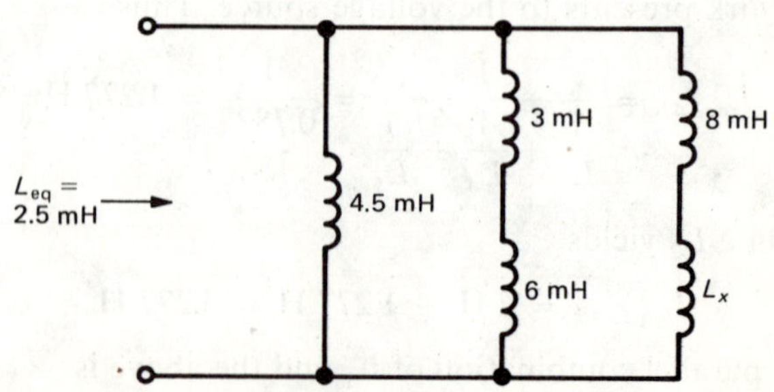

ANSWER $L_x = 7$ mH

PA 5-8 Prove that the inductive current divider relationship

$$\frac{i_{L2}}{i} = \frac{(L_1 \| L_3)}{L_2 + (L_1 \| L_3)}$$

may be used to determine the current in L_2 in Fig. PA 5-8.

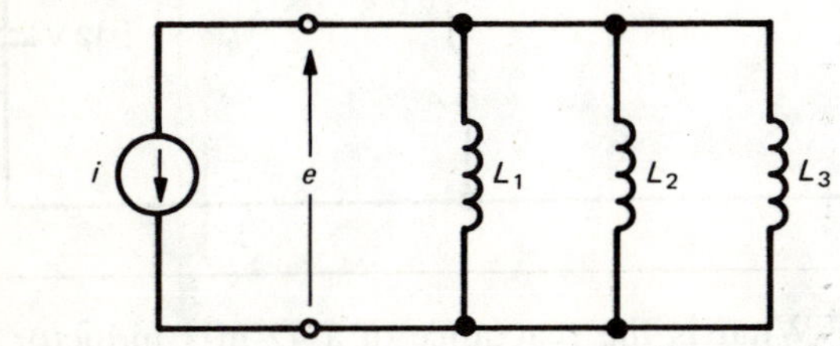

PA 5-9 What is the reactance and susceptance of a 12-mH inductor at 400 kHz?

ANSWER

$X_L = 30.2 \text{ k}\Omega$, $B_L = 3.32 \times 10^{-5}$ mhos

PA 5-10 Determine E_x in Fig. PA 5-10. Assume $\mathbf{E}_A = 18/\underline{0^\circ}$ volts, $\mathbf{I}_B = 0.06/\underline{30^\circ}$ amps, $L_1 = 6$ H, $L_2 = 3$ H, and $\omega = 100$ rad/sec.

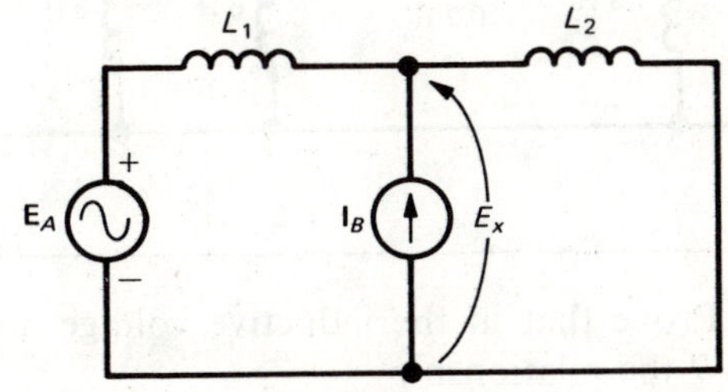

ANSWER $15.9/\underline{-40.9^\circ}$ V

PROBLEMS WITHOUT ANSWERS

P 5-1 What voltage is developed across a 3-henry inductor if the current through it is changing at a rate of 1.5 mA/1 msec?

P 5-2 In Fig. P 5-2, determine I_{bb} 2.5 sec after the switch closes.

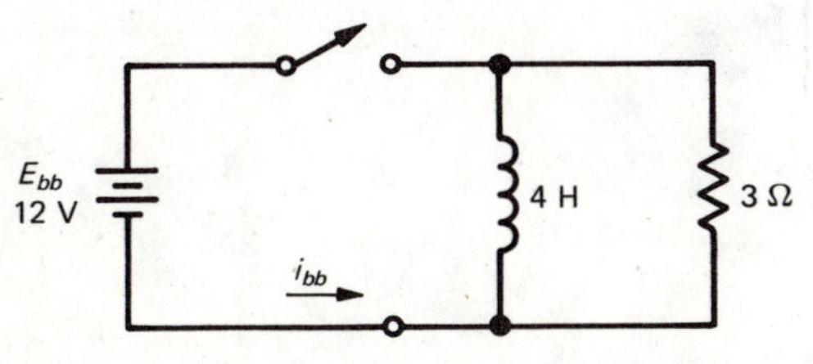

P 5-3 How much energy has the battery in the previous problem delivered at the end of the 2.5 sec? How much energy is stored in the network that the battery is supplying i_{bb} to?

P 5-4 A coil has a 3/8-in. radius and 2.5-in. length. How many turns of wire are required for an inductance of 18 μH?

P 5-5 Inductors of 1, 2, and 3.5 henry are connected in series. What is the equivalent inductance?

P 5-6 Determine L_{eq} in Fig. P 5-6.

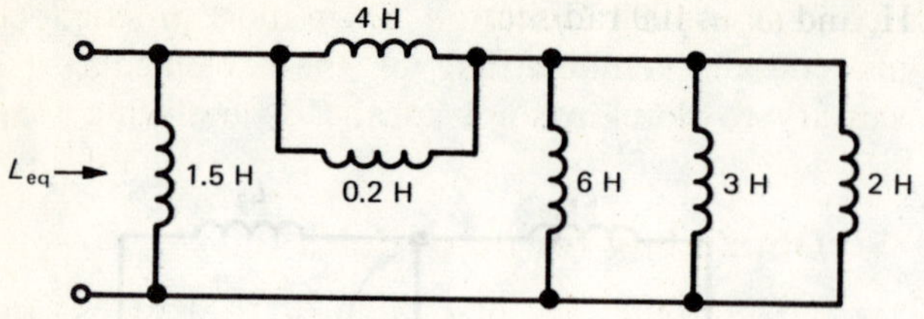

P 5-7 Prove that in the inductive voltage divider of Fig. P 5-7 the relationship

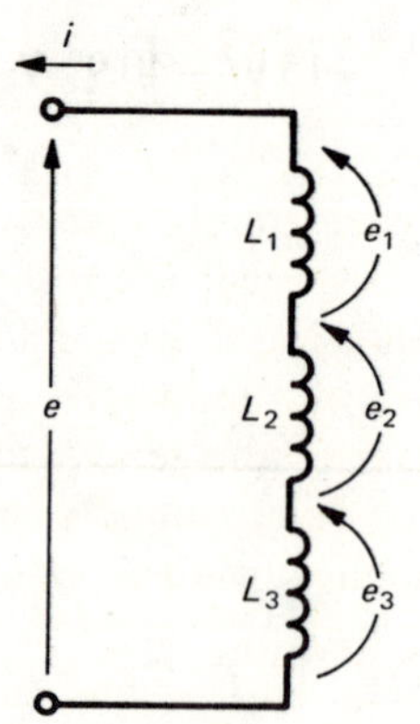

$$\frac{e_2}{e} = \frac{L_2}{L_1 + L_2 + L_3}$$

holds true for any e or i waveform.

P 5-8 Determine V_{AB} in Fig. P 5-8 two seconds after the switches close simultaneously.

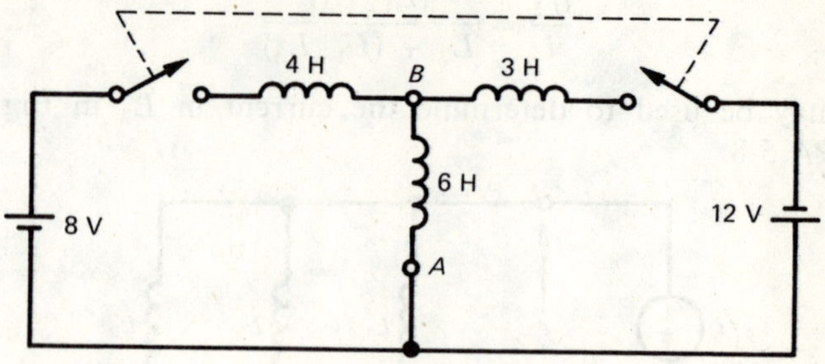

P 5-9 What is the reactance of a 42-mH inductor at 20 kHz?

P 5-10 Determine the current if a voltage given by $e = 72 \sin 377t$ volts is impressed across a 2.1-H inductor?

P 5-11 What is the susceptance at 60 Hz seen looking into the input terminals of Fig. P 5-6?

P 5-12 Determine E_{AB} in Fig. P 5-12.

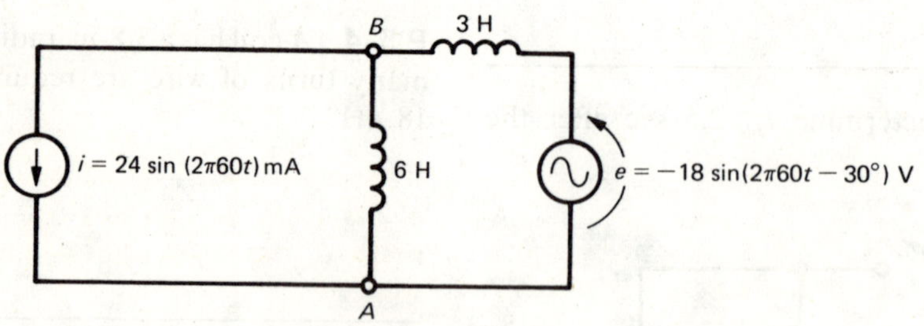

6 *RLC* circuits

Up to this point we have studied the response of a circuit containing either pure resistance, capacitance, or inductance to sinusoidal excitation. Since most practical circuits contain combinations of these elements, it is necessary to develop some means of analyzing them.

6-1 Ohm's law for ac circuits

As a starting point the first thing to recognize is that Ohm's law in its basic form is applicable to ac as well as dc circuits. That is, the current is directly proportional to the applied voltage and inversely proportional to something analogous to resistance which opposes the flow of current. In equation form,

1) $$\mathbf{I} = \frac{\mathbf{E}}{(?)}$$

If the circuit were purely resistive, the unknown (?) factor would be resistance; and if the circuit were purely capacitive or inductive, it would be the capacitive reactance $(-jX_C)$ or the inductive reactance $(+jX_L)$ respectively. However, in the general case where the circuit opposition may be due to any combination of R, L, and C, we call the net effect of these elements the impedance of the circuit. Impedance is given the symbol Z, it is measured in ohms, and it has both magnitude and direction. Hence, the denominator of (1) would be written as Z.

6-2 The impedance concept

How then do we determine the impedance Z of circuit so that we may use Ohm's law in the following relationships?

2) $$\mathbf{I} = \frac{\mathbf{E}}{\mathbf{Z}} \qquad (6\text{-}1a)$$

3) $$\mathbf{E} = \mathbf{ZI} \qquad (6\text{-}1b)$$

4) $$\mathbf{Z} = \frac{\mathbf{E}}{\mathbf{I}} \qquad (6\text{-}1c)$$

If we can use the impedance concept in any of the forms of Eq. (6-1), we can just as well use it in applying the network theorems and Kirchhoff's laws to ac circuit analysis.

Intuitively we sense that impedance for an *RLC* circuit must somehow be the combined effect of R, L, and C, since from equation (1) $\mathbf{Z}$ must $= R\angle 0°$ in a purely resistive circuit, or $\mathbf{Z} = -jX_C = X_C\angle -90°$ in a purely capacitive circuit, or $\mathbf{Z} = jX_L = X_L\angle 90°$ in a purely inductive circuit. The question is how do we get the net effect of all these, or any combination, together? The answer lies in what is called an impedance diagram, as shown in Fig. 6-1. Note that resistance is plotted along

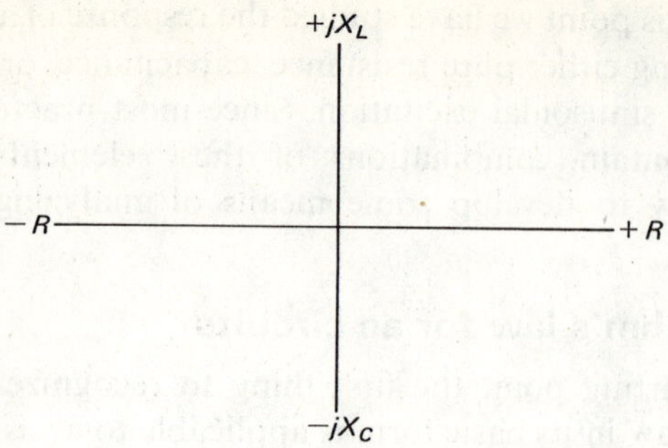

FIGURE 6-1

the real (horizontal) axis and reactance along the imaginary (vertical) axis.

You will recall that, in the purely resistive circuit, I is in phase with E so that if E is at a reference angle of zero degrees, I is also at zero degrees. Thus from Eq. (6-1c),

$$\mathbf{Z} = \frac{E\underline{/0^\circ}}{I\underline{/0^\circ}} = R\underline{/0^\circ} \tag{6-2a}$$

However, in the purely capacitive circuit, the current led the voltage by 90°. Hence, if we consider E at a reference angle of zero degrees, we have from Eq. (6-1c)

$$\mathbf{Z} = \frac{E\underline{/0^\circ}}{I\underline{/90^\circ}} = X_C\underline{/-90^\circ} = -jX_C \tag{6-2b}$$

And in the purely inductive circuit where I lags E by 90°, we have

$$\mathbf{Z} = \frac{E\underline{/0^\circ}}{I\underline{/-90^\circ}} = X_L\underline{/90^\circ} = jX_L \tag{6-2c}$$

Inspecting Eq. (6-2a), (6-2b), and (6-2c), we see that resistance has zero degrees associated with it, while X_C and X_L have -90° and $+90^\circ$ respectively. Since these quantities have magnitude and direction, we will treat them as vector quantities and plot them on the impedance diagram along their respective axes. It should be borne in mind, however, that impedance (**Z**) is not a rotating vector similar to **E** and **I**.

Since the coordinate system of Fig. 6-1 defines the impedance plane, we can locate any resistance, reactance, or impedance as a point in that plane. For example, assume it is desired to locate the impedance $6\underline{/0^\circ}$ ohms. We simply move out 6 units along a line of 0° (which means along the R axis) and make a mark at $R = 6$. If it were desired to express this point in terms of rectangular rather than polar coordinates, we would locate $6 + j0$, which is of course identical to $6\underline{/0^\circ}$ in polar form.

If we had a capacitive reactance of $-j9$ ohms $= 9\underline{/-90^\circ}$ ohms, we simply move down the $-jX_C$ axis 9 units. Similarly, to locate an inductive reactance of $j4$ ohms $= 4\underline{/90^\circ}$ ohms, we move 4 units up the jX_L axis. These points are located in Fig. 6-2.

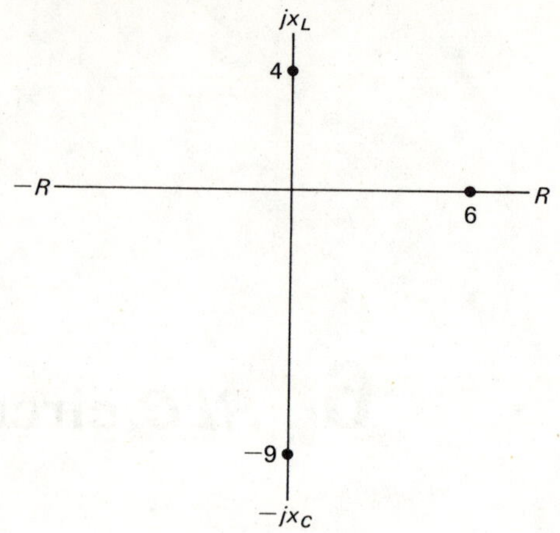

FIGURE 6-2

Now suppose you somehow had to locate an impedance given by $\mathbf{Z} = 6 + j4$ ohms. Intuitively we might suspect that this impedance is somehow related to a resistance and inductance since it contains a real part (6 ohms) and a plus j component ($j4$ ohms). However, that is just a passing thought. The problem is how to locate this impedance on our impedance plane and the manner in which it is done is to simply move 6 units along the $+R$ axis and then straight up in the $j4$ direction. This is shown in Fig. 6-3a. The same point could have

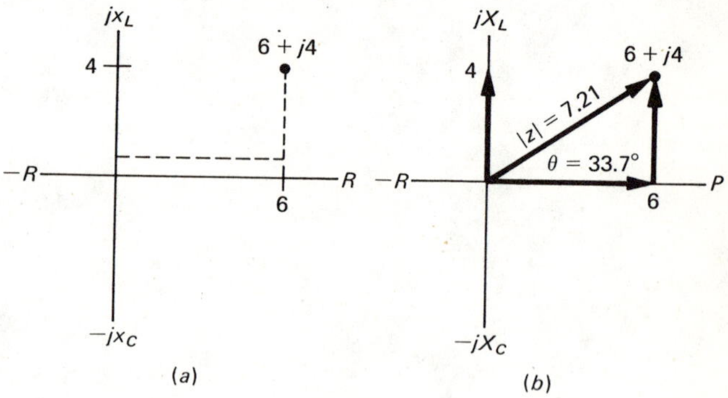

FIGURE 6-3

been located by drawing a vector of $6\underline{/0^\circ}$ ohms and another at $4\underline{/90^\circ}$ ohms and then vectorially adding them as shown in Fig. 6-3b. The tip of the resultant (Z) would lie on the point $6 + j4$.

If it is desired to express this resultant in polar form, we merely apply the pythagorean theorem (see Fig. 6-4)

$$Z = \sqrt{R^2 + X^2} \tag{6-3a}$$

where Z = the magnitude of the impedance (ohms)
R = net resistive component
X = net reactive component

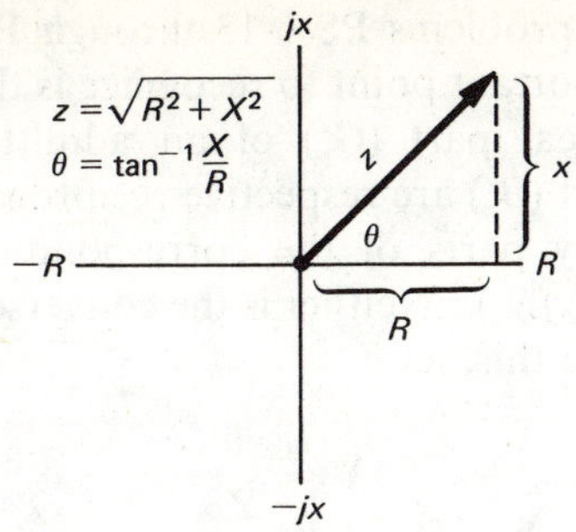

FIGURE 6-4

to determine the magnitude of Z. The phase angle is given by

$$\theta = \tan^{-1}\frac{X}{R} \tag{6-3b}$$

Thus converting $\mathbf{Z} = 6 + j4$ ohms from rectangular to polar form yields

$$Z = \sqrt{6^2 + 4^2} = 7.21\ \Omega$$

and

$$\theta = \tan^{-1}\tfrac{4}{6} = 33.7^\circ$$

It is just as often true that you may have to convert from polar to rectangular form. Thus if you are given $Z\underline{/\theta}$ and wish to resolve it into its resistive (R) and reactive (X) components, we use the relationships (see Fig. 6-5)

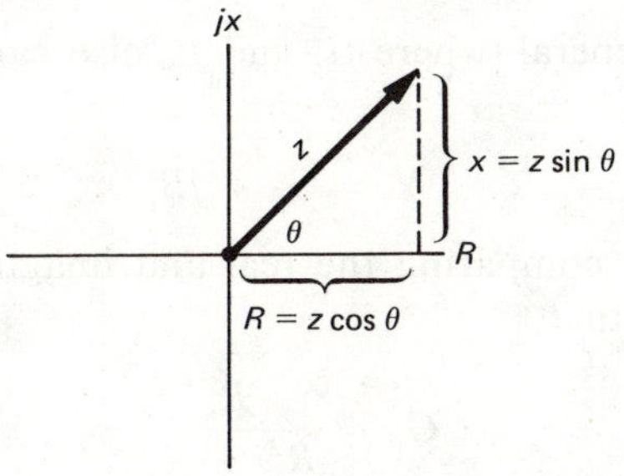

FIGURE 6-5

$$R = Z\cos\theta \tag{6-4a}$$

$$X = Z\sin\theta \tag{6-4b}$$

Thus $\mathbf{Z} = 18\underline{/120^\circ}\ \Omega$ resolves into

$$R = 18\cos 120^\circ = 18\,(-0.5) = -9\ \Omega$$

$$X = 18\sin 120^\circ = 18\,(0.866) = +15.6\ \Omega$$

Hence $\mathbf{Z} = 18\underline{/120^\circ}$ ohms $= -9 + j15.6\ \Omega$

Note that the resistive component of Z was negative and the reactive part was positive and hence inductive in nature.

Manipulations involving rectangular to polar and vice versa conversions may be conveniently performed on a suitable slide rule, and it is highly recommended the reader possess one and be familiar with this type of operation.

If you know your vector (phasor) algebra and can convert from rectangular to polar and vice versa form with ease, you've got it made, because from here on most of the circuit analysis involves the analytic tools of the previous text.[1] The major difference is that you will be working with phasors and using the related math (complex number algebra); but Kirchhoff's laws, the network theorems, etc., still hold.

Now carefully work through problems PS 6-1 through PS 6-8 in order to be sure you are ready to proceed. These problems will review many concepts of vector algebra.

6-3 The series *RLC* circuit

In dc circuit analysis the equivalent resistance of resistors in series is the sum of the individual resistors. In ac circuits we may have an analogous situation except that now we speak of impedances connected in series, and the equivalent impedance is the sum of the individual impedances. Thus (see Fig. 6-6)

$$\mathbf{Z}_{eq} = \mathbf{Z}_1 + \mathbf{Z}_2 + \mathbf{Z}_3 + \cdots + \mathbf{Z}_N \tag{6-5}$$

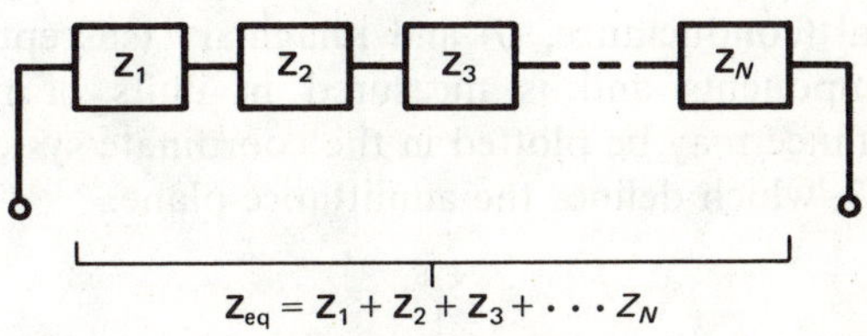

FIGURE 6-6

Any of the impedances may be a pure resistance or a pure reactance. Some may be expressed as complex numbers in rectangular or polar form. All we have to worry about is being able to form $\mathbf{Z}_{eq}$ by the correct application of vector addition. The phasor addition may be performed graphically, but this is usually time-consuming and inherently inaccurate due to plotting errors. A much more desirable approach is the analytic one in which all the individual impedances are first converted to rectangular form. Then we can algebraically add all the real (resistive) components and all the imaginary (reactive) components. The result is a complex number which is $\mathbf{Z}_{eq}$. If desired, this rectangular form of $\mathbf{Z}_{eq}$ can easily be converted to polar form by Eq. (6-3) or by suitable procedures on an appropriate slide rule.

Once $\mathbf{Z}_{eq}$ is determined, we can plug into Eq. (6-1) to apply Ohm's law or whatever.

[1] The terms *vector* and *phasor* are used interchangeably in this text to provide a gradual transition between the older, commonly used (vector) nomenclature and the newer (phasor) terminology.

Now work problems PS 6-9 through PS 6-12 so that the nitty-gritty techniques are completely clear to you.

6-4 Admittance and the parallel *RLC* circuit

The impedance concept can be used to solve parallel as well as series circuits. For example, any two parallel-connected impedances, say $\mathbf{Z}_1$ and $\mathbf{Z}_2$, may be simplified to an equivalent impedance $\mathbf{Z}_{eq}$ given by

1) $$\mathbf{Z}_{eq} = \frac{\mathbf{Z}_1\mathbf{Z}_2}{\mathbf{Z}_1 + \mathbf{Z}_2} \qquad (6\text{-}6)$$

The form of Eq. (1) is identical with that for two resistors in parallel.

Equation (1) may be successively applied to a string of parallel-connected resistors, or we may write

2) $$\frac{1}{\mathbf{Z}_{eq}} = \frac{1}{\mathbf{Z}_1} + \frac{1}{\mathbf{Z}_2} + \frac{1}{\mathbf{Z}_3} + \cdots + \frac{1}{\mathbf{Z}_N} \qquad (6\text{-}7)$$

However, just as conductance is the reciprocal of resistance, we have an analogous quantity called admittance (**Y**), which is the reciprocal of impedance.

3) $$\mathbf{Y} = \frac{1}{\mathbf{Z}}$$

The admittance is also a vector quantity which in general has real (conductance, *G*) and imaginary (susceptance, *B*) components and is measured in units of mhos. Admittance may be plotted in the coordinate system of Fig. 6-7, which defines the admittance plane.

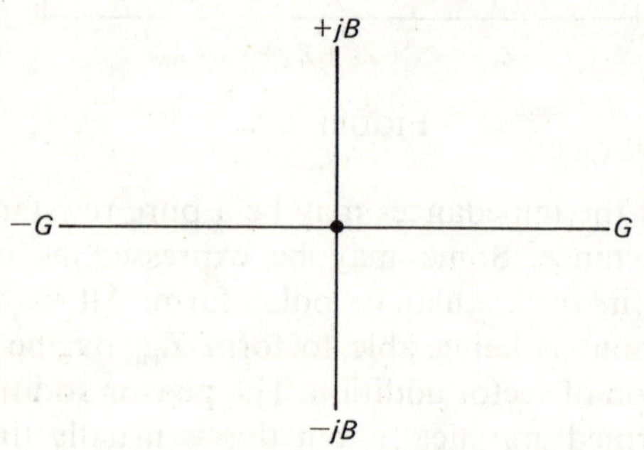

FIGURE 6-7

Hence, with this definition of admittance, we can rewrite (2) as (see Fig. 6-8)

4) $$\mathbf{Y}_{eq} = \mathbf{Y}_1 + \mathbf{Y}_2 + \mathbf{Y}_3 + \cdots + \mathbf{Y}_N \qquad (6\text{-}8)$$

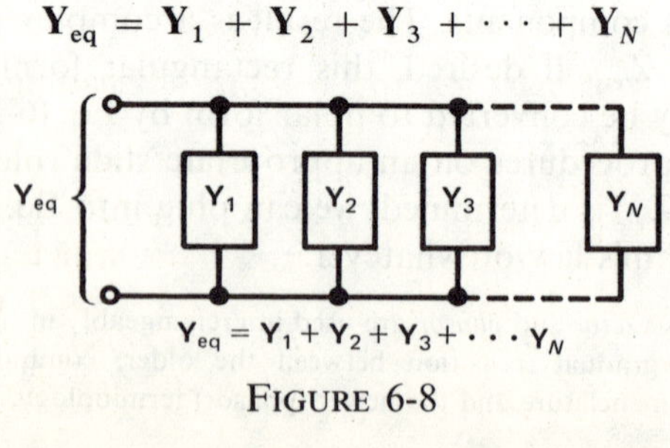

FIGURE 6-8

Now work problems PS 6-13 through PS 6-16.

A most important point to recognize is that in general neither the real part (G_p) of an admittance nor its imaginary part (B_p) are respective reciprocals of the real and imaginary parts of the corresponding series impedance ($R_s + jX_s$).[2] Neither is the converse relationship true. To prove this, let

1) $$\mathbf{Y} = \frac{1}{\mathbf{Z}}$$

where

2) $$\mathbf{Z} = R_s + jX_s$$

and we will let R_s and X_s carry their own signs. In other words, R_s and/or X_s may actually be negative. Substituting (2) into (1) yields

3) $$\mathbf{Y} = \frac{1}{R_s + jX_s} \qquad (6\text{-}9a)$$

Using the complex conjugate of the denominator in (3) to perform the indicated division in rectangular form, we have

4) $$\mathbf{Y} = \frac{1}{R_s + jX_s} \times \frac{R_s - jX_s}{R_s - jX_s} = \frac{R_s - jX_s}{R_s^2 + X_s^2}$$

or separating (4) into its real and imaginary components

5) $$\mathbf{Y} = \frac{R_s}{R_s^2 + X_s^2} - j\frac{X_s}{R_s^2 + X_s^2} \qquad (6\text{-}9b)$$

Since in general (where G_p and B_p also carry their own signs)

6) $$\mathbf{Y} = G_p + jB_p \qquad (6\text{-}9c)$$

We see by comparing the real and imaginary parts of (5) and (6) that

7) $$G_p = \frac{R_s}{R_s^2 + X_s^2} \qquad (6\text{-}9d)$$

and

8) $$B_p = \frac{-X_s}{R_s^2 + X_s^2} \qquad (6\text{-}9e)$$

Note that $G_p \neq 1/R_s$, and that $B_p \neq 1/X_s$. Also observe the change of sign in equation (8).

Now work through problem PS 6-17.

While we are at it, let us develop the conversion from an admittance to an impedance. Thus

1) $$\mathbf{Z} = \frac{1}{\mathbf{Y}}$$

[2]The "*p*" subscript is used to indicate that an admittance may be considered as a conductance in *parallel* with a susceptance. The "*s*" subscript is used to indicate that an impedance may be considered as a resistance in *series* with a reactance.

or

2) $$R_s + jX_s = \frac{1}{G_p + jB_p} \qquad (6\text{-}10a)$$

Again, we will let the components of Z and Y carry their own signs. Applying the complex conjugate to (2) yields

3) $$R_s + jX_s = \frac{1}{G_p + jB_p} \times \frac{G_p - jB_p}{G_p - jB_p}$$

$$= \frac{G_p}{G_p^2 + B_p^2} - j\frac{B_p}{G_p^2 + B_p^2} \qquad (6\text{-}10b)$$

from which we see that

4) $$R_s = \frac{G_p}{G_p^2 + B_p^2} \qquad (6\text{-}10c)$$

5) $$X_s = \frac{-B_p}{G_p^2 + B_p^2} \qquad (6\text{-}10d)$$

Again note that $R_s \neq 1/G_p$, and that $X_s \neq 1/B_p$. Also observe the change of sign in (5). The conversions of Eqs. (6-9*c*), (6-9*d*), (6-10*c*), and (6-10*d*) are useful when going from a series to parallel or vice versa configuration.

Now work through problem PS 6-18.

6-5 Power in the *RLC* circuit

It was demonstrated in the chapters pertaining to capacitors and inductors that a pure C or L dissipates zero power. Therefore, any power dissipated in an *RLC* circuit must be due to the resistance. What we would like to do is develop some relation between the applied voltage and current or the circuit resistance and the voltage or current. In other words, we are looking for relationships analogous to $P = EI = E^2/R = I^2R$ that we were exposed to in dc circuit analysis.

One approach is to simply reason that when looking into the terminals of some *RLC* network as in Fig. 6-9*a*, the power dissipated in the impedance **Z** will be given by

1) $$P = I^2R \qquad (6\text{-}11)$$

where R is the resistive component of Z and I is the magnitude of the rms current. The phase angle associated with I is of no matter in determining P in (1) and hence is ignored. Now **Z** will in general be of the form $R + jX$ as shown in Fig. 6-9*b*, where X can be positive or negative depending upon whether the circuit has an inductive or capacitive component of Z. In any event, we are letting X carry its own sign. The exact nature of the *RLC* network is of no concern either, since whether it is a series, parallel, series-parallel, or whatever configuration, it can at any given frequency be replaced by the equivalent form of Fig. 6-9*b*.

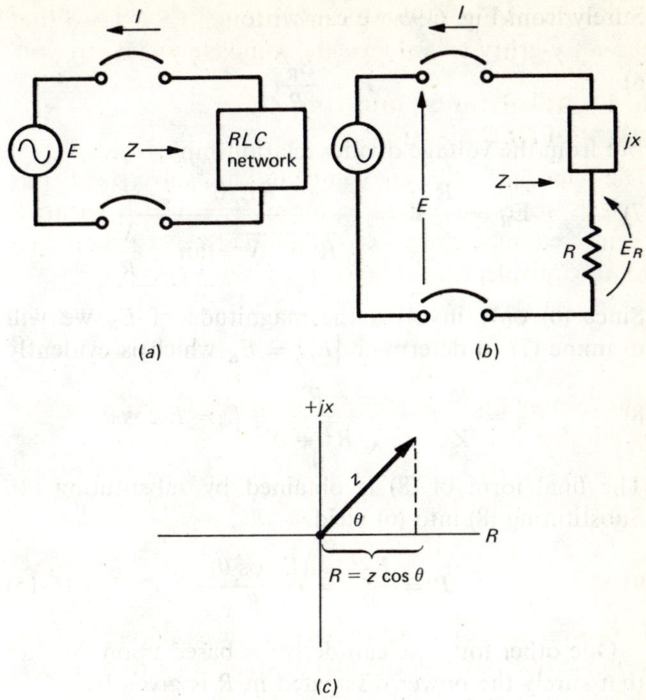

FIGURE 6-9

By directing our attention to the impedance diagram of Fig. 6-9*c*, we note that

2) $$R = Z\cos\theta$$

Substituting (2) into (1), we have

3) $$P = I^2Z\cos\theta \qquad (6\text{-}12)$$

where I and Z are the current and impedance magnitudes and the factor $\cos\theta$ is called the power factor (PF).

From equation (2) we may define the power factor as

4) $$PF = \cos\theta = \frac{R}{Z} = \frac{R}{\sqrt{R^2 + X^2}} \qquad (6\text{-}13)$$

If Z has an inductive component, I will lag E, and in that case we speak of a lagging power factor. Conversely, if Z has a capacitive component, I leads E and the power factor is said to be a leading PF.

The power factor may also be defined as

5) $$PF = \frac{\text{true power}}{\text{apparent power}}$$

$$= \frac{\text{wattmeter reading}}{\text{voltmeter reading} \times \text{ammeter reading}} \qquad (6\text{-}14)$$

where true power is the power read by a wattmeter connected in the circuit of Fig. 6-9*a*, whereas the denominator of (5) is the product of an ac voltmeter which directly reads E and an ac ammeter which directly reads I.

Let us derive some other expressions for the power.

Surely from Fig. 6-9*b* we can write

6) $$P = \frac{E_R^2}{R}$$

but from the voltage divider relationship

7) $$\mathbf{E}_R = \frac{R}{Z}\mathbf{E} = \frac{RE/0^\circ}{\sqrt{R^2 + X^2}\Big/\tan^{-1}\dfrac{X}{R}}$$

Since (6) only involves the magnitude of E_R we will examine (7) to determine $|E_R| = E_R$ which is evidently

8) $$E_R = \frac{R}{Z}E = \frac{R}{\sqrt{R^2 + X^2}}E = E\cos\theta$$

The final form of (8) is obtained by substituting (4). Substituting (8) into (6) yields

9) $$P = \frac{E_R^2}{R} = \frac{(E\cos\theta)^2}{R} \qquad (6\text{-}15)$$

One other form we can derive is based upon the fact that surely the power dissipated in R is given by

10) $$P = IE_R$$

Substituting (8) into (10) yields

11) $$P = IE\cos\theta \qquad (6\text{-}16)$$

If the resistive component (R) of the impedance ($\mathbf{Z}$) is known, then it is preferable to use Eq. (6-11), as it avoids the power factor angle.

Now work through problem PS 6-19.

6-6 Maximum power transfer

As you can see in any of the power expressions we have derived, a $PF = 1$ maximizes the power delivered to the network of Fig. 6-9*a*. Hence, unity power factor is generally a desirable condition and various circuit modifications or additions are attempted to bring it about. The essential thought is to make purely resistive the load that the generator or voltage source sees. In that case, $\theta = 0^\circ$, $PF = 1$, and the generator current and voltage are in phase.

To make the circuit look purely resistive, it is only necessary to somehow cancel any reactive component seen by the generator, and this is easily done by adding the opposite kind of reactance in series with the circuit. Thus if the load is inductive in nature, we could add a capacitor in series (or shunt) with it and choose its value so that at the operating frequency its reactance cancels that of the load. In many cases where electromechanical generators are used, the generator internal impedance is inductive in nature, and that too may be compensated for (if necessary) in a similar manner.

If the voltage source has a resistive component also, then, after achieving unity power factor, the generator resistance (R_G) should, if possible, be matched to the load resistance (R_L) for maximum power transfer. This is an extension of the maximum power transfer theorem we studied in the previous text.

Now work through problem PS 6-20.

PROBLEMS WITH SOLUTIONS

PS 6-1 Sketch the impedance $\mathbf{Z} = 4 - j0$ ohms in the impedance plane. Express it in polar form.

SOLUTION See Fig. PS 6-1 for the sketch. Converting $4 - j0$ to polar form we have

$$|Z| = \sqrt{R^2 + X^2} = \sqrt{(4)^2 + (0)^2} = 4$$

$$\theta = \tan^{-1}\frac{X}{R} = \tan^{-1}\frac{0}{4} = 0^\circ$$

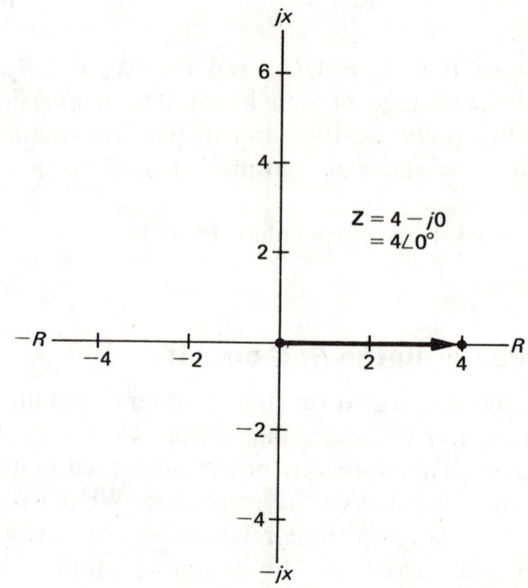

FIGURE PS 6-1

PS 6-2 Sketch the impedance $\mathbf{Z} = -3 + j0$ ohms in the impedance plane (an impedance diagram) and also convert it to polar form.

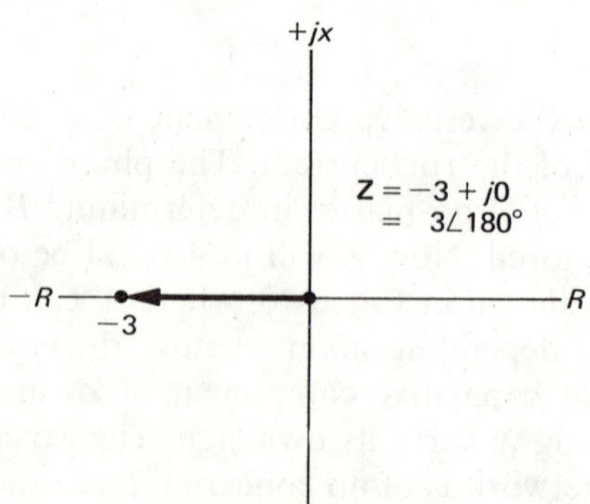

FIGURE PS 6-2

SOLUTION See Fig. PS 6-2 for the sketch. Converting $-3 + j0$ to polar form we have

$$|Z| = \sqrt{(-3)^2 + (0)^2} = 3$$

$$\theta = \tan^{-1}\frac{0}{-3} = \tan^{-1}(-0) = 180°$$

Think of (-0) as being equal to $0\underline{/180°}$.

PS 6-3 Sketch the impedance $\mathbf{Z} = 2\underline{/0°}$ ohms in the impedance plane and convert it to rectangular form.

SOLUTION See Fig. PS 6-3 for the sketch. To convert $2\underline{/0°}$ ohms to rectangular form,

$$R = Z\cos\theta = 2\cos 0° = 2\,\Omega$$

$$X = Z\sin\theta = 2\sin 0° = 0\,\Omega$$

Therefore $2\underline{/0°}$ ohms $= 2 + j0\,\Omega$

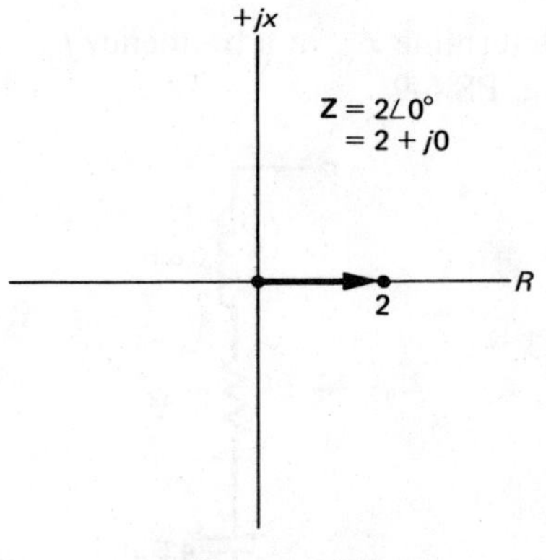

FIGURE PS 6-3

PS 6-4 Sketch the impedance $\mathbf{Z} = 3 - j5$ ohms in the impedance plane and convert it to polar form.

SOLUTION See Fig. PS 6-4.

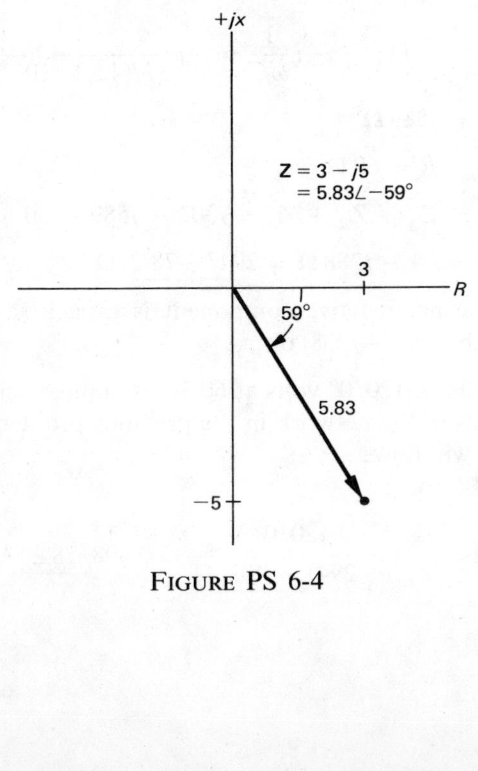

FIGURE PS 6-4

$$|Z| = \sqrt{R^2 + X^2} = \sqrt{(3)^2 + (-5)^2} = 5.83$$

$$\theta = \tan^{-1}\frac{-5}{3} = -59 \text{ or } 301°$$

PS 6-5 Sketch the impedance $\mathbf{Z} = (4 - j9) + (2 + j5)$ in the impedance plane and also convert it to polar form.

SOLUTION Prior to making any sketch let us first simplify $\mathbf{Z}$ by adding the real (resistive) components together, and the imaginary (reactive) components also.

$$\begin{array}{r} 4 - j9 \\ +\; 2 + j5 \\ \hline \mathbf{Z} = \quad 6 - j4 \end{array}$$

Converting $6 - j4$ to polar form yields

$$|Z| = \sqrt{(6)^2 + (-4)^2} = 7.21\,\Omega$$

$$\theta = tan^{-1}\frac{-4}{6} = -33.7°$$

See Fig. PS 6-5.

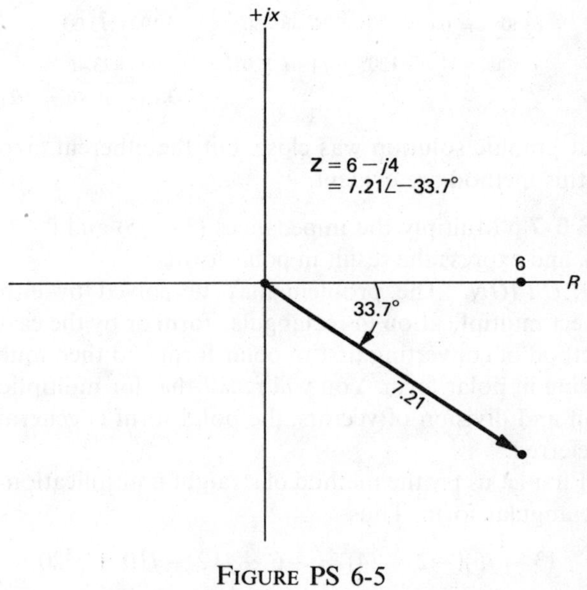

FIGURE PS 6-5

PS 6-6 Using vector addition by graphic methods, determine the sum of the impedances $\mathbf{Z} = \mathbf{Z}_1 + \mathbf{Z}_2$ where $\mathbf{Z}_1 = 6\underline{/30°}$ and $\mathbf{Z}_2 = 4\underline{/120°}$. Check your result by converting these vectors to rectangular form.

SOLUTION The graphic solution is shown in Fig. PS 6-6. First we construct $\mathbf{Z}_1$ and $\mathbf{Z}_2$. Then we translate $\mathbf{Z}_2$ (dashed line) so that its tail sits on the head of $\mathbf{Z}_1$. The resultant $\mathbf{Z}$ is then constructed by drawing a line from the origin to the head of the translated vector $\mathbf{Z}_2$. Measuring the length and angle of the resultant indicates that $\mathbf{Z} \cong 7\underline{/64.5°}$ ohms.

An easier and more accurate way to obtain a solution is to recognize that we have a problem in vector addition.

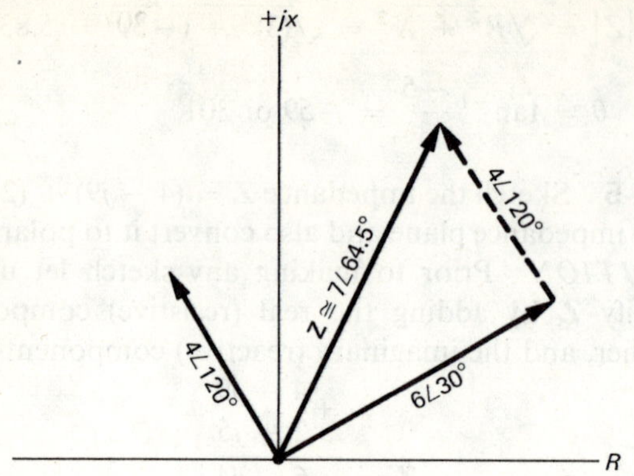

FIGURE PS 6-6

Since vector addition (or subtraction) is most conveniently performed in rectangular form, let us convert $\mathbf{Z}_1$ and $\mathbf{Z}_2$ to rectangular form, add the respective components, and then convert back to polar form to check our solution. Thus

$$\begin{array}{lllll} & \mathbf{Z}_1 = 6\underline{/30^\circ} & = 6\cos 30^\circ + j6\sin 30^\circ = & 5.19 + j3.00 & \\ + & \mathbf{Z}_2 = 4\underline{/120^\circ} & = 4\cos 120^\circ + j4\sin 120^\circ = & -2.00 = j3.46 & \\ \hline & \mathbf{Z} = & & 3.19 + j6.46 = 7.21\underline{/63.7^\circ} & \end{array}$$

Our graphic solution was close, but the inherent errors in this method are evident.

PS 6-7 Multiply the impedances $(3 - j5)$ and $(-2 - j4)$, and express the result in polar form.

SOLUTION The problem may be solved by either direct multiplication in rectangular form or by the easier method of converting first to polar form and then multiplying in polar form. You will recall that for multiplication and division of vectors, the polar form is generally preferred.

First let us try the method of straight multiplication in rectangular form. Thus

$$\begin{aligned}(3 - j5)(-2 - j4) &= -6 - j12 + j10 + j^2 20 \\ &= -6 - j2 - 20 \\ &= -26 - j2 = 26.1\underline{/-176^\circ}\end{aligned}$$

By the second method,

$$3 - j5 = 5.83\underline{/-59^\circ} \quad \text{and} \quad -2 - j4 = 4.47\underline{/-116^\circ}$$

Therefore

$$(5.83\underline{/-59^\circ})(4.47\underline{/-116^\circ}) = 26.1\underline{/-176^\circ}$$

which checks.

PS 6-8 Divide the impedance $3 - j4$ by $6 - j5$ and express the result in polar form.

SOLUTION Division, you will recall, may be performed directly in rectangular form, with the aid of the complex conjugate, or much more easily in polar form. Using the first method,

$$\begin{aligned}\frac{3 - j4}{6 - j5} &= \frac{3 - j4}{6 - j5} \times \frac{6 + j5}{6 + j5} \\ &= \frac{18 + j15 - j24 - j^2 20}{6^2 + 5^2} = \frac{38 - j9}{61} \\ &= 0.623 - j0.148 = 0.64\underline{/-13.3^\circ}\end{aligned}$$

Using the second method,

$$\frac{3 - j4}{6 - j5} = \frac{5\underline{/-53.1^\circ}}{7.81\underline{/-39.8^\circ}} = 0.64\underline{/-13.3^\circ}$$

Aside from some slight round-off error in the angle, the results are in agreement. Note how much easier the latter method is. Nevertheless there are occasions where we specifically prefer to work with the complex conjugate, so be sure you are familiar with it.

PS 6-9 Determine $\mathbf{Z}_{eq}$ at a frequency $f = 60$ Hz in the circuit of Fig. PS 6-9.

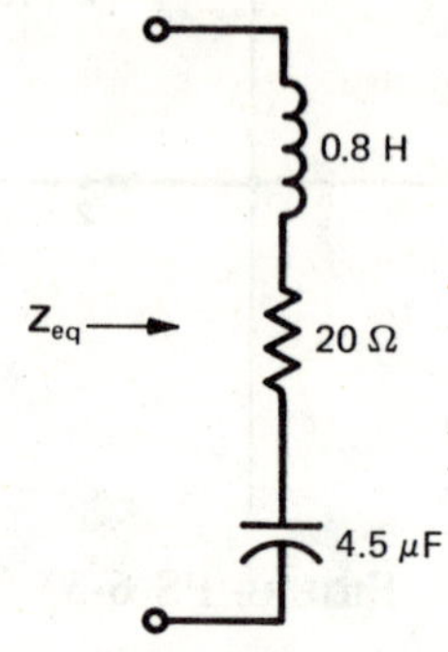

FIGURE PS 6-9

SOLUTION

$$\omega = 2\pi f = 377 \text{ rad/sec}$$

$$\mathbf{Z}_1 = jX_L = j\omega L = j377(0.8) = j302\ \Omega$$

$$\mathbf{Z}_2 = -jX_C = -j\frac{1}{\omega C} = -j\frac{1}{377(4.5 \times 10^{-6})}$$

$$= j589\ \Omega$$

$$\mathbf{Z}_3 = R = 60\ \Omega$$

$$\begin{aligned}\mathbf{Z}_{eq} &= \mathbf{Z}_1 + \mathbf{Z}_2 + \mathbf{Z}_3 = j302 - j589 + 60 \\ &= 60 - j288\ \Omega = 294\underline{/-78.2^\circ}\ \Omega\end{aligned}$$

Note the net reactive component is capacitive as evidenced by the $(-j288)$ ohms.

PS 6-10 If $120\underline{/0^\circ}$ volts at 60 Hz is applied across the terminals of the network in the previous problem, what current will flow?

SOLUTION

$$\mathbf{I} = \frac{\mathbf{E}}{\mathbf{Z}_{eq}} = \frac{120\underline{/0^\circ}\text{ V}}{294\underline{/-78.2^\circ}\ \Omega} = 0.408\underline{/78.2^\circ}\text{ A}$$

Note the current leads the voltage by $\angle 78.2^\circ$ as might be expected, since $\mathbf{Z}_{eq}$ had a capacitive component.

PS 6-11 Determine the capacitor voltage E_C in Fig. PS 6-11 by utilizing the voltage divider relationship. Compare its magnitude with that of the applied voltage E_{in}.

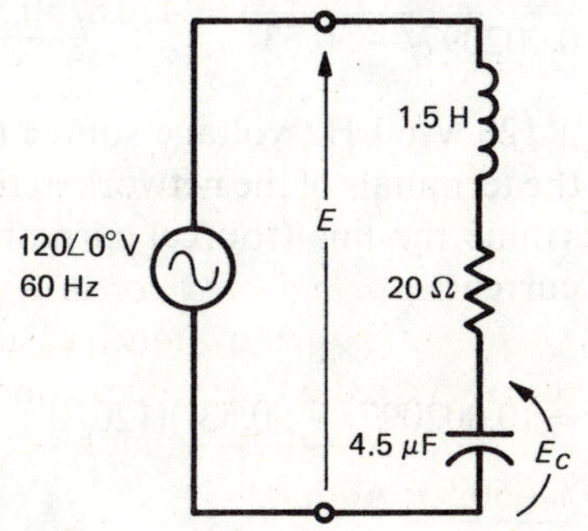

FIGURE PS 6-11

SOLUTION

$$\omega = 2\pi f = 377 \text{ rad/sec}$$

$$\mathbf{Z}_1 = jX_L = j\omega L = j565\ \Omega$$

$$\mathbf{Z}_2 = -jX_C = -j\frac{1}{\omega C} = -j589\ \Omega$$

$$\mathbf{Z}_3 = R = 20\ \Omega$$

$$\mathbf{E}_C = \frac{\mathbf{Z}_2\mathbf{E}}{\mathbf{Z}_1 + \mathbf{Z}_2 + \mathbf{Z}_3} = \frac{-j589(120\angle 0^\circ)}{j565 - j589 + 20}$$

$$= \frac{(589\angle -90^\circ)(120\angle 0^\circ)}{20 - j23.5} = \frac{(589\angle -90^\circ)(120\angle 0^\circ)}{30.9\angle -49.6^\circ}$$

$$E_C = 2290\angle -40.4^\circ \text{ V}$$

Comparing the magnitude of E_C with E, the applied voltage, we find a rather astonishing result—E_C is actually 2,290 volts despite the fact that we are only applying 120 volts. This is not an error. It is a direct result of a phenomenon called resonance which we will explore in a later chapter. However, in case you cannot stand the suspense, work through the next problem, which will shed a little light on the subject.

PS 6-12 Using the data of the previous problem, determine, also by the voltage divider relationship, E_R and E_L. Then check to see if the sum of the vector voltages across C, R, and L equals the applied voltage E. Construct a vector diagram to illustrate this.

SOLUTION

$$\mathbf{E}_L = \frac{\mathbf{Z}_1\mathbf{E}}{\mathbf{Z}_1 + \mathbf{Z}_2 + \mathbf{Z}_3} = \frac{(565\angle 90^\circ)(120\angle 0^\circ)}{30.9\angle -49.6^\circ}$$

$$= 2194\angle 140^\circ \text{ V}$$

$$\mathbf{E}_R = \frac{\mathbf{Z}_3\mathbf{E}}{\mathbf{Z}_1 + \mathbf{Z}_2 + \mathbf{Z}_3} = \frac{(20\angle 0^\circ)(120\angle 0^\circ)}{30.9\angle -49.6^\circ}$$

$$= 77.7\angle 49.6^\circ \text{ V}$$

and from the previous problem

$$\mathbf{E}_C = 2290\angle -40.4^\circ \text{ V}$$

Checking to see if

$$\mathbf{E} = \mathbf{E}_C + \mathbf{E}_R + \mathbf{E}_L$$

$$= 2290\angle -40.4^\circ + 77.7\angle 49.6^\circ + 2194\angle 140^\circ$$

$$= (1744 - j1484) + (50.4 + j59.2) + (-1681 + j1410)$$

$$= 113.6 - j14.75 = 114.6\angle -7.398^\circ$$

which is reasonably close to the applied $120\angle 0^\circ$ voltage. Round-off errors tend to be aggravated when performing operations involving the difference between two large numbers, as was our case here.

Figure PS 6-12 is a vector sketch of the circuit voltages.

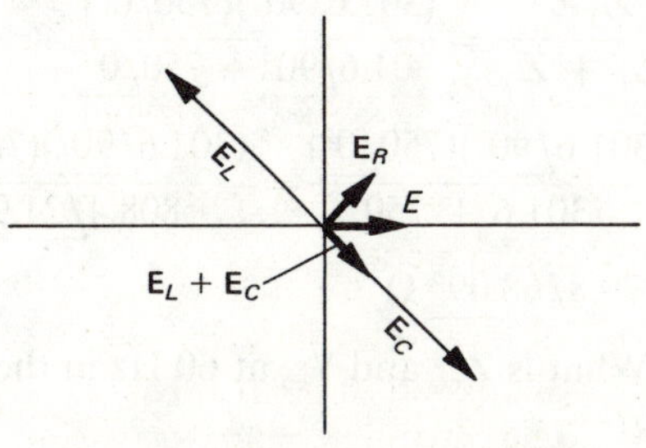

FIGURE PS 6-12

If we add these vectors, say, by first adding $\mathbf{E}_L$ to $\mathbf{E}_C$, the result is a small vector at a negative angle almost equal to E_R in magnitude and opposite sign in the angle. You can see that adding E_R to it would yield a resultant which is essentially equal to E. From a physical viewpoint we see that the inductive and capacitive reactances, although quite large, almost cancel, and hence $\mathbf{E}_L$ and $\mathbf{E}_C$ likewise also almost cancel. Therefore the equivalent impedance presented to E at 60 Hz is not very large. Hence, a substantial current would flow through the relatively large reactances, which results in voltage drops that can exceed the applied voltage. We will discuss this in the chapter on resonance phenomena.

PS 6-13 Convert the impedance $\mathbf{Z} = 6 + j8$ ohms to an admittance and sketch it in the admittance plane.

SOLUTION

$$\mathbf{Y} = \frac{1}{\mathbf{Z}} = \frac{1}{6 + j8} = \frac{1}{10\angle 53.1^\circ}$$

$$= 0.1\angle -53.1^\circ \text{ mhos} = 0.0600 - j0.08 \text{ mhos}$$

See Fig. PS 6-13 for a polar and rectangular plot of $\mathbf{Y}$.

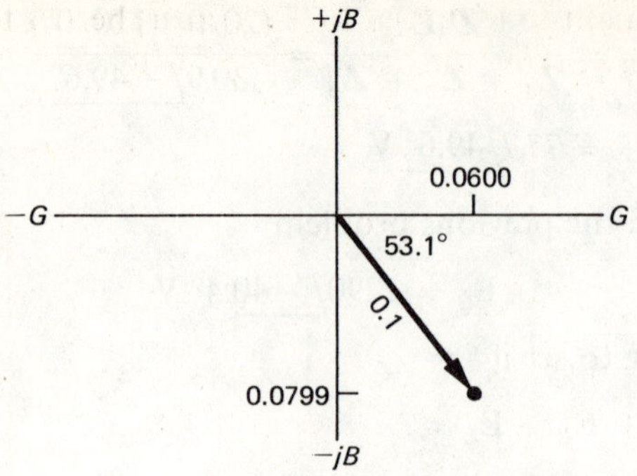

FIGURE PS 6-13

PS 6-14 What is the equivalent impedance of a 0.8 H inductor in parallel with a 750-ohm resistor at 60 Hz?

SOLUTION

$$\omega = 2\pi f = 377 \text{ rad/sec}$$

$$\mathbf{Z}_1 = jX_L = j\omega L = j301.6 = 301.6\angle 90^\circ\ \Omega$$

$$\mathbf{Z}_2 = 750\angle 0^\circ\ \Omega$$

$$\mathbf{Z}_{eq} = \frac{\mathbf{Z}_1\mathbf{Z}_2}{\mathbf{Z}_1 + \mathbf{Z}_2} = \frac{(301.6\angle 90^\circ)(750\angle 0^\circ)}{301.6\angle 90^\circ + 750\angle 0^\circ}$$

$$= \frac{(301.6\angle 90^\circ)(750\angle 0^\circ)}{j301.6 + 750} = \frac{(301.6\angle 90^\circ)(750\angle 0^\circ)}{808.4\angle 21.91^\circ}$$

$$= 279.8\angle 68.09^\circ\ \Omega$$

PS 6-15 What is $\mathbf{Z}_{eq}$ and $\mathbf{Y}_{eq}$ at 60 Hz in the circuit of Fig. PS 6-15?

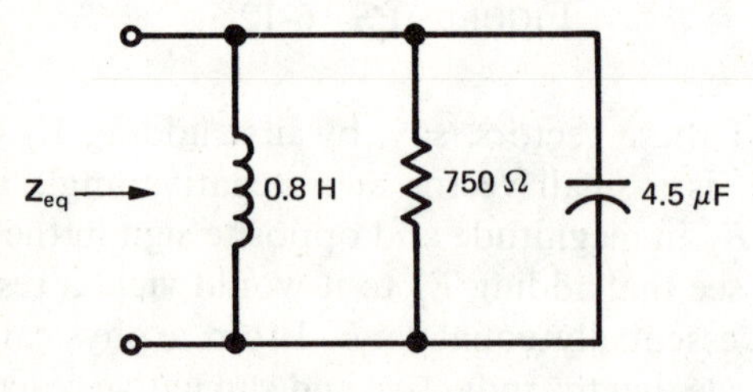

FIGURE PS 6-15

SOLUTION

$$\omega = 377 \text{ rad/sec}$$

$$\mathbf{Z}_1 = j\omega L = 301.6\angle 90^\circ\ \Omega$$

$$\mathbf{Z}_2 = 750\ \Omega$$

$$\mathbf{Z}_3 = -jX_C = -j\frac{1}{\omega C} = 589.4\angle -90^\circ\ \Omega$$

$$\mathbf{Y}_{eq} = \frac{1}{\mathbf{Z}_{eq}} = \frac{1}{\mathbf{Z}_1} + \frac{1}{\mathbf{Z}_2} + \frac{1}{\mathbf{Z}_3}$$

$$= \frac{1}{301.6\angle 90^\circ} + \frac{1}{750\angle 0^\circ} + \frac{1}{589.4\angle -90^\circ}$$

$$= \underbrace{0.003316\angle -90^\circ}_{\mathbf{Y}_L} + \underbrace{0.001333\angle 0^\circ}_{\mathbf{Y}_R} + \underbrace{0.001697\angle 90^\circ}_{\mathbf{Y}_C}$$

or

$$\mathbf{Y}_{eq} = 0.001333 - j0.001619 = 0.002097\angle -50.53^\circ \text{ mhos}$$

$$\mathbf{Z}_{eq} = \frac{1}{\mathbf{Y}_{eq}} = \frac{1}{0.002097\angle -50.53^\circ} = 476.8\angle 50.53^\circ\ \Omega$$

PS 6-16 If a 120 V/60 Hz voltage source ($\mathbf{E}_s$) is connected across the terminals of the network in the previous problem, determine the line (source) current $\mathbf{I}_s$ and the three branch currents.

SOLUTION

$$\mathbf{I}_s = \mathbf{Y}_{eq}\mathbf{E}_s = (0.002097\angle -50.53^\circ)(120\angle 0^\circ)$$

$$= 0.2517\angle -50.53^\circ \text{ A}$$

$$\mathbf{I}_C = \mathbf{Y}_C\mathbf{E}_s = (0.001697\angle 90^\circ)(120\angle 0^\circ) = 0.2036\angle 90^\circ \text{ A}$$

$$\mathbf{I}_R = \mathbf{Y}_R\mathbf{E}_s = (0.001333\angle 0^\circ)(120\angle 0^\circ) = 0.1600\angle 0^\circ \text{ A}$$

$$\mathbf{I}_L = \mathbf{Y}_L\mathbf{E}_s = (0.003316\angle -90^\circ)(120\angle 0^\circ) \text{ A}$$

$$= 0.3979\angle -90^\circ \text{ A}$$

As an exercise, the reader may verify that the vector sum of the branch currents equals the line current.

PS 6-17 Determine the conductance and susceptance of the admittance seen looking into the *AB* terminals of Fig. PS 6-17*a* at a frequency of 500 Hz. Use more than one method to solve the problem and sketch the circuit and components of the equivalent admittance.

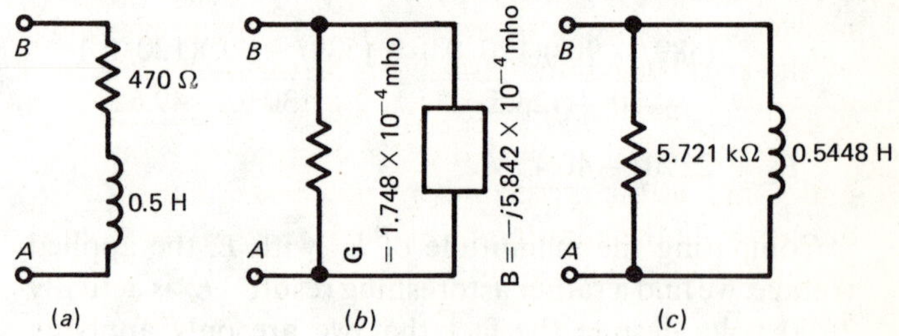

FIGURE PS 6-17

$$\omega = 2\pi(500) \text{ rad/sec} = 3142 \text{ rad/sec}$$

$$X_L = \omega L = 3142(0.5) = 1571\ \Omega$$

1) $$\mathbf{Z}_{AB} = 470 + j1571 = 1639\angle 73.34^\circ\ \Omega$$

2) $$\mathbf{Y}_{AB} = \frac{1}{\mathbf{Z}_{AB}} = 6.098 \times 10^{-4}\angle -73.34 \text{ mhos}$$

or

3) $$\mathbf{Y}_{AB} = \underbrace{1.748 \times 10^{-4}}_{G} - \underbrace{j5.842 \times 10^{-4}}_{B} \text{ mhos}$$

The above equation is the most obvious approach to finding G and B. An alternative is to use the series to parallel conversions of Eqs. (6-9*d*) and (6-9*e*) which yield

$$4)\quad \mathbf{Y}_{AB} = \frac{R}{R^2 + X^2} + j\frac{j(-X)}{R^2 + X^2} = \frac{470}{(470)^2 + (1571)^2}$$

$$-j\frac{1571}{(470)^2 + (1571)^2} = \frac{470}{2.689 \times 10^6} = \frac{1571}{2.689 \times 10^6}$$

or

$$5)\quad \mathbf{Y}_{AB} = \underbrace{1.748 \times 10^{-4}}_{G} - j\underbrace{5.842 \times 10^{-4}}_{B} \text{ mhos}$$

Fortunately we arrive at the same solution by either method.

To determine the equivalent parallel circuit, we reason that since in the general case $\mathbf{Y}_{eq} = \mathbf{Y}_1 + \mathbf{Y}_2 +$, etc., where $\mathbf{Y}_1$, $\mathbf{Y}_2$, etc., are parallel connected elements, Eq. (5), which is of the same form, exhibits $\mathbf{Y}_1 = G = 1.748 \times 10^{-4}$ mhos and $\mathbf{Y}_2 = -jB = -j5.842 \times 10^{-4}$ mhos. Hence, the equivalent circuit to this extent would appear as shown in Fig. PS 6-17*b*. We know that since the G branch has no phase angle associated with it, this branch may be represented by a pure conductance whose equivalent resistance is

$$R = \frac{1}{G} = \frac{1}{1.748 \times 10^{-4} \text{ mhos}} = 5721\ \Omega$$

The next step is to determine what exhibits a susceptance of $-j5.842 \times 10^{-4}$ mhos at 500 Hz. If we can answer this question we can model the remaining branch of the parallel configuration. Since the susceptance is negative, it must be inductive in nature. You should recall this from Chap. 5, but if not, just remember that jX_L and $-jX_C$ in the impedance plane correspond to $-jB_L$ and $+jB_C$ in the admittance plane. Hence we are looking for an inductor whose susceptance is given by

$$-jB_L = -j5.842 \times 10^{-4} \text{ mhos} = \frac{1}{jX_L}$$

$$= \frac{1}{j2\pi(500)L} = \frac{-j}{2\pi(500)L}$$

Solving for L yields

$$5.842 \times 10^{-4} = \frac{1}{2\pi(500)L}$$

$$L = \frac{1}{2\pi(500)5.842 \times 10^{-4}} = 0.5448 \text{ H}$$

Thus the equivalent circuit would appear as shown in Fig. PS 6-17*c*.

PS 6-18 Determine the elements of a series circuit that would present an input impedance at 500 Hz that is equivalent to that seen looking into the AB terminals of the network on the left in Fig. PS 6-18.

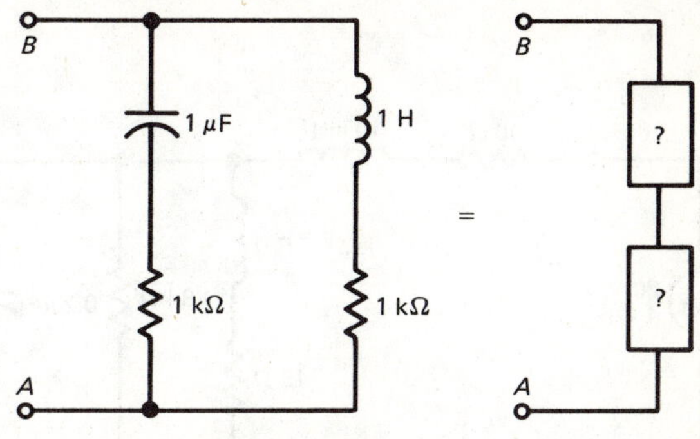

FIGURE PS 6-18

SOLUTION

$$\omega = 2\pi(500) = 3142 \text{ rad/sec}$$

The impedance of the capacitive branch is

$$\mathbf{Z}_C = R - j\frac{1}{\omega C} = 1000 - j318.3\ \Omega = 1049\underline{/-17.66^\circ}\ \Omega$$

The impedance of the inductive branch is

$$\mathbf{Z}_L = R + j\omega L = 1000 + j3142\ \Omega = 3297\underline{/72.34^\circ}\ \Omega$$

The admittance of the capacitive branch is

$$\mathbf{Y}_C = \frac{1}{\mathbf{Z}_C} = \frac{1}{1049\underline{/-17.66^\circ}} = 9.533 \times 10^{-4}\underline{/17.66^\circ} \text{ mhos}$$

$$= (9.084 + j2.892)10^{-4} \text{ mhos}$$

The admittance of the inductive branch is

$$\mathbf{Y}_L = \frac{1}{\mathbf{Z}_L} = \frac{1}{3297\underline{/72.34^\circ}} = 3.033 \times 10^{-4}\underline{/-72.34} \text{ mhos}$$

$$= 9.202 \times 10^{-5} - j2.890 \times 10^{-4} \text{ mhos}$$

The admittance looking into the AB terminals is then

$$\mathbf{Y}_{AB} = \mathbf{Y}_C + \mathbf{Y}_L$$

$$\begin{array}{rl} & 9.084 \times 10^{-4} + j2.892 \times 10^{-4} \\ & 0.9202 \times 10^{-4} - j2.890 \times 10^{-4} \\ \hline \mathbf{Y}_{AB} = & 10.004 \times 10^{-4} + j0.002 \times 10^{-4} \end{array}$$

or

$$\mathbf{Y}_{AB} \cong 10 \times 10^{-4}\underline{/0^\circ} \text{ mhos}$$

and

$$\mathbf{Z}_{AB} = \frac{1}{\mathbf{Y}_{AB}} = 1000\underline{/0^\circ}\ \Omega$$

Hence there is no reactive component and the equivalent series circuit consists solely of a 1 kΩ resistor. As an exercise, you might solve this problem for different frequencies. The results are surprising.

PS 6-19 Determine the power dissipated in the network of Fig. PS 6-19. Use more than one method and compare the results.

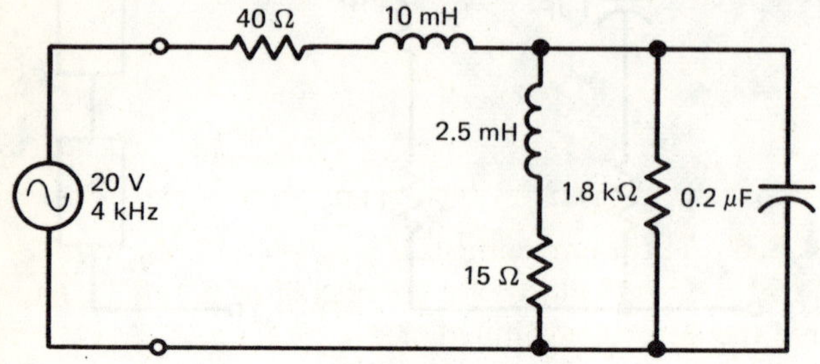

FIGURE PS 6-19

SOLUTION First we simplify the network to determine the impedance the voltage source looks into. Starting with the three parallel branches,

For the 0.2μF capacitor and noting ω = 25,130 rad/sec,

$$\mathbf{Y}_1 = j\omega C = j5.027 \text{ mmhos}$$

For the 1800-ohm resistor,

$$\mathbf{Y}_2 = \frac{1}{1.8 \text{ k}\Omega} = 0.5555 \text{ mmhos}$$

For the 2.5-mH, 15-ohm branch,

$$\mathbf{Z}_3 = 15 + j\omega(2.5 \text{ mH}) = 15 + j62.83\,\Omega$$
$$= 64.59\underline{/76.57^\circ}\,\Omega$$

Therefore

$$\mathbf{Y}_3 = \frac{1}{\mathbf{Z}_3} = 15.48\underline{/-76.57^\circ} \text{ mmhos}$$
$$= 3.595 - j15.06 \text{ mmhos}$$

The equivalent admittance of the three parallel branches is then

$$\mathbf{Y} = \mathbf{Y}_1 + \mathbf{Y}_2 + \mathbf{Y}_3$$
$$= j5.027 + 0.5555 + 3.595 - j15.06$$
$$= 4.151 - j10.03 \text{ mmhos}$$
$$\mathbf{Y} = 10.85\underline{/-67.52^\circ} \text{ mmhos}$$

The corresponding impedance is

$$\mathbf{Z}_4 = \frac{1}{\mathbf{Y}} = \frac{1}{10.85\underline{/-67.52^\circ} \text{ mmhos}} = 92.12\underline{/67.52^\circ}\,\Omega$$
$$\mathbf{Z}_4 = 35.22 + j85.12\,\Omega$$

This impedance will appear in series with that of the 40-ohm, 10-mH inductor whose impedance is

$$\mathbf{Z}_5 = 40 + j\omega(10\text{mH}) = 40 + j251.3\,\Omega$$

The total impedance presented to the voltage source is then

$$\mathbf{Z} = \mathbf{Z}_4 + \mathbf{Z}_5 = (35.22 + j85.12) + (40 + j251.3)$$
$$\mathbf{Z} = 75.22 + j336.4\,\Omega = 344.8\underline{/77.39^\circ}\,\Omega$$

The power factor by Eq. (6-13) is

$$PF = \cos\theta = \frac{R}{Z} = \frac{75.22}{344.8} = 0.2182 \text{ (lagging)}$$

The source current is (taking the applied voltage at a reference angle of 0°)

$$\mathbf{I} = \frac{\mathbf{E}}{\mathbf{Z}} = \frac{20\underline{/0^\circ} \text{ V}}{344.8\underline{/77.39^\circ}\,\Omega} = 0.0580\underline{/-77.39^\circ}\text{A}$$

The network power dissipation is, by Eq. (6-16),

$$P = EI\cos\theta = EI\,(P.F.) = (\text{ 20 V})\,(0.580 \text{ A})\,(0.2182)$$
$$= 0.253 \text{ W}$$

An alternative solution is

$$P = I^2R = (0.0580)^2\, 75.22 = 0.253 \text{ W}$$

which verifies the previous solution.

PS 6-20 How would you achieve unity power factor in the previous problem, and what is the new power dissipation?

SOLUTION The network impedance presented to the voltage source in the previous problem was determined as

$$\mathbf{Z} = 75.22 + j336.4\,\Omega$$

This means we must cancel the $j336.4$ ohms to make the network purely resistive. To cancel this inductive component we will insert a capacitor in series with the voltage source and the network so that

$$-jX_C = -j336.4\,\Omega$$

Therefore

$$X_C = 336.4\,\Omega = \frac{1}{2\pi(4000)C}$$

Solving for C yields

$$C = \frac{1}{2\pi(4000)\, 336.4} = 0.1183\,\mu\text{F}$$

The voltage source now sees only 75.22 ohms and hence

$$P = \frac{E^2}{R} = \frac{(20 \text{ V})^2}{75.22\,\Omega} = 5.3 \text{ W}$$

PROBLEMS WITH ANSWERS

PA 6-1 Express the impedance $\mathbf{Z} = 4 - j8$ ohms in polar form and sketch its location in the impedance plane.

ANSWER $8.944\angle -63.43°$

PA 6-2 Express the impedance $\mathbf{Z} = -2 + j4$ ohms in polar form and locate it in the impedance plane.

ANSWER $4.47\angle 117°$

PA 6-3 Express the impedance $Z = (3 - j4.5) - (-3 + j6)$ in polar form. Check your solution by graphic methods.

ANSWER $12.1\angle -60.3°$

PA 6-4 Express the impedance $Z = 12\angle -120°$ in rectangular form.

ANSWER $(-6 - j10.4)$

PA 6-5 Evaluate $Z = \dfrac{(8\angle 60°)(3 - j4)}{5\angle 20° - j5} + 6\angle -40°$

ANSWER $9.82\angle 4.67°$

PA 6-6 Two impedances, $\mathbf{Z}_1 = 5 - j8$ and $\mathbf{Z}_2 = 9\angle 20°$, are connected in series. What is the equivalent impedance in polar form?

ANSWER $14.3\angle -20.1°$

PA 6-7 Two impedances, $\mathbf{Z}_1 = -5 + j3$, and $\mathbf{Z}_2 = 10\angle 30°$, are connected in parallel. Determine the equivalent impedance.

ANSWER $(-2.65 + j6.07)$

PA 6-8 Three impedances, $\mathbf{Z}_1 = 4 - j12$, $\mathbf{Z}_2 = 6\angle 18°$, and $\mathbf{Z} = 5 + j6$ ohms, are connected in parallel. Determine the equivalent impedance.

ANSWER $3.63\angle 15.75\ \Omega$

PA 6-9 Determine E_x in Fig. PA 6-9.

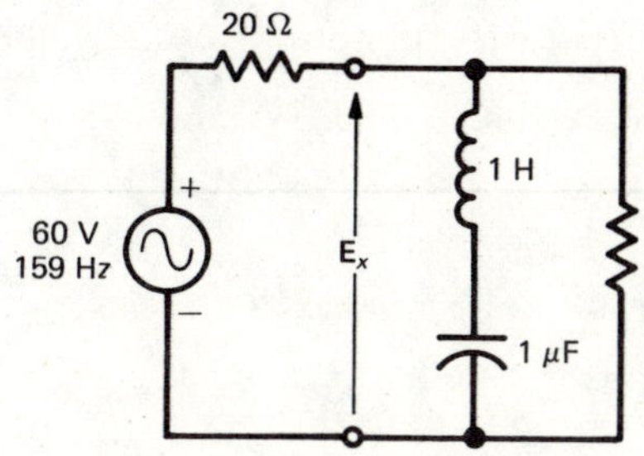

ANSWER $5.795\angle -82.6°$

PA 6-10 Determine $\mathbf{I}_{AB}$ in Fig. PA 6-10.

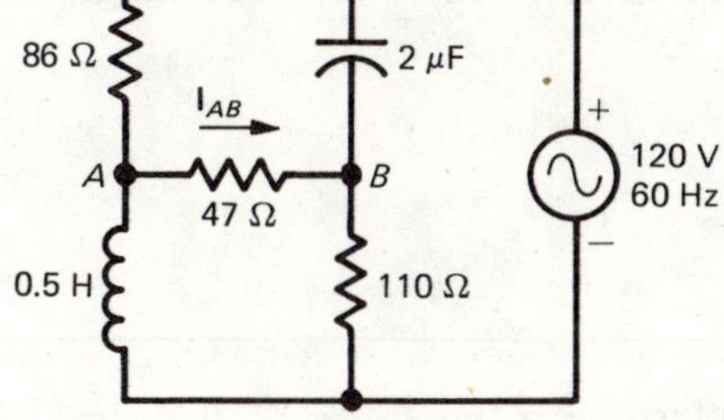

ANSWER $0.458\angle 194°$ A

PA 6-11 What is the admittance $\mathbf{Y}_{AB}$ seen at a frequency of 28 MHz in Fig. PA 6-11? Determine the components of the equivalent parallel circuit.

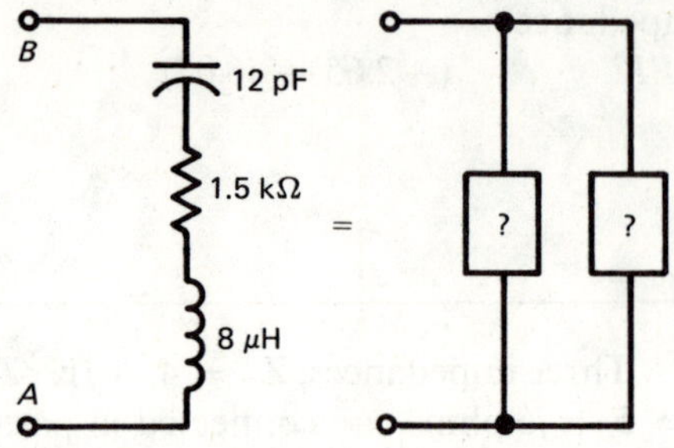

ANSWER
$Y_{AB} = (0.4807 - j0.2989)$mmhos, $R = 2.08$ kΩ, $L = 19\ \mu$H

PA 6-12 Determine the power delivered to the load in Fig. PA 6-12. What is the power factor?

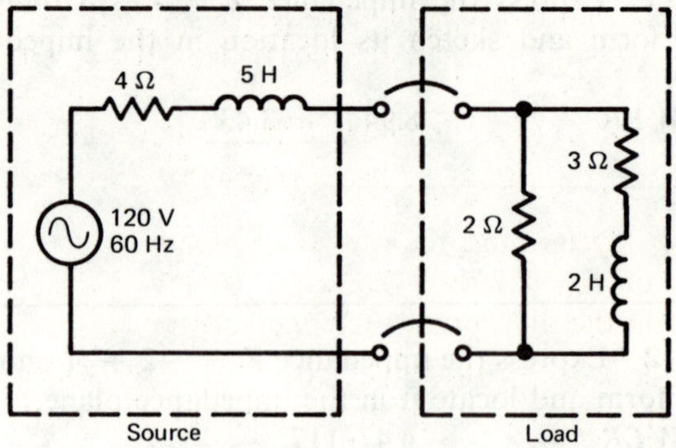

ANSWER $P = 8.1$ mW, $PF = 90°$ lagging

PA 6-13 How could you modify the circuit of Fig. PA 6-12 so as to obtain unity power factor? Completely specify any components used.
ANSWER Insert a capacitor whose reactance is 1885 ohms at 60 Hz in series between the source and the load. $C = 1.41\ \mu$F.

PROBLEMS WITHOUT ANSWERS

P 6-1 Express $\mathbf{Z} = 4 - j3$ in polar form.

P 6-2 Express $\mathbf{Z} = 18\underline{/-75°}$ in rectangular form.

P 6-3 Determine $\mathbf{Z} = \mathbf{Z}_1 + \mathbf{Z}_2$ if $\mathbf{Z}_1 = 3 + j4$, and $\mathbf{Z}_2 = -6\underline{/45°}$.

P 6-4 Determine $\mathbf{Z} = \mathbf{Z}_1 - \mathbf{Z}_2$ if $\mathbf{Z}_1 = 4\underline{/30°}$, and $\mathbf{Z}_2 = 5\underline{/-90°}$.

P 6-5 Determine $\mathbf{Z}^2 = \mathbf{Z}_1\mathbf{Z}_2$ if $\mathbf{Z}_1 = 6 - j12$, and $\mathbf{Z}_2 = 8\underline{/50°}$.

P 6-6 Determine $\mathbf{A} = \mathbf{Z}_1/\mathbf{Z}_2$ if $\mathbf{Z}_1 = 6 - j12$ and $\mathbf{Z}_2 = 5 + j7$.

P 6-7 What is the complex conjugate of $(-2 + j3)$?

P 6-8 Determine the resistive and reactive components of $\mathbf{A} = \mathbf{Z}_1/\mathbf{Z}_2$ if $\mathbf{Z}_1 = 5 - j8$ and $\mathbf{Z}_2 = 4 - j3$. Perform the division in rectangular form.

P 6-9 If $\mathbf{Z} = 12 - j3$ ohms, what is the corresponding admittance?

P 6-10 If $\mathbf{Y} = 0.5 - j0.2$ mhos, what is the corresponding impedance?

P 6-11 If $\mathbf{Z} = 6 + j4$ ohms at 60 Hz, draw the equivalent circuit and determine the series-equivalent components.

P 6-12 If $\mathbf{Y} = 0.3 - j0.1$ mhos at 60 Hz, draw the equivalent circuit and determine the parallel-equivalent components.

P 6-13 Determine the components of the simplest series circuit that will yield an input impedance identical to that offered by the parallel circuit of Fig. P 6-13 at 1500 Hz.

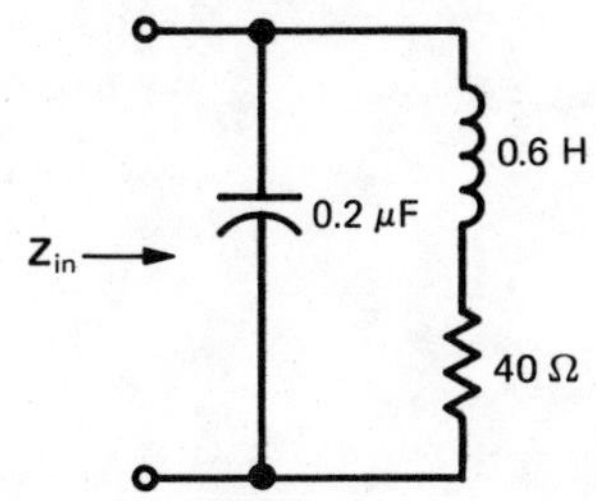

P 6-14 Determine $\mathbf{E}_x$ in Fig. P 6-14.

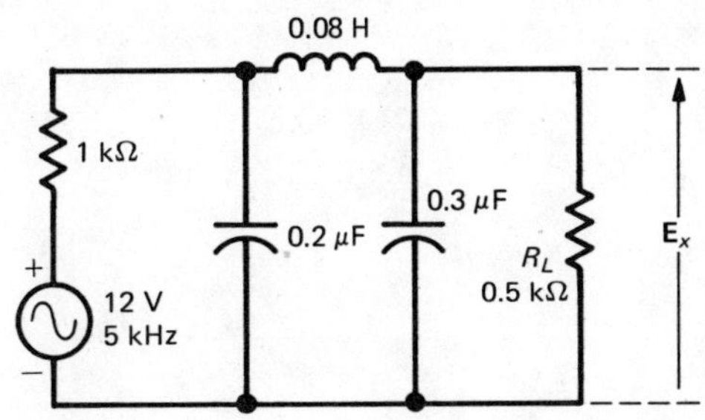

P 6-15 What is the power delivered to the 0.5 kΩ load resistor in Fig. P 6-14?

P 6-16 What circuit element could be introduced in the previous problem so that R_L receives maximum power?

7 Q and series-parallel conversions

7-1 Q

When dealing with sinusoidal or even other waveforms, the term Q is frequently encountered. The meaning of Q may be expressed in several ways, but its broadest interpretation is that it is a measure of the energy-storing ability of a circuit or component. For example, a pure inductor or capacitor can only store energy and hence dissipates zero power. Any circuit resistance, however, dissipates energy as heat. Thus Q is a figure of merit which compares the energy-storing ability of a circuit or component to that dissipated as heat. In this sense Q is a measure of the pureness of an inductor or capacitor, since any losses must be due to a resistive circuit component.

Specifically, for ac circuits we may define Q as

$$Q = \frac{\text{energy stored per cycle}}{\text{energy dissipated per cycle}} = \frac{\text{reactive power}}{\text{real power}} \tag{7-1}$$

where the reactive power is the power stored in the net reactive component and the real power is the power dissipated in the net resistive component.

We will examine Q again in the next chapter, but the definition provided by Eq. (7-1) will do for now.

At a given frequency most circuits, parts of circuits, or even components can be represented as a net reactance in series with a resistance, as shown in Fig. 7-1, or

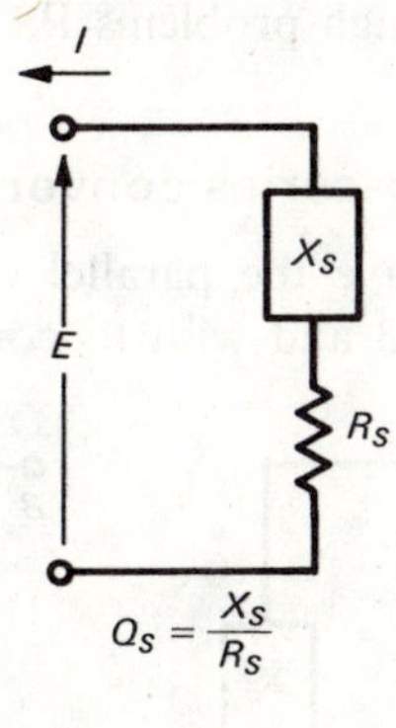

FIGURE 7-1

as a net reactance in parallel with a resistance, as shown in Fig. 7-2. In general $R_S \neq R_P$ and $X_S \neq X_P$. Just how the series and parallel configurations are interrelated will be examined shortly. For the moment our goal is to develop expressions for the Q of the series and parallel circuits.

Since I in the series circuit of Fig. 7-1 is common to R_S and X_S, we may write

$$Q_S = \frac{\text{reactive power}}{\text{real power}} = \frac{I^2 X_S}{I^2 R_S} = \frac{X_S}{R_S} \tag{7-2}$$

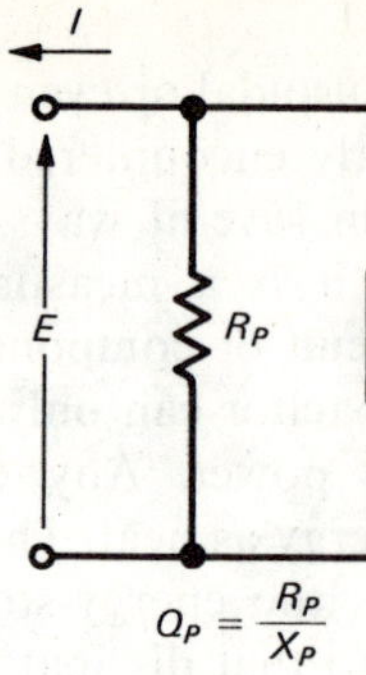

FIGURE 7-2

For the parallel circuit of Fig. 7-2, E is common and hence

$$Q_P = \frac{\text{reactive power}}{\text{real power}} = \frac{E^2/X_P}{E^2/R_P} = \frac{R_P}{X_P} \tag{7-3}$$

It is advisable to memorize the final forms of Eqs. (7-2) and (7-3).

Since the reactance X is in general a function of frequency, Q will also vary with frequency. However, R may also vary with frequency, since it may depend upon skin effect, core losses in inductors, and dielectric losses in capacitors. Hence, the manner in which Q varies with frequency is not always obvious. Measurements or manufacturers' data must be consulted so that Q at the frequency of interest is known.

Now work through problems PS 7-1 and PS 7-2.

7-2 Parallel-to-series conversions

Quite often we have the parallel configuration in the left side of Fig. 7-3 and wish to convert it to its series equivalent as shown on the right at some particular frequency or range of frequencies. Reactance X_P represents the equivalent parallel reactance and R_P is the equivalent parallel resistance.

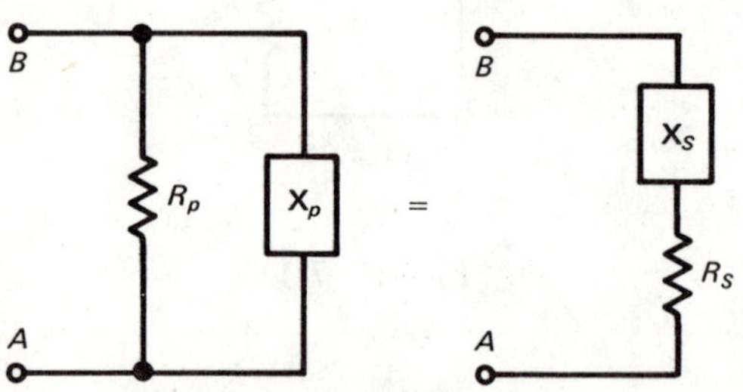

FIGURE 7-3

If numerical values are specified for the various circuit parameters, you can take your slide rule, calculator, or whatever and determine $\mathbf{Y}_P$ for the parallel circuit in polar form. Then take the reciprocal of $\mathbf{Y}_P$ and obtain $\mathbf{Z}_P$, still in polar form. Resolve $\mathbf{Z}_P$ from polar form into its resistive and reactive components as a next step. Now if the series and parallel networks are to be equivalent, the resistive component of $\mathbf{Z}_P$ must equal R_S and the reactive component must equal X_S. This is one approach, and we have actually solved this type of problem in previous chapters.

Work through problems PS 7-3 and PS 7-4 at this time.

Although we have just seen in problems PS 7-3 and PS 7-4 that a mathematical solution to a specific problem is readily obtained when sufficient data is given, what we often seek is the general analysis of some circuit involving these conversions. Hence, let us develop a purely general approach, as follows:

For the parallel arrangement in Fig. 7-3, we have

1) $$\mathbf{Z}_{ABP} = \frac{R_P(jX_P)}{R_P + jX_P}$$

where $\mathbf{Z}_{ABP}$ is the impedance seen looking into the AB terminals, R_P is the shunt resistive component, and X_P is the equivalent reactive component. We will let X_P carry its own sign, so that X_P will be positive for a net inductive reactance and negative for a net capacitive reactance.

Multiplying numerator and denominator of (1) by the complex conjugate of the denominator yields, after some simplification,

2) $$\mathbf{Z}_{ABP} = \frac{R_P(jX_P)}{R_P + jX_P} \times \frac{R_P - jX_P}{R_P - jX_P}$$

$$= \frac{R_P X_P^2}{R_P^2 + X_P^2} + j\frac{R_P^2 X_P}{R_P^2 + X_P^2} \tag{7-4}$$

Inspection of equation (2) indicates that, with the aid of the complex conjugate, we have transformed the parallel circuit to a series equivalent with a resistive component

3) $$R_S = \frac{R_P X_P^2}{R_P^2 + X_P^2} \tag{7-5}$$

and a reactive component

4) $$X_S = \frac{R_P^2 X_P}{R_P^2 + X_P^2} \tag{7-6}$$

Since there was no sign change in the reactive term of 7-4, X_S and X_P are the same kind of reactance.

It is often convenient to express (3) and (4) in terms of Q_P, which you will recall from Eq. (7-3) is given by

5) $$Q_P = \frac{R_P}{X_P} \tag{7-7a}$$

Thus, dividing numerator and denominator of (3) by X_P^2, we obtain

6) $$R_S = \frac{\dfrac{R_P X_P^2}{X_P^2}}{\dfrac{R_P^2}{X_P^2} + \dfrac{X_P^2}{X_P^2}} = \frac{R_P}{Q_P^2 + 1} \qquad (7\text{-}7b)$$

And similarly, treating (4), we obtain

7) $$X_S = \frac{\dfrac{R_P^2 X_P}{X_P^2}}{\dfrac{R_P^2}{X_P^2} + \dfrac{X_P^2}{X_P^2}} = \frac{Q_P^2 X_P}{Q_P^2 + 1} = \frac{X_P}{1 + \dfrac{1}{Q_P^2}} \qquad (7\text{-}7c)$$

You might note that if Q_P is large, say $Q_P \geqq 10$, R_S becomes much less than R_P, and $X_S \cong X_P$.

Now do problem PS 7-5.

7-3 Series-to-parallel conversions

You now know the parallel-to-series conversion. It is just as important to go the other way and change the series arrangement of Fig. 7-4 to its parallel equivalent.

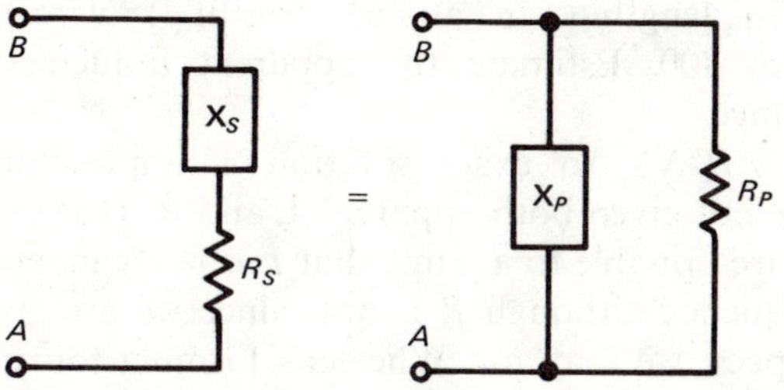

FIGURE 7-4

As a starting point in the development of this conversion, we may reason that if the two networks are to be equivalent, their admittances at the *AB* terminals must be identical.

1) $$\mathbf{Y}_{ABS} = \mathbf{Y}_{ABP}$$

2) $$\frac{1}{R_S + jX_S} = \mathbf{Y}_{ABP}$$

Applying the complex conjugate to the left side of (2),

3) $$\frac{1}{R_S + jX_S} \times \frac{R_S - jX_S}{R_S - jX_S} = \underbrace{\frac{R_S}{R_S^2 + X_S^2}}_{G_P} \underbrace{- j\frac{X_S}{R_S^2 + X_S^2}}_{B_P} \qquad (7\text{-}8)$$

Now the real or purely conductive term of (3) must be the conductance of $\mathbf{Y}_{ABS}$ which, if the networks are to be identical, must equal G_P, the conductance of $\mathbf{Y}_{ABP}$. Thus

4) $$R_P = \frac{1}{G_P} = \frac{R_S^2 + X_S^2}{R_S} \qquad (7\text{-}9)$$

To introduce

5) $$Q_S = \frac{X_S}{R_S} \qquad (7\text{-}10a)$$

we divide numerator and denominator of (4) by R_S^2 to obtain

6) $$R_P = R_S(1 + Q_S^2) \qquad (7\text{-}10b)$$

Similarly the susceptance of $\mathbf{Y}_{ABS}$, which is the reactive term of (3), must equal B_P, the susceptance of $\mathbf{Y}_{ABP}$. Thus, and again introducing Q_S, we have

7) $$X_P = \frac{1}{B_P} = \frac{R_S^2 + X_S^2}{X_S} = X_S\left(\frac{1}{Q_S^2} + 1\right) \qquad (7\text{-}10c)$$

Reactance X_P is the same kind of reactance as X_S. In other words, if X_S is capacitive (inductive), so is X_P capacitive (inductive). This was also true for the parallel-to-series conversions.

Now work through problems PS 7-6 and PS 7-7.

The series-to-parallel and parallel-to-series conversions are summarized in Table 7-1.

Table 7-1

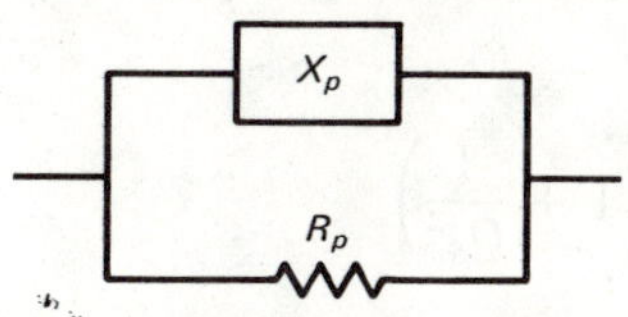	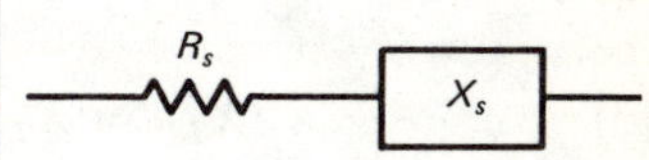
$Q_P = \dfrac{R_P}{X_P}$	$Q_S = \dfrac{X_S}{R_S}$
$R_P = R_S(1 + Q_S^2)$	$R_S = \dfrac{R_P}{1 + Q_P^2}$
$X_P = X_S\left(1 + \dfrac{1}{Q_S^2}\right)$	$X_S = \dfrac{X_P}{1 + \dfrac{1}{Q_P^2}}$

7-4 Inductor *Q*

We shall see in the next chapter that the frequency response of *LCR* networks may, at least over a certain frequency range, be highly dependent upon inductor *Q*. Since an inductor may be characterized by the equivalent circuit of Fig. 7-5*a*, we will use this as a starting point for evaluating inductor *Q*.

In Fig. 7-5*a*, *L* is the inductance of the coil winding, including any core material effects; R_L is the net resistance due to winding resistance, skin effect, and core

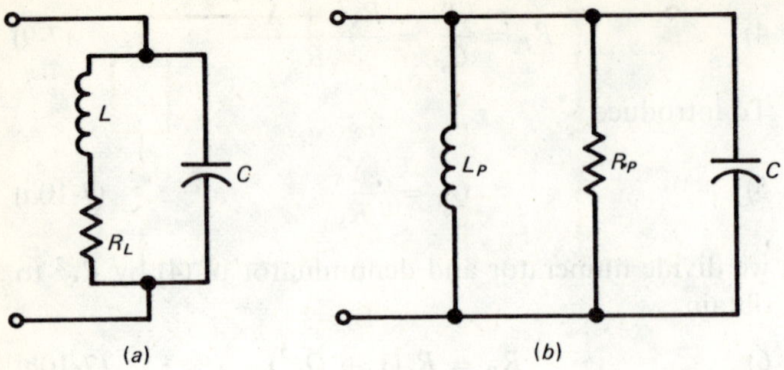

FIGURE 7-5

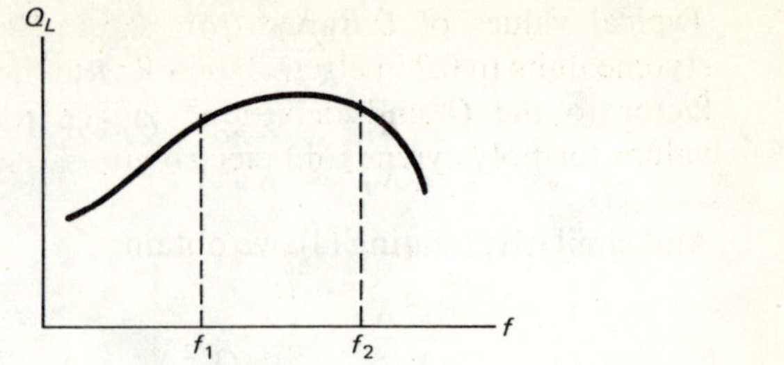

FIGURE 7-6

losses if applicable; and C is the equivalent capacitance across the coil due to capacitance between the various turns of which the coil is wound plus any stray capacitances which may exist due to the manner in which the coil is constructed. As a first step, let us convert the series $L - R_L$ branch to its parallel equivalent, which is shown in parallel with C in Fig. 7-5b.

The Q of the series $L - R_L$ branch in Fig. 7-5a is

1) $$Q_S = \frac{X_L}{R_L}$$

Hence, in Fig. 7-5b,

2) $$R_P = R_L(Q_S^2 + 1)$$

and

3) $$X_{LP} = X_L\left(1 + \frac{1}{Q_S^2}\right)$$

or

4) $$L_P = L\left(1 + \frac{1}{Q_S^2}\right)$$

Fig. 7-5b may be somewhat more readily analyzed than Fig. 7-5a. The most striking observation is that we can find some frequency, called the self-resonant frequency, at which the susceptance of the L_P branch will be canceled by the susceptance of the C branch so that the coil actually looks like a pure resistor R_P. At frequencies above the self-resonant frequency, the coil looks like a capacitor shunted by some resistance. However, inductors are normally operated at frequencies far below their self-resonant frequencies, so the capacitor C in Fig. 7-5a or 7-5b may be neglected. In that case, Eq. (1) actually defines the inductor Q as

5) $$Q_L = \frac{X_L}{R_L} \tag{7-11}$$

Typically the Q of an inductor may vary with frequency in a manner somewhat resembling Fig. 7-6. As you can see, for this coil at least, Q_L is substantially constant over the frequency range f_1 to f_2, and it is also a maximum in this region. Ideally this inductor would be used between these frequencies.

The reason Q_L is fairly constant between f_1 and f_2 is that skin effect and other losses cause the resistive component to increase with frequency just as fast as X_L increases. Hence, the ratio Q_L is relatively constant.

PROBLEMS WITH SOLUTIONS

PS 7-1 At 300 kHz a single-layer coil with a diameter of 6.3 in., length 6.1 in., and 61 turns of #18 wire exhibits a Q of 400. Estimate the apparent inductance and resistance.

SOLUTION An exact solution is impossible since we are not given both apparent L and R. However, it is not unreasonable to assume that L is fairly independent of frequency although R is not. Since we are given the coil specs, we may use Wheeler's formula for approximating the inductance. Thus

$$L \cong \frac{r^2N^2}{9r + 10l}\,\mu\text{H} = \frac{\left(\frac{6.3}{2}\right)^2 (61)^2}{9\left(\frac{6.3}{2}\right) + 10(6.1)} = 413\ \mu\text{H}$$

You will recall that this relationship is accurate to within 1 percent if $l > 0.8r$, which is satisfied here. Thus

$$X_S = X_L = 2\pi\ 300\ \text{kHz}\ (413\ \mu\text{H}) = 779\ \Omega$$

Since $Q = Q_S = 400$

$$R_S = \frac{X_S}{Q_S} = \frac{779}{400} \cong 2\ \Omega$$

PS 7-2 A practical capacitor exhibits dielectric losses and shunt leakage resistance. Over a wide frequency range both these effects can be represented by the equivalent circuit of Fig. PS 7-2 where G, a conductance, represents the net resistive losses, and C represents the apparent capacity. A commonly used figure of merit for capacitors is a dimensionless factor called the dissipation factor D, which is given by

$$D = \frac{G}{B_C}$$

Typical values of D range from 0.5×10^{-4} in polystyrene units to 0.2 in electrolytics. Relate the dissipation factor to the Q and determine the corresponding Q values for polystyrene and electrolytic capacitors.

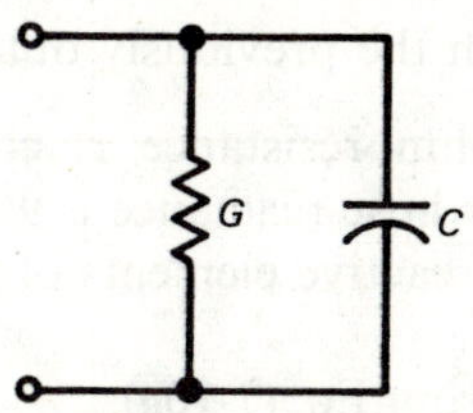

FIGURE PS 7-2

SOLUTION For a parallel circuit, Q is given by Eq. (7-3) as follows:

1) $$Q = Q_P = \frac{R_P}{X_P}$$

Since

2) $$D = \frac{G}{B_C}$$

and

3) $$G = \frac{1}{R_P}$$

4) $$B_C = \frac{1}{X_P}$$

we can substitute (3) and (4) into (2) to obtain

5) $$D = \frac{1/R_P}{1/X_P} = \frac{X_P}{R_P}$$

Substituting (5) into (1) yields the capacitor Q in terms of D. Thus

6) $$Q = Q_P = \frac{1}{D}$$

For polystyrene capacitors, Q will be in the neighborhood of

7) $$Q = \frac{1}{D} = \frac{1}{0.5 \times 10^{-4}} = 2 \times 10^4$$

whereas for electrolytics

8) $$Q = \frac{1}{D} = \frac{1}{0.2} = 5$$

Of course this is only true over the frequency range that the equivalent circuit of Fig. PS 7-2 holds true and where G and C are substantially constant.

PS 7-3 Convert the parallel circuit on the left of Fig. PS 7-3 to its series equivalent shown on the right.

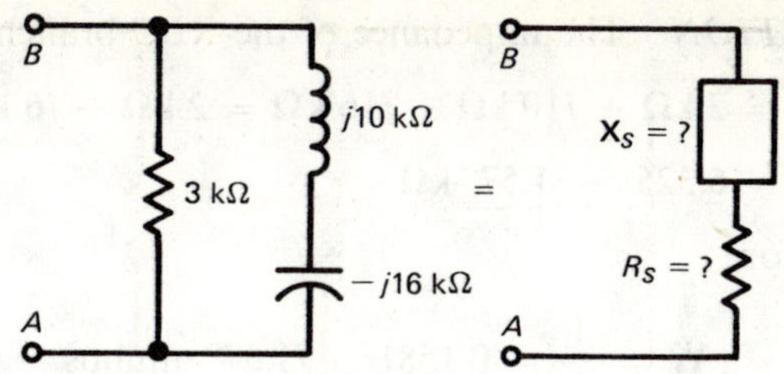

FIGURE PS 7-3

SOLUTION The net reactance in the LC branch of the parallel circuit is

$$\mathbf{X}_P = j10\text{ k}\Omega - j16\text{ k}\Omega = -j6\text{ k}\Omega$$

Therefore

$$\mathbf{B}_P = \frac{1}{\mathbf{X}_P} = \frac{1}{-j6\text{ k}\Omega} = j0.16667\text{ mmhos}$$

For the resistive branch,

$$G_P = \frac{1}{R_P} = \frac{1}{3\ k\Omega} = 0.3333\text{ mmhos}$$

Hence

$$\mathbf{Y}_P = G_P + j B_P = 0.3333 + j0.16667\text{ mmhos}$$
$$= 0.3727\underline{/26.57^\circ}\text{ mmhos}$$

And

$$\mathbf{Z}_P = \frac{1}{\mathbf{Y}_P} = 2.683\underline{/-26.57^\circ}\text{ k}\Omega = 2.4\text{ k}\Omega - j1.200\text{ k}\Omega$$

Hence if $\mathbf{Z}_{ABP} = \mathbf{Z}_{ABS}$

$$R_S = 2.4\text{ k}\Omega$$
$$\mathbf{X}_S = -j1.200\text{ k}\Omega$$

PS 7-4 Convert the parallel circuit on the left of Fig. PS 7-4 to its series equivalent shown on the right.

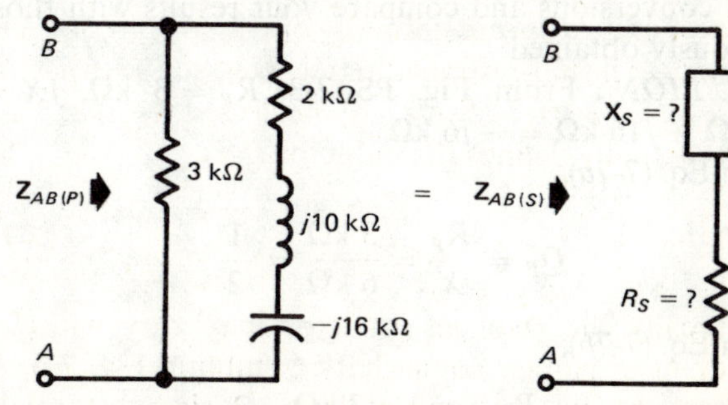

FIGURE PS 7-4

SOLUTION The impedance of the *RLC* branch is

$$\mathbf{Z}_1 = 2\ \text{k}\Omega + j10\ \text{k}\Omega - j16\ \text{k}\Omega = 2\ \text{k}\Omega - j6\ \text{k}\Omega$$
$$= 6.325\underline{/-71.57^\circ}\ \text{k}\Omega$$

Therefore

$$\mathbf{Y}_1 = \frac{1}{\mathbf{Z}_1} = 0.1581\underline{/+71.57^\circ}\ \text{mmhos}$$
$$= 0.05 + j0.15\ \text{mmhos}$$

For the 3 kΩ branch,

$$\mathbf{Y}_2 = \frac{1}{3\ \text{k}\Omega} = 0.3333\ \text{mmhos}$$

Thus looking into the *AB* terminals of the parallel configuration,

$$\mathbf{Y}_{ABP} = \mathbf{Y}_2 + \mathbf{Y}_1 = 0.3833 + j0.15\ \text{mmhos}$$
$$= 0.4116\underline{/+21.37^\circ}\ \text{mmhos}$$

Hence

$$\mathbf{Z}_{ABP} = \frac{1}{\mathbf{Y}_{ABP}} = 2.43\underline{/-21.37^\circ}\ \text{k}\Omega$$
$$= 2.262\ \text{k}\Omega - j0.8853\ \text{k}\Omega$$

Equating the resistive and reactive components of $\mathbf{Z}_{ABP}$ and $\mathbf{Z}_{ABS}$ yields

$$R_S = 2.262\ \text{k}\Omega$$
$$\mathbf{X}_S = -j0.8853$$

As an alternative approach, we might determine $\mathbf{Z}_{AB}$ from

$$\mathbf{Z}_{ABP} = \frac{3\ \text{k}\Omega\,(2\ \text{k}\Omega - j6\ \text{k}\Omega)}{3\ \text{k}\Omega + 2\ \text{k}\Omega - j6\ \text{k}\Omega} = 2.429\underline{/-21.37}\ \text{k}\Omega$$
$$= \underbrace{2.262\ \text{k}\Omega}_{R_S} - \underbrace{j0.8853\ \text{k}\Omega}_{X_S}$$

which is probably a shorter method if the parallel network contains only two branches.

PS 7-5 Repeat problem PS 7-3 using the parallel-to-series conversions and compare your results with those previously obtained.
SOLUTION From Fig. PS 7-3, $R_P = 3\ \text{k}\Omega$, $jX = j10\ \text{k}\Omega - j16\ \text{k}\Omega = -j6\ \text{k}\Omega$.
From Eq. (7-7*a*),

$$Q_P = \frac{R_P}{X} = \frac{3\ \text{k}\Omega}{6\ \text{k}\Omega} = \frac{1}{2}$$

From Eq. (7-7*b*),

$$R_S = \frac{R_P}{Q_P^2 + 1} = \frac{3\ \text{k}\Omega}{(0.5)^2 + 1} = 2.4\ \text{k}\Omega$$

From Eq. (7-7*c*),

$$X_S = \frac{X_P}{1 + \dfrac{1}{Q_P^2}} = \frac{6\ \text{k}\Omega}{1 + \dfrac{1}{(0.5)^2}} = 1.2\ \text{k}\Omega$$

which checks with the previously obtained results.

PS 7-6 A 16-ohm resistance is connected in series with a capacitor whose reactance is 96 ohms. Determine the resistive and reactive elements of the parallel equivalent circuit.
SOLUTION From Eq. (7-10*a*),

1) $$Q_S = \frac{X_S}{R_S} = \frac{96}{16} = 6$$

From Eq. (7-10*b*),

2) $$R_P = R_S(Q_S^2 + 1) = 16(37) = 592\ \Omega$$

From Eq. (7-10*c*),

3) $$X_P = X_S\left(1 + \frac{1}{Q_S^2}\right) = 96\left(1 + \frac{1}{36}\right) = 98.7\ \Omega$$

The elements of the parallel circuit consist of a 592-ohm resistor shunted by a capacitor whose reactance is 98.7 ohms at the operating frequency.

PS 7-7 Determine the series equivalent circuit of the network in Fig. PS 7-7*a*. Use the series-parallel conversion table in this chapter.

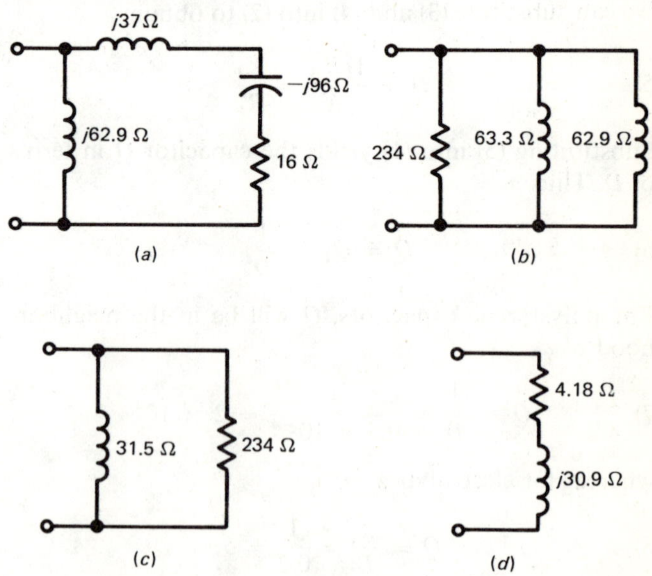

FIGURE PS 7-7

SOLUTION First we will take the branch containing the series-connected elements and convert it to its parallel form.

Thus

1) $X_S = 96\ \Omega - 37\ \Omega = 59\ \Omega$ (capacitive)

2) $$Q_S = \frac{X_S}{R_S} = \frac{59}{16} = 3.69$$

Therefore

3) $$R_P = 16\,[1 + (3.69)^2] = 234\ \Omega$$

4) $$X_P = 59\left[1 + \frac{1}{(3.69)^2}\right] = 63.3\ \Omega$$

The network is now equivalent to Fig. PS 7-7*b*, which reduces to Fig. PS 7-7*c*, since

5) $$X_{Leq} = \frac{(63.3)(62.9)}{63.3 + 62.9} = 31.5\ \Omega$$

Converting Fig. PS 7-7*c* to its series equivalent of Fig. PS 7-7*d*, we have

6) $$Q_P = \frac{R_P}{X_P} = \frac{234}{31.5} = 7.416 \text{ and } Q_P^2 = 55$$

7) $$R_S = \frac{234}{55 + 1} = 4.18\ \Omega$$

8) $$X_S = \frac{31.5}{1 + 1/55} = 30.9\ \Omega$$

PROBLEMS WITH ANSWERS

PA 7-1 Determine Q_S at 1.6 MHz in Fig. PS 7-1.

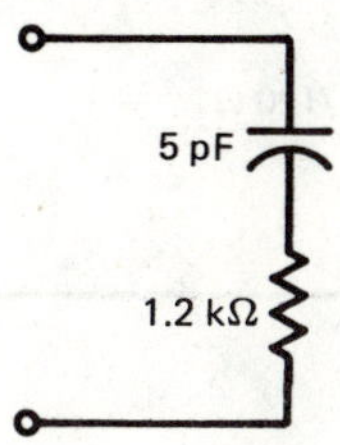

ANSWER $Q_S = 16.6$

PA 7-2 Determine Q_P at 12 kHz in Fig. PA 7-2.

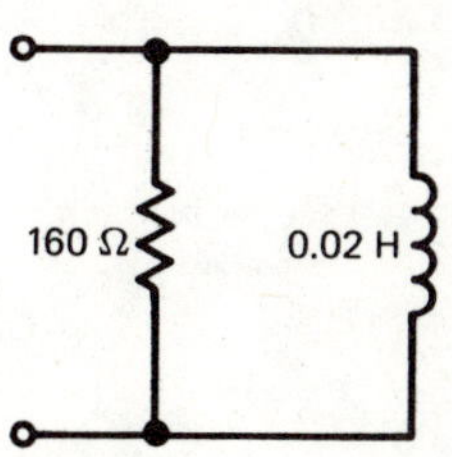

ANSWER $Q_P = 0.106$

PA 7-3 Construct the parallel equivalent of the series circuit of Fig. PA 7-3*a* for the frequency at which $R_S = 0.6$ kΩ and $X_L = 2.4$ kΩ.

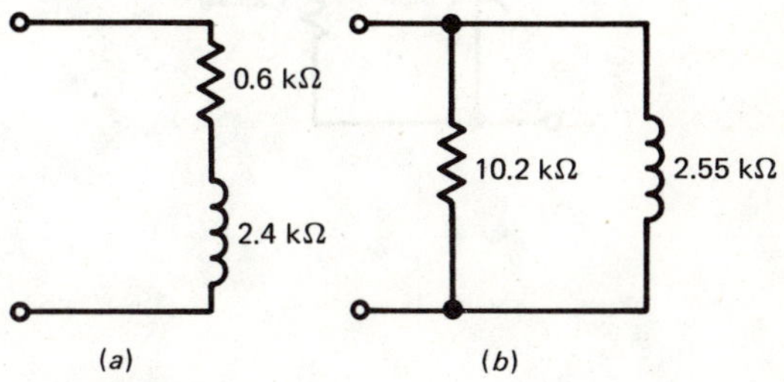

ANSWER See Fig. PA 7-3*b*.

PA 7-4 Estimate the impedance **X** required in Fig. PA 7-4 so that $\mathbf{Z}_{AB}$ is purely resistive at 500 Hz.

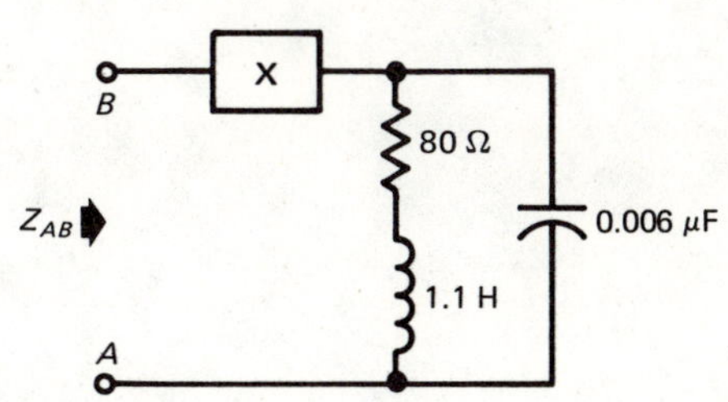

ANSWER $\mathbf{X} \cong -j3.46$ kΩ

PROBLEMS WITHOUT ANSWERS

P 7-1 At what frequency is $Q_S = 1$ in Fig. P 7-1?

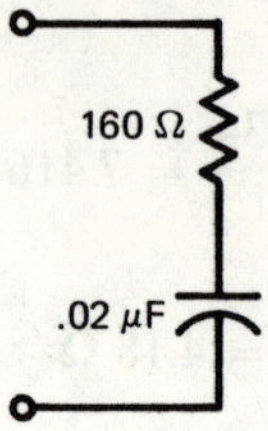

P 7-2 Determine Q_S of the series equivalent of Fig. P 7-2 at 1.6 MHz.

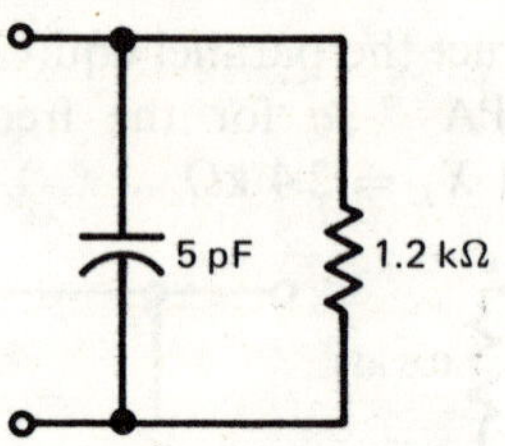

P 7-3 Determine the elements of the equivalent series circuit of Fig. P 7-3 at 18 kHz.

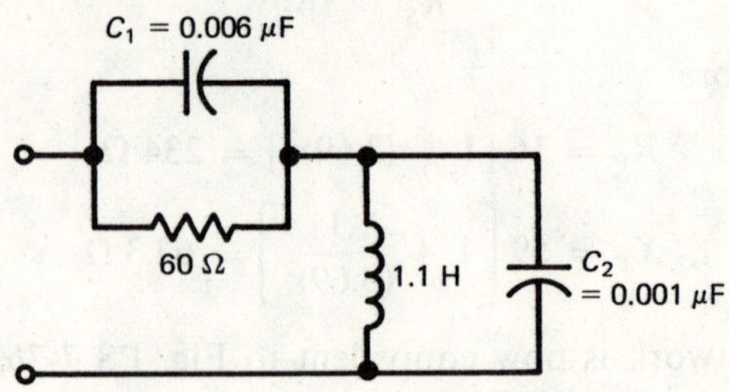

P 7-4 Express $\mathbf{Z}_{AB}$ in rectangular form for the circuit of Fig. P 7-4. Use the conversion techniques developed in this chapter and then check yourself using some other method.

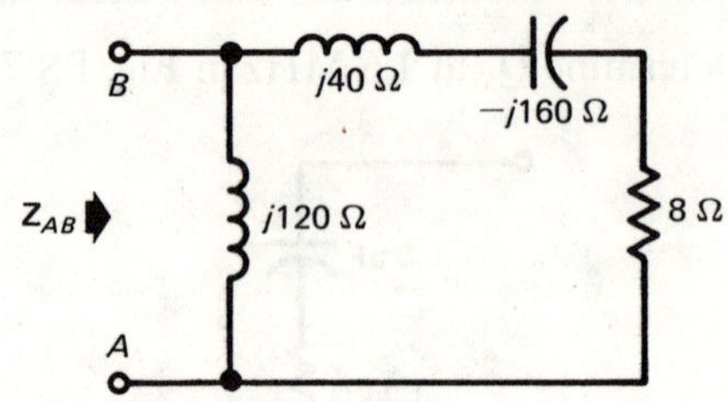

8 resonance phenomena

In preceding chapters we occasionally encountered problems with seemingly wild solutions. For example, there was a case in which the voltage across a reactive element in a series *LRC* circuit was much greater than the applied voltage over a certain frequency range. Circuits which exhibit a marked variation in their response characteristics over a particular frequency range are called tuned circuits, and this phenomenon is called resonance.

Tuned (resonant) circuits are usually utilized whenever there exists a need to discriminate between signals of different frequencies. In your radio or TV, tuned circuits are used to separate one transmitting station's signal from another.

8-1 Series resonance

In this section we shall investigate the well-known phenomenon of series resonance. In the material which follows, the most fundamental concepts which are introduced in basic electronics will be expanded.

Consider the series circuit of Fig. 8-1*a*. The impedance of that part of the circuit to the right of the *AB* terminals is

$$\mathbf{Z}_{AB} = R_L + j\left(\omega L - \frac{1}{\omega C}\right) \tag{8-1a}$$

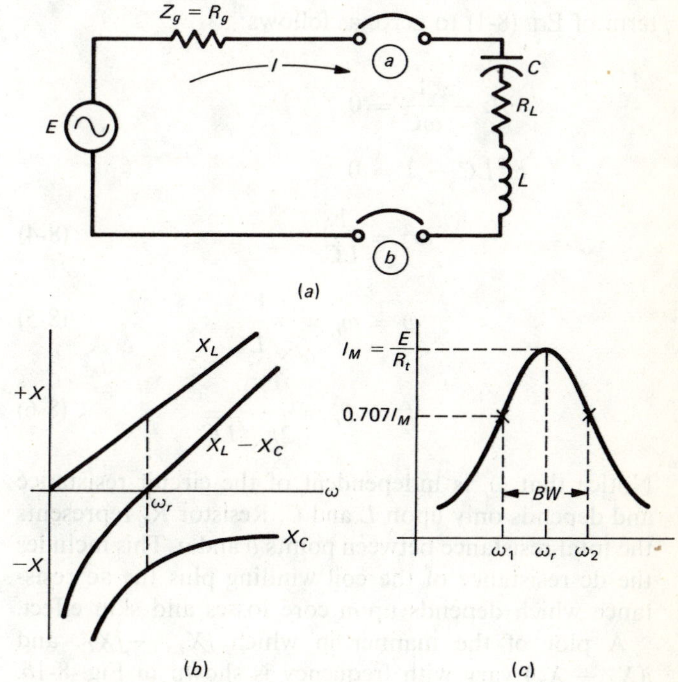

FIGURE 8-1

At some angular frequency ω_r, the reactive term, will equal zero and the impedance will be purely resistive. This condition is known as *series resonance*, and ω_r or f_r is the series-resonant frequency.

In polar form, the general expression for the impedance seen looking into terminals AB is

$$\mathbf{Z}_{AB} = \sqrt{R_L^2 + \left(\omega L - \frac{1}{\omega C}\right)^2} \Bigg/ \tan^{-1}\frac{\omega L - (1/\omega C)}{R_L} \qquad (8\text{-}1b)$$

and the current is

$$\mathbf{I} = \frac{\mathbf{E}}{\mathbf{Z}_g + \mathbf{Z}_{AB}} = \frac{\mathbf{E}}{R_g + R_L + j[\omega L - (1/\omega C)]} \qquad (8\text{-}2)$$

If the generator resistance ($R_g = 0$), then

$$\mathbf{I} = \frac{\mathbf{E}}{\mathbf{Z}_{AB}}$$

From Eq. (8-1) we can see that $\mathbf{Z}_{AB}$ will exhibit a minimum impedance equal to R_L ohms. If the source impedance R_g is purely resistive as indicated, then the current is in phase with the applied voltage.

If $R_g \neq 0$, it must be added to R_L to give a total equivalent circuit resistance referred to as R_t.

$$R_t = R_g + R_L \qquad (8\text{-}3)$$

The series-resonant frequency may be expressed in terms of the circuit parameters by equating the reactive term of Eq. (8-1) to zero, as follows:

$$\omega L - \frac{1}{\omega C} = 0$$

$$\omega^2 LC - 1 = 0$$

$$\omega^2 = \frac{1}{LC} \qquad (8\text{-}4)$$

$$\omega = \omega_r = \frac{1}{\sqrt{LC}} \qquad (8\text{-}5)$$

$$f = f_r = \frac{1}{2\pi\sqrt{LC}} \qquad (8\text{-}6)$$

Notice that ω_r is independent of the circuit resistance and depends only upon L and C. Resistor R_L represents the total resistance between points b and a. This includes the dc resistance of the coil winding plus the ac resistance which depends upon core losses and skin effect.

A plot of the manner in which jX_L, $-jX_C$, and $j(X_L - X_C)$ vary with frequency is shown in Fig. 8-1b. At ω_r, the positive X distance is equal to the negative X distance, and the net reactance is zero. The manner in which the current varies with frequency is the well-known resonance curve shown in Fig. 8-1c. The current is maximum at ω_r because $\mathbf{Z}_{AB}$ is a minimum and equal to R_L.

8-2 Bandwidth of a series-resonant circuit

It would be very convenient to have some means of describing the steepness of the resonance curve, since this would indicate how sharply we can select the desired frequency from adjacent frequencies. The method used is based upon the following considerations. At resonance, the power dissipated in the resonant circuit is a maximum, since the current is maximum. There will then be two frequencies, one on each side of f_r, where the power dissipated is one-half the power at resonance. These frequencies are designated as the upper (f_2) and lower (f_1) half-power frequencies. Let us now proceed to locate these half-power frequencies on the resonance curve of Fig. 8-1c. Remember that when we speak of power here, we mean real power that is dissipated in resistive elements. At f_r,

$$P_r = I_r^2 R_t$$

At f_2,

$$P_2 = 1/2\, P_r$$

Therefore,

$$I_2^2 R_t = 1/2\, I_r^2 R_t$$

$$I_2 = \frac{I_r}{\sqrt{2}} = 0.707 I_r$$

In a similar manner we can show that for f_1,

$$I_1 = \frac{I_r}{\sqrt{2}} = 0.707 I_r$$

Now it might be desirable to determine the bandwidth of the tuned circuit by inspection of the parameters, instead of by direct measurements upon an actual circuit. We can surely set up the following proportion, since we have developed the relationship between I at resonance and I at the half-power frequencies (I_{12}). Subscript 12 is used to designate a half-power point occurring at ω_1 and ω_2.

$$\frac{|I_{12}|}{|I_r|} = \frac{1}{\sqrt{2}} = \frac{E/\sqrt{R_t^2 + X_{12}}}{E/R_t}$$

$$\frac{1}{\sqrt{2}} = \frac{R_t}{\sqrt{R_t^2 + X_{12}^2}}$$

$$\frac{1}{2} = \frac{R_t^2}{R_t^2 + X_{12}^2} \qquad (8\text{-}7)$$

Solving for the relationship between X_{12} and R_t, we obtain

$$X_{12} = \pm R_t$$

Notice that the net reactance is equal to the net resistance at the half-power points. This also tells us that the phase angle is plus or minus 45°. At ω_2 the circuit appears inductive and the phase angle is $-45°$, while at ω_1 the net reactance is capacitive and the current leads the voltage by 45°. The net reactance may be expressed in terms of L, C, and ω as follows:

$$X_{12} = \omega_{12}L - \frac{1}{\omega_{12}C} = \pm R_t$$

$$\omega_{12}^2 LC - 1 = \pm\omega_{12}R_t C$$

$$LC\omega_{12}^2 \mp R_t C\omega_{12} - 1 = 0$$

which is analogous to the quadratic formula

$$ax^2 + bx + c = 0$$

whose solution is

$$x = \frac{-b}{2a} \pm \frac{\sqrt{b^2 - 4ac}}{2a}$$

Therefore

$$\pm R_t C \pm \frac{\sqrt{R_t^2 C^2 + 4LC}}{2LC}$$

Since $\sqrt{R_t^2 C^2 + 4LC}$ is surely greater than $R_t C$, the case where the radical is preceded by a negative sign will result in negative frequencies. A negative frequency is meaningless to us, and this case is discarded. With just the positive sign before the radical, we have two possible frequencies:

$$\omega_{12} = \frac{\pm R_t C + \sqrt{R_t^2 C^2 + 4LC}}{2LC}$$

The two roots are then

$$2\pi f_1 = \omega_1 = \frac{-R_t C + \sqrt{R_t^2 C^2 + 4LC}}{2LC}$$

and

$$2\pi f_2 = \omega_2 = \frac{+R_t C + \sqrt{R_t^2 C^2 + 4LC}}{2LC}$$

We now have developed three formulas that will enable us to determine the resonant and half-power frequencies in terms of circuit parameters. The frequency range between ω_1 and ω_2 is termed the *bandwidth*, *B*. That is, $B = \omega_2 - \omega_1$. A word of caution at this point: the quantity R_t, includes both generator and coil resistance. The coil resistance varies with frequency, because of skin effect, etc., which means that R_t due to R_L also varies with frequency. The value of R_L will not necessarily be the same at f_1, f_r, or f_2.

Fortunately, even though ac coil resistance varies with frequency, the ratio of coil reactance to coil resistance stays reasonably constant over the bandwidth in most cases. As R_L increases with frequency, so does X_L, and the ratio of X_L to R_L stays fairly constant. The quantity X_L/R_L is referred to as the coil Q, or Q_L, and it affords us a convenient way of analyzing the tuned circuit. While coil manufacturers do not usually have charts of R_L versus frequency, curves of Q_L versus f are readily available (Fig. 8-2).

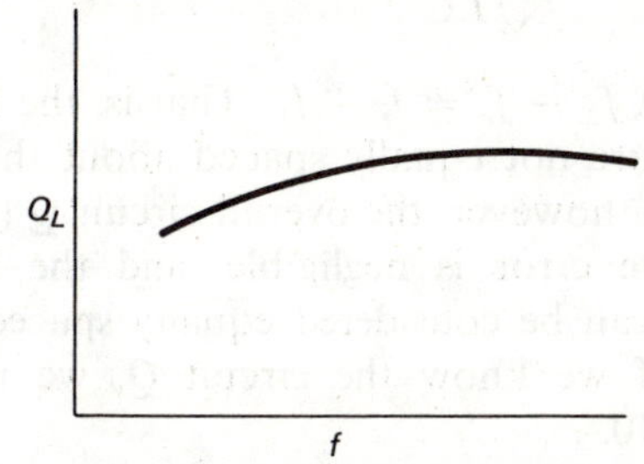

FIGURE 8-2

Later we shall develop a relation between Q_L and the half-power frequencies. But first let us see if we can relate the resonant and half-power frequencies directly to the circuit parameters. If we multiply ω_1 by ω_2, the result is

$$\omega_1\omega_2 = \frac{R_t^2C^2 + 4LC - R_t^2C^2}{4L^2C^2} = \frac{1}{LC}$$

but $\omega_r = 1/\sqrt{LC}$; therefore,

$$\omega_1\omega_2 = \omega_r^2 \qquad (8\text{-}8a)$$

or $f_1 f_2 = f_r^2$. This is the same as writing

$$\frac{f_1}{f_r} = \frac{f_r}{f_2} \qquad (8\text{-}8b)$$

The term bandwidth, as we have been using it up to now, does not really tell us much unless the resonant frequency is also specified. For example, if you were told that the bandwidth of a series resonant circuit is 100 Hz, could you tell if it were a sharply tuned circuit? Of course not. If the f_r is 500 Hz, 100 Hz would be a large percentage of f_r, resulting in a broad response curve. If f_r were 1 MHz, the tuning would be very sharp. Therefore, what we really need as a figure of merit for judging the selectivity of a tuned circuit is the ratio of the bandwidth to the resonant frequency. This ratio is sometimes referred to in the literature as the per unit bandwidth or just bandwidth (*B*). We shall, however, define

$$\frac{\omega_2 - \omega_1}{\omega_r} = \frac{f_2 - f_1}{f_r} = \text{per unit } B \qquad (8\text{-}9a)$$

and $\omega_2 - \omega_1 = \Delta\omega$ as the bandwidth.

Therefore, let us now develop a simple relationship between the per unit *B* and the circuit parameters. We have already shown that

$$\omega_2 = \frac{R_t C + \sqrt{R_t^2 C^2 + 4LC}}{2LC} \qquad (8\text{-}9b)$$

$$\omega_1 = \frac{-R_tC + \sqrt{R_t^2C^2 + 4LC}}{2LC} \quad (8\text{-}9c)$$

$$\omega_r = \frac{1}{\sqrt{LC}} \quad (8\text{-}9d)$$

In general, $f_2 - f_r \neq f_r - f_1$. That is, the half-power frequencies are not equally spaced about the resonant frequency. If, however, the overall circuit Q (Q_t, that is) is ≥ 10, the error is negligible, and the half-power frequencies can be considered equally spaced about f_r. Therefore, if we know the circuit Q, we may write, when $Q_t \geq 10$,

$$\omega_2 = \omega_r + \frac{B}{2} = \omega_r + \frac{\omega_r}{2Q_t} \quad (8\text{-}9e)$$

$$\omega_1 = \omega_r - \frac{B}{2} = \omega_r - \frac{\omega_r}{2Q_t} \quad (8\text{-}9f)$$

If the circuit Q is about 10 or more, the voltage across either L or C will also essentially be a maximum at ω_r and exhibits a frequency-response curve similar to that of the current. The same bandwidth Q and other relationships may be used. For example, the circuit Q could be evaluated by measuring 0.707 points of coil voltage. More about this later.

Let us summarize the results of our efforts up to this point.

1. The resonant frequency of a series RLC circuit is

$$\omega_r = \frac{1}{\sqrt{LC}} \quad (8\text{-}10a)$$

2. The current at resonance is in phase with the applied voltage and has a maximum value equal to E/R_t.
3. We have defined two half-power frequencies, ω_1 and ω_2, at which the current is 70.7 percent of the resonant current and the phase angle between the current and the applied voltage is plus or minus 45°.
4. The resonant frequency is the geometric mean of ω_1 and ω_2. That is,

$$\omega_r = \sqrt{\omega_1\omega_2} \quad (8\text{-}10b)$$

5. The bandwidth, or spacing between the half-power frequencies, is

$$B = \omega_2 - \omega_1 \quad (8\text{-}10c)$$

6. The per unit bandwidth is a measure of the circuit selectivity, since it expresses what percentage of the resonant frequency is occupied by the bandwidth. The per unit bandwidth is related to the circuit Q by

$$\text{Per unit } B = \frac{\omega_2 - \omega_1}{\omega_r} = \frac{1}{Q_t} \quad (8\text{-}10d)$$

7. The above relationship (6) tells us that, if we know the circuit Q and ω_r, we can determine the bandwidth from

$$B = \omega_2 - \omega_1 = \frac{\omega_r}{Q_t} \quad (8\text{-}10e)$$

8. Conversely, the total Q of a series-resonant circuit can be determined by performing measurements to evaluate ω_1, ω_r, and ω_2.

$$Q_t = \frac{\omega_r}{\omega_2 - \omega_1} \quad (8\text{-}10f)$$

9. If the source which drives the resonant circuit is purely resistive in nature, the circuit Q can be related to the coil Q by

$$Q_t = \frac{1}{(R_g/\omega_rL) + 1/Q_L} \quad (8\text{-}10g)$$

10. If the circuit Q is known and it is desired to evaluate the coil Q, we can solve for Q_L and obtain

$$Q_L = \frac{1}{(1/Q_t) - (R_g/\omega_rL)} \quad (8\text{-}10h)$$

11. If the circuit Q is known and is in the neighborhood of 10 or more, the half-power frequencies may be considered evenly spaced about ω_r. That is,

$$\omega_2 = \omega_r + \frac{B}{2} = \omega_r + \frac{\omega_r}{2Q_t} \quad (8\text{-}10i)$$

$$\omega_1 = \omega_r - \frac{B}{2} = \omega_r - \frac{\omega_r}{2Q_t} \quad (8\text{-}10j)$$

Now work through problems PS 8-1, PS 8-2, and PS 8-3.

8-3 Analysis with Q not constant

Up to this point we have restricted ourselves to cases in which the coil Q is essentially constant over a range of frequencies. While this is the usual case, there are times, particularly when dealing with iron-core inductors at audio frequencies, when Q_L itself varies considerably with a slight change in frequency, as indicated in Fig. 8-3. To analyze this situation, let us go back to fundamentals.

In a series RLC circuit the current at any frequency is given by

1) $$\mathbf{I} = \frac{\mathbf{E}}{R + jX} \quad (8\text{-}11)$$

where $\mathbf{E}$ = the applied voltage
$R = R_t$, the total circuit resistance
$X = X_L - X_C$, the net reactance

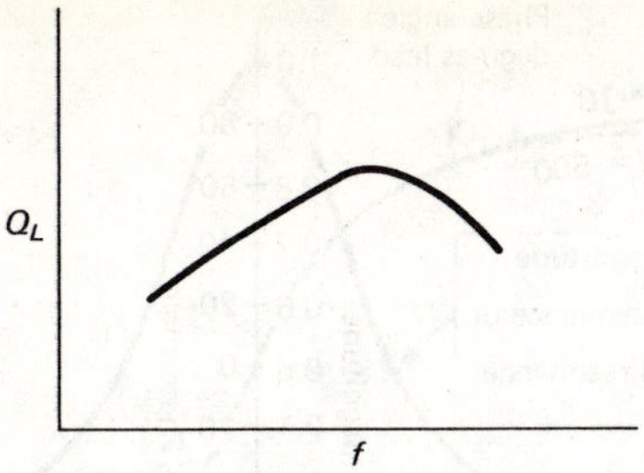

FIGURE 8-3

At resonance, $\mathbf{I} = \mathbf{E}/R_r$, where R_r = total circuit resistance at resonance. Therefore,

2) $$\frac{\mathbf{I}}{\mathbf{I}_r} = \frac{\mathbf{E}/(R + jX)}{\mathbf{E}/R_r} = \frac{R_r}{R + jX}$$

Dividing numerator and denominator by R_r,

3) $$\frac{\mathbf{I}}{\mathbf{I}_r} = \frac{1}{(R/R_r) + j(X/R_r)}$$

Since we wish to express I/I_r as a function of ω, we can write, for X,

4) $$X = X_L - X_C = \omega L - \frac{1}{\omega C} = \frac{\omega^2 LC - 1}{\omega C}$$

but $\omega_r = 1/\sqrt{LC}$; therefore $LC = 1/\omega_r{}^2$, and $C = 1/\omega_r{}^2 L$. Therefore,

5) $$X = \frac{\omega^2(1/\omega_r{}^2) - 1}{\omega/\omega_r{}^2 L} = \frac{(\omega/\omega_r)^2 - 1}{(\omega/\omega_r)1/\omega_r L}$$

Substituting in (3), we have

6) $$\frac{\mathbf{I}}{\mathbf{I}_r} = \frac{1}{\dfrac{R}{R_r} + j\dfrac{(\omega/\omega_r)^2 - 1}{(\omega/\omega_r)R_r/\omega_r L}}$$

But $Q_r = \omega_r L/R_r$, where Q_r is the total circuit Q at resonance. Hence,

7) $$\frac{\mathbf{I}}{\mathbf{I}_r} = \frac{1}{\dfrac{R}{R_r} + j\dfrac{(\omega/\omega_r)^2 - 1}{(\omega/\omega_r)1/Q_r}} = \frac{1}{\dfrac{R}{R_r} + jQ_r\dfrac{(\omega/\omega_r)^2 - 1}{\omega/\omega_r}}$$

$$= \frac{1}{\dfrac{R}{R_r} + jQ_r\left(\dfrac{\omega}{\omega_r} - \dfrac{1}{\omega/\omega_r}\right)}$$

8) $$\frac{\mathbf{I}}{\mathbf{I}_r} = \frac{1}{\dfrac{R}{R_r} + jQ_r\left(\dfrac{\omega}{\omega_r} - \dfrac{\omega_r}{\omega}\right)} \qquad (8\text{-}12)$$

Usually coil Q as a function of frequency is available from manufacturers' data. Therefore, it may be more convenient to express R/R_r in terms of Q.

$$R = \frac{\omega L}{Q}$$

$$R_r = \frac{\omega_r L}{Q_r}$$

and Q does not necessarily equal Q_r. Remember, R is the total circuit resistance; therefore,

$$\frac{R}{R_r} = \frac{\omega L/Q}{\omega_r L/Q_r} = \frac{\omega}{\omega_r}\frac{Q_r}{Q}$$

$$\frac{\mathbf{I}}{\mathbf{I}_r} = \frac{1}{\dfrac{\omega}{\omega_r}\dfrac{Q_r}{Q} + jQ_r\left(\dfrac{\omega}{\omega_r} - \dfrac{\omega_r}{\omega}\right)}$$

$$= \frac{1}{\dfrac{f}{f_r}\dfrac{Q_r}{Q} + jQ_r\left(\dfrac{f}{f_r} - \dfrac{f_r}{f}\right)} \qquad (8\text{-}13)$$

Equations (8-12) and (8-13) are exact and involve no assumptions other than L remaining constant. Let us see how Eq. (8-13) would be used in a typical problem. Study problem PS 8-4.

8-4 The universal resonance curve

At times we are faced with the problem of determining what frequency is required to give a specific current response. This is a difficult problem, since it means solving for ω in Eq. (8-13), repeated here as Eq. (8-14*a*).

1) $$\frac{\mathbf{I}}{\mathbf{I}_r} = \frac{1}{\dfrac{\omega}{\omega_r}\dfrac{Q_r}{Q} + jQ_r\left(\dfrac{\omega}{\omega_r} - \dfrac{\omega_r}{\omega}\right)} \qquad (8\text{-}14a)$$

Furthermore, when ω is the unknown, we cannot determine Q, since it depends on ω. Possibly Q can be expressed as a function of ω for a given coil, but this is a major problem in itself.[1] It would be much easier to develop a chart that plots the current ratio versus frequency. This, then, is our next undertaking.

To reduce the complexity of our problem, we shall first make the assumption that Q does not appreciably vary in the region about resonance. That is, $Q = Q_r$. Equation (8-14*a*) therefore reduces to

2) $$\frac{\mathbf{I}}{\mathbf{I}_r} = \frac{1}{\dfrac{\omega}{\omega_r} + jQ_r\left(\dfrac{\omega}{\omega_r} - \dfrac{\omega_r}{\omega}\right)} \qquad (8\text{-}14b)$$

Let $\omega = \omega_r + k\omega_r$, where k may be positive or negative. This substitution may be simplified to

3) $$\omega = \omega_r(1 + k) \qquad (8\text{-}14c)$$

[1] An excellent treatment of tuned circuits may be found in F. E. Terman, "Electronic and Radio Engineering," 4th ed., McGraw-Hill Book Company, New York, 1955.

The parameter k carries its own sign. From Eq. (8-14c) we see that

$$4)\qquad k = \frac{\omega - \omega_r}{\omega_r} \qquad (8\text{-}14d)$$

Substituting (8-14c) in (8-14b), we have

$$5)\quad \frac{\mathbf{I}}{\mathbf{I}_r} = \frac{1}{\dfrac{\omega_r(1+k)}{\omega_r} + jQ_r\left[\dfrac{\omega_r(1+k)}{\omega_r} - \dfrac{\omega_r}{\omega_r(1+k)}\right]}$$

Simplifying,

$$\frac{\mathbf{I}}{\mathbf{I}_r} = \frac{1}{1 + k + jQ_r\left(1 + k - \dfrac{1}{1+k}\right)}$$

$$= \frac{1}{1 + k + jQ_r\left[\dfrac{(1+k)^2 - 1}{1+k}\right]}$$

$$= \frac{1}{1 + k + jQ_r k\dfrac{2+k}{1+k}} \qquad (8\text{-}14e)$$

Equation (8-14e) may be plotted as a universal resonance curve in the following manner. Assume values of $Q_r k$ along the horizontal axis of a graph and plot the corresponding $\mathbf{I}/\mathbf{I}_r$ values along the vertical axis. The range of values that $Q_r k$ should encompass will be determined by the Q values expected and by deviation from the resonant frequency. Usually Q will lie between 10 and 500, so these two cases will be considered. Even for the case where $Q = 10$, we are seldom interested in the current response at frequencies 30 percent higher or lower than ω_r. This fixes the range of k, which is determined as follows.

From Eq. (8-14c) $\omega = \omega_r(1 + k)$. When ω is 30 percent higher than ω_r,

$$\omega = \omega_r + 0.3\omega_r = 1.3\omega_r$$

Substituting in Eq. (8-14d) yields

$$k = \frac{1.3\omega_r - \omega_r}{\omega_r} = 0.3$$

When ω is 30 percent lower than ω_r

$$\omega = \omega_r - 0.3\omega_r = 0.7\omega_r$$

and for this case we find the lower limit of k, which is

$$k = -0.3$$

A satisfactory range along the horizontal axis can be determined by taking $Q_r k$ equal to the product of the lowest Q considered, which is 10, and $k = \pm 0.3$, which is -3 to $+3$.

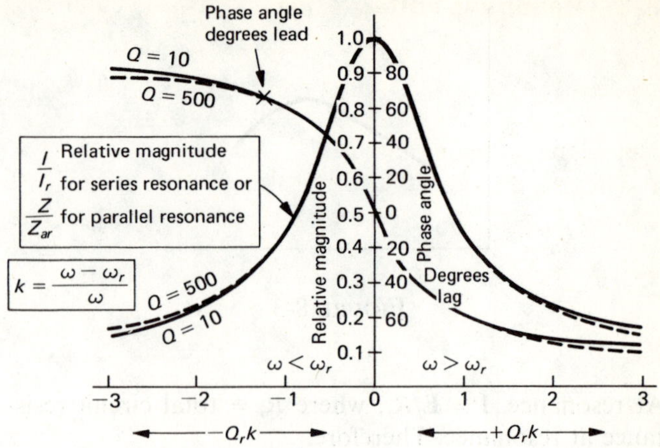

FIGURE 8-4

Let us go through a few illustrative calculations using (8-14e) to show how the universal resonance curve of Fig. 8-4 is sketched. If $Q_r k = -3$, and $Q_r = 10$, $k = -0.3$,

$$\frac{\mathbf{I}}{\mathbf{I}_r} = \frac{1}{1 - 0.3 + j10(-0.3)\left(\dfrac{2 - 0.3}{1 - 0.3}\right)} = 0.137\underline{/84.5^\circ}$$

If $k = +0.3$, and $Q_r = 10$, $Q_r k = 3$,

$$\frac{\mathbf{I}}{\mathbf{I}_r} = \frac{1}{1 + 0.3 + j10(0.3)\left(\dfrac{2 + 0.3}{1 + 0.3}\right)} = \frac{1}{1.3 + j3\dfrac{2.3}{1.3}}$$

$$= \frac{1}{1.3 + j5.3} = \frac{1}{5.5\underline{/76.2^\circ}} = 0.18\underline{/-76.2^\circ}$$

Note that the curve is almost symmetrical about ω_r for these values of $Q_r k$. The higher the Q, the better the symmetry.

Now suppose $Q_r k = -3$, but $Q_r = 500$. This means $k = -0.006$:

$$\frac{\mathbf{I}}{\mathbf{I}_r} = \frac{1}{1 - 0.006 + j500(-0.006)\left(\dfrac{2 - 0.006}{1 - 0.006}\right)}$$

$$\approx \frac{1}{1 - j3(2)} = \frac{1}{1 - j6} = \frac{1}{6.1\underline{/-80.6^\circ}} = 0.16\underline{/80.6^\circ}$$

If we were to continue the process and obtain sufficient points to sketch two resonance curves, one for $Q = 10$ and the other for $Q = 500$, we should find them virtually superimposed. The curve for $Q = 500$ may be used for values of Q ranging to infinity with very little error, as evidenced in Fig. 8-4. We shall see later that Fig. 8-4 may also be used for parallel-resonant circuits.

Now work through problems PS 8-5, PS 8-6, and PS 8-7.

8-5 Resonant voltage rise

An interesting and useful phenomenon about series-resonant circuits is the large rise in voltage that occurs across L or C at ω_r when Q_t is large. We can prove this fact in the following manner.

The magnitude of the voltage across the capacitor is $E_C = IX_C$, but at resonance $I = I_r = E/R$. Therefore, $E_{C_r} = \frac{E}{R}X_{C_r}$. But at ω_r, $X_{C_r} = X_{L_r}$, or

$$E_{C_r} = \frac{EX_{L_r}}{R} = EQ_{t_r} \qquad (8\text{-}15a)$$

Q_{t_r} is the circuit Q at resonance.

Remarkably, the capacitor or inductor voltage at resonance may be Q_{t_r} times larger than the applied voltage. If 10 volts is applied to a series-resonant circuit having a $Q_{t_r} = 100$, the capacitor or inductor voltage will be 1,000 volts. When circuits of this type are designed, the capacitor working voltage must be determined on this basis. Actually ω_r is not the exact frequency at which E_L or E_C is a maximum, but the difference is slight if $Q_{t_r} \geqq 10$. The exact frequency at which E_c is a maximum is

$$\omega = \omega_r\sqrt{1 - \frac{1}{2Q_{t_r}^{\ 2}}} \qquad (8\text{-}15b)$$

which lies slightly below ω_r. If $Q_{t_r} = 10$, this frequency is essentially the same as ω_r and the maximum capacitor voltage will approximately equal the capacitor voltage at resonance.

The voltage across the coil at resonance (E_{L_r}) is complicated by the fact that L has a series resistance (R_L) associated with it. Therefore, we shall start with Z_L rather than X_L.

$$Z_L = \sqrt{R_L^{\ 2} + (\omega L)^2}$$

Since R_L is usually unspecified but Q_L is specified, we can write

$$Z_L = \sqrt{(\omega L)^2\left[\frac{R_L^{\ 2}}{(\omega L)^2} + \frac{(\omega L)^2}{(\omega L)^2}\right]} = \omega L\sqrt{\frac{1}{Q_L^{\ 2}} + 1}$$

$$E_{L_r} = I_r Z_{L_r} = \frac{E}{R}\omega_r L\sqrt{\frac{1}{Q_{L_r}^{\ 2}} + 1}$$

$$= EQ_{t_r}\sqrt{\frac{1}{Q_{L_r}^{\ 2}} + 1} \approx EQ_{t_r} \qquad \text{when } Q_{L_r} \geq 10$$

$$(8\text{-}16a)$$

The exact frequency at which the coil voltage is a maximum is slightly above ω_r and is given by

$$\omega = \frac{\omega_r}{\sqrt{1 - 1/2Q_{t_r}^{\ 2}}} \qquad (8\text{-}16b)$$

Again, if $Q_{t_r} \geq 10$, the coil voltage may be considered maximum at ω_r.

A note of caution—whenever you make any calculations involving the coil Q in the vicinity of ω_r, be sure to use the value of Q corresponding to ω_r. Q_L may vary widely over a broad frequency range, and therefore it is best either to measure the coil Q at the frequency of interest or to use manufacturer's data, which should plot Q_L versus f.

Let us apply the material we have covered to some illustrative problems. In all cases, rms values of voltage and current are understood unless otherwise stated. Work through problems PS 8-8 and PS 8-9.

8-6 Parallel antiresonance

We shall now investigate the phenomenon of parallel resonance, or antiresonance, as it is sometimes called. The circuit of Fig. 8-5*a* illustrates a completely general

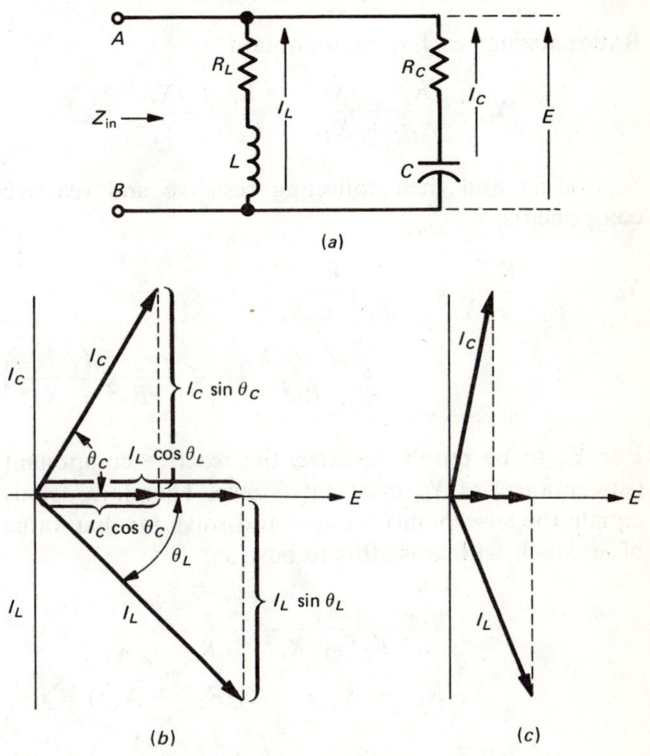

FIGURE 8-5

antiresonant circuit. The impedance seen looking into the AB terminals may vary widely, depending upon the Q of the inductive and capacitive branches. At frequencies below the resonant frequency, the impedance of the inductive branch is low and a large current will flow through the coil. The current through the capacitive branch will be small because X_C is large at low frequencies. The line current flowing into the input terminals is therefore large. At high frequencies, the inductive branch offers a high impedance but the capacitive branch has low impedance. Again the line current is relatively large. At some in-between frequency, the input impedance will be largest and the line current will be minimum. This is not necessarily the same frequency at which the line current is in phase with the applied voltage. If Q is even as low as 5, however, the error is in the neighborhood of 1 percent, and therefore maximum impedance will be considered as occurring at the same frequency, which results in unity power factor. Thus at some frequency which we shall define as the antiresonant frequency (f_{ar}), the impedance seen looking into terminals AB is purely resistive. Our first objective is to determine how this frequency is related to the circuit parameters.

Since we are dealing with a parallel circuit, it is more convenient to work with admittances.

$$\mathbf{Y}_{in} = \frac{1}{Z_{in}} = \frac{1}{Z_L} + \frac{1}{Z_C} = \frac{1}{R_L + jX_L} + \frac{1}{R_C - jX_C}$$

Rationalizing[2] each term, we obtain

$$\mathbf{Y}_{in} = \frac{R_L - jX_L}{R_L^2 + X_L^2} + \frac{R_C + jX_C}{R_C^2 + X_C^2}$$

Separating and then collecting resistive and reactive components,

$$\mathbf{Y}_{in} = \frac{R_L}{R_L^2 + X_L^2} + \frac{R_C}{R_C^2 + X_C^2} + j\left(\frac{X_C}{R_C^2 + X_C^2} - \frac{X_L}{R_L^2 + X_L^2}\right)$$

For $\mathbf{Y}_{in}$ to be purely resistive, the reactive component (susceptance) of $\mathbf{Y}_{in}$ must equal zero. Therefore, let us equate the susceptance to zero and solve for that value of ω which will cause this to be true:

$$\frac{X_C}{R_C^2 + X_C^2} - \frac{X_L}{R_L^2 + X_L^2} = 0$$

$$X_C(R_L^2 + X_L^2) - X_L(R_C^2 + X_C^2) = 0$$

$$\frac{1}{\omega C}(R_L^2 + \omega^2 L^2) - \omega L\left(R_C^2 + \frac{1}{\omega^2 C^2}\right) = 0$$

$$R_L^2 + \omega^2 L^2 - \omega^2 LC\frac{\omega^2 C^2 R_C^2 + 1}{\omega^2 C^2} = 0$$

$$\omega^2 C^2 R_L^2 + \omega^4 L^2 C^2 - \omega^2 LC(\omega^2 C^2 R_C^2 + 1) = 0$$

Factoring an $\omega^2 C$ out of each term, we have

$$CR_L^2 + \omega^2 L^2 C - L(\omega^2 C^2 R_C^2 + 1) = 0$$

[2] Another way of expressing the fact that numerator and denominator are multiplied by the conjugate of the denominator.

Expanding and collecting terms,

$$\omega^2 L^2 C - \omega^2 LC^2 R_C^2 + CR_L^2 - L = 0$$

$$\omega^2 = \frac{L - CR_L^2}{L^2 C - LC^2 R_C^2} = \frac{L - CR_L^2}{LC(L - CR_C^2)}$$

$$\omega = \omega_{ar} = \frac{1}{\sqrt{LC}}\sqrt{\frac{L - CR_L^2}{L - CR_C^2}} \tag{8-17}$$

Notice that the parallel antiresonant frequency is actually dependent upon the branch resistances. In series circuits the resonant frequency was independent of the circuit resistances. Equation (8-17) is quite interesting. It indicates that resonance can be established not only by varying ω, L, or C but also by controlling R_L or R_C. This, however, is rarely done in practice, since R_L and R_C tend to deteriorate the selectivity of the circuit.

In most communications circuits, the resistance in the capacitive branch is negligible and that of the inductive branch is small if the coil Q is reasonably high. Then L will usually be much larger than CR_L^2 or CR_C^2, and Eq. (8-17) reduces to $\omega_{ar} = 1/\sqrt{LC}$, which is the same as the series-resonance case. If either CR_L^2 or CR_C^2 should be greater than L, the quantity under the radical will be negative, which results in an imaginary value of ω_{ar}. This is something we cannot physically generate and therefore has no meaning other than that no resonant condition will exist at any frequency. If $R_L = R_C$, the quantity under the radical is equal to 1, and therefore $\omega_{ar} = 1/\sqrt{LC}$ for this case. Should R_L equal R_C and also equal $\sqrt{L/C}$, ω_{ar} is indeterminate, and the circuit appears resistive at all frequencies.

In practical antiresonant circuits, the resistance in the capacitive branch is usually negligible, and Eq. (8-17) reduces to

1) $$\omega_{ar} = \frac{1}{\sqrt{LC}}\sqrt{\frac{L - CR_L^2}{L}}$$

which may be manipulated into

2) $$\omega_{ar} = \frac{1}{\sqrt{LC}}\sqrt{\frac{L^2 - LCR_L^2}{L^2}}$$

Squaring both sides and substituting the series-resonant frequency ω_r for $1/\sqrt{LC}$, we obtain

3) $$\omega_{ar}^2 = \omega_r^2\frac{L^2 - (1/\omega_r^2)R_L^2}{L^2} = \omega_r^2\left(1 - \frac{R_L^2}{\omega_r^2 L^2}\right)$$

and since $\omega_r L/R_L = Q_L$, we have

4) $$\omega_{ar} = \omega_r\sqrt{1 - \frac{1}{Q_L^2}} \tag{8-18}$$

which indicates that the series- and parallel-resonant

frequencies are almost identical when Q_L is large in the vicinity of resonance.

Note that the value of Q_L in Eq. (8-18) is based upon the coil Q at ω_r and not at ω_{ar}. A slightly different expression for ω_{ar} is obtained if the coil Q at ω_{ar} is worked into (2). Squaring both sides of (2), we have

$$5)\qquad \omega_{ar}{}^2 = \frac{1}{LC}\left[\frac{L^2 - LCR^2}{L^2}\right]$$

Substituting $\omega_r{}^2$ for $1/LC$ and separating the bracketed term into two terms,

$$6)\qquad \omega_{ar}{}^2 = \omega_r{}^2\left(\frac{L^2}{L^2} - \frac{1}{\omega_r{}^2}\frac{R^2}{L^2}\right)$$

Multiplying both numerator and denominator of the R^2/L^2 term by $\omega_{ar}{}^2$,

$$7)\qquad \omega_{ar}{}^2 = \omega_r{}^2\left(1 - \frac{\omega_{ar}{}^2R^2}{\omega_r{}^2\omega_{ar}{}^2L^2}\right) = \omega_r{}^2\left(1 - \frac{\omega_{ar}{}^2}{\omega_r{}^2Q_L{}^2}\right)$$

where Q_L this time is the coil Q determined at ω_{ar}. Solving for ω_{ar}, we obtain

$$8)\qquad \omega_{ar} = \frac{\omega_r}{\sqrt{1 + 1/Q_L{}^2}} \qquad (8\text{-}19)$$

Comparing Eqs. (8-18) and (8-19), we see that even though the coil Q at ω_r may differ from that at ω_{ar}, the antiresonant frequency ω_{ar} is still essentially equal to ω_r if Q_L is about 10 or more.

A physical interpretation of the circuit conditions at ω_{ar} may be obtained from Fig. 8-5*b*. The current in each branch is determined by the impedance of that branch. The current in the capacitive branch (I_C) will lead the applied voltage by an angle θ_C. The component of I_C which is in phase with the applied voltage (E) is $I_C \cos \theta_C$, while the quadrature (reactive) component is $I_C \sin \theta_C$. In the inductive branch, the current I_L will lag the applied voltage by an angle θ_L. We can also resolve I_L into an in-phase and a quadrature component, $I_L \cos \theta_L$ and $I_L \sin \theta_L$ respectively. At ω_{ar}, the magnitudes and phase angles of I_L and I_C need not be the same, since R_L and R_C might be different. The quadrature components, $I_C \sin \theta_C$ and $I_L \sin \theta_L$, cancel, which leaves a total in-phase current of $I_C \cos \theta_C$ plus $I_L \cos \theta_L$. The impedance seen looking into the *AB* terminals of Fig. 8-5*a* at ω_{ar} is then a finite quantity equal to the applied voltage divided by the net in-phase current.

When the Q of each branch is high, so that the branch reactance is much larger than the branch resistance, the quadrature currents will be much larger than the in-phase currents. This is illustrated in Fig. 8-5*c*. The net in-phase current at ω_{ar} is therefore lower, which means that the input impedance at antiresonance is higher for the higher Q circuit. At ω_{ar}, the branch currents may be quite large, but their vector sum, if Q is high, results in a small line current.

Now work problems PS 8-10 and PS 8-11.

8-7 A practical antiresonant circuit

Figure 8-6 illustrates a practical antiresonant circuit commonly used in communications work. We have already developed Eq. (8-18), which expresses the resonant frequency of Fig. 8-5*a*. This, with $R_C = 0$, is the same as Fig. 8-6. Since one of our primary objectives is to gain skill in manipulating and interpreting complex-number equations, let us start from scratch in our analysis of Fig. 8-6. Let us first verify Eq. (8-18).

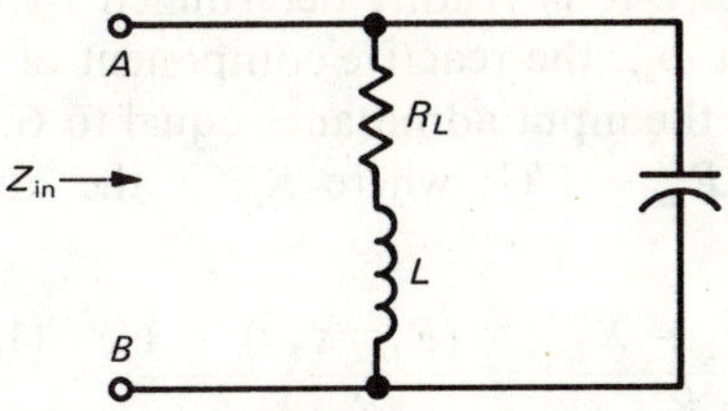

FIGURE 8-6

$$1)\qquad \mathbf{Y}_{in} = \frac{1}{\mathbf{Z}_{in}} = \frac{1}{R_L + jX_L} + \frac{1}{-jX_C}$$

$$= \frac{R_L - jX_L}{R_L{}^2 + X_L{}^2} + j\frac{1}{X_C}$$

$$2)\qquad = \frac{R_L}{R_L{}^2 + X_L{}^2} + j\left(\frac{1}{X_C} - \frac{X_L}{R_L{}^2 + X_L{}^2}\right)$$

$$3)\qquad = G + jB = \frac{R_L}{R_L{}^2 + X_L{}^2} + j\frac{R_L{}^2 + X_L{}^2 - X_LX_C}{X_C(R_L{}^2 + X_L{}^2)} \qquad (8\text{-}20)$$

For Z_{in} to be purely resistive, the reactive component of Y_{in} must equal zero. That is,

$$4)\qquad \frac{R_L{}^2 + X_L{}^2 - X_LX_C}{X_C(R_L{}^2 + X_L{}^2)} = 0$$

The above will be true when the numerator is zero or

$$5)\qquad R_L{}^2 + X_L{}^2 - X_LX_C = 0$$

Solving for the value of ω which makes the expression zero,

$$6)\qquad R_L{}^2 + \omega^2L^2 - \frac{\omega L}{\omega C} = 0$$

$$7)\qquad \omega^2 = \frac{L}{L^2C} - \frac{R_L{}^2}{L^2} = \frac{1}{LC} - \frac{R_L{}^2}{L^2}$$

$$8)\qquad \omega = \omega_{ar} = \sqrt{\frac{1}{LC} - \frac{R_L{}^2}{L^2}}$$

This can be manipulated into the form of Eq. (8-18) if we again let $\omega_r^2 = 1/LC$ and $Q_L = \omega L/R_L$.

9) $$\omega_{ar} = \sqrt{\omega_r^2 - \frac{\omega_r^2 R_L^2}{\omega_r^2 L^2}}$$

$$= \sqrt{\omega_r^2 - \frac{\omega_r^2}{Q_L^2}} = \omega_r \sqrt{1 - \frac{1}{Q_L^2}}$$

which verifies Eq. (8-18).

8-8 Input impedance at antiresonance

The impedance at ω_{ar}, seen looking into terminals *a* and *b* of Fig. 8-6, is readily determined by examining Eq. (8-20). At ω_{ar}, the reactive component of Y_{in} is zero, which leaves the input admittance equal to G. Therefore $\mathbf{Z}_{in} = Z_{ar} = R_{ar} = 1/G$; where R_{ar} is the antiresonant impedance.

1) $$R_{ar} = \frac{R_L^2 + X_L^2}{R_L} = \frac{(R_L^2/X_L^2) + 1}{R_L/X_L^2} = \frac{(1/Q_L^2) + 1}{1/Q_L X_L}$$

All reactances are taken at ω_{ar}.

2) $$R_{ar} = X_L Q_L\left(1 + \frac{1}{Q_L^2}\right) \qquad (8\text{-}21a)$$

and if Q_L is large, say 10 or more,

3) $$R_{ar} \approx X_L Q_L \qquad (8\text{-}21b)$$

We can express R_{ar} in terms of X_C by solving step 5 in Sec. 8-7 for X_C in terms of X_L and can thereby obtain

$$R_L^2 + X_L^2 - X_L X_C = 0$$

4) $$X_C = \frac{X_L^2 + R_L^2}{X_L} = \frac{1 + R_L^2/X_L^2}{1/X_L}$$

$$= X_L\left(1 + \frac{1}{Q_L^2}\right) \qquad (8\text{-}22a)$$

From Eq. (8-22*a*) we see that X_C does not quite equal X_L at ω_{ar}, but the difference is slight if Q_L is large. Solving Eq. (8-22*a*) for X_L, we obtain

5) $$X_L = X_C \frac{1}{1 + 1/Q_L^2} \qquad (8\text{-}22b)$$

and substituting in Eq. (8-21*a*) yields

6) $$R_{ar} = X_C Q_L \qquad (8\text{-}21c)$$

Equation (8-21*c*) can be used to express R_{ar} directly in terms of the circuit parameters as follows:

7) $$R_{ar} = X_C Q_L = \frac{1}{\omega_{ar} C}\frac{\omega_{ar} L}{R_L} = \frac{L}{R_L C} \qquad (8\text{-}21d)$$

From (8-21*d*), additional useful forms of expressing R_{ar} may be derived. For example,

$$R_{ar} = \frac{X_C X_L}{R_L} \approx \frac{X_C^2}{R_L} \approx \frac{X_L^2}{R_L} = Q_L^2 R_L \qquad (8\text{-}21e)$$

If resistance is present in the capacitive branch, it should be added to R_L when determining R_{ar}. For example, Eqs. (8-21*d*) and (8-21*e*) would now be

$$R_{ar} = \frac{L}{(R_L + R_C)C} = \frac{X_C X_L}{R_L + R_C}$$

$$= \frac{X_L^2}{R_L + R_C} = \frac{X_C^2}{R_L + R_C} \qquad (8\text{-}21f)$$

Equation (8-21*f*) is not exact, but it is sufficiently accurate when the branch *Q*'s are about 10 or more.

The total *Q* of the parallel circuit when resistance is present in both branches may be taken as

$$Q_T = \frac{X_L}{R_L + R_C} \qquad (8\text{-}23)$$

when Q_T is 10 or more. An upper-case subscript (T) will be used to differentiate the total *Q* of the antiresonant circuit itself from the total *circuit Q*, designated Q_t. The total circuit *Q*, which includes the effect of source impedance, is developed in Sec. 8-9.

With the aid of the various expressions for R_{ar}, it may be shown that the branch currents are approximately Q_T times the line current at ω_{ar}. Let I_g equal the line current forced by some source E_g into the *AB* terminals of Fig. 8-6. We may write the following proportion:

$$\frac{E_g}{I_g} = R_{ar}$$

$$\frac{E_g}{I_L} = Z_L$$

If Q_T is high, therefore,

$$\frac{I_L}{I_g} = \frac{R_{ar}}{Z_L} \approx \frac{Q_T X_L}{X_L} = Q_T \qquad (8\text{-}24a)$$

Since $X_L \approx X_C$ at ω_{ar}, when Q_T is large, we see that

$$\frac{I_C}{I_g} = Q_T \qquad (8\text{-}24b)$$

Now study problem PS 8-12 carefully.

8-9 Selectivity of an antiresonant circuit

The selectivity of a series-resonant circuit was described in terms of the bandwidth relative to the resonant frequency. This in turn was shown to be a function of the circuit *Q*, as expressed in the equation

$$\frac{\omega_2 - \omega_1}{\omega_r} = \frac{1}{Q_t}$$

It would be desirable to develop an analogous relationship for the selectivity of an antiresonant circuit.

Our first consideration will be how to represent the source which drives the antiresonant circuit. That is, shall a current or voltage source be used? The series-resonant circuit had to be driven from a low-impedance or constant-voltage source if the circuit was to be selective. If the series circuit were driven from a constant-current (high-impedance) source, the current would be independent of conditions in the tuned circuit and no resonant current rise would manifest itself. An antiresonant circuit, on the other hand, should be driven from a constant current rather than from a constant voltage source. When the coil Q is high, the antiresonant circuit exhibits its maximum impedance at essentially the frequency of antiresonance.[3] If a constant-voltage source is connected across the antiresonant circuit, there will be no detectable change in voltage across it, since the voltage is held constant. The line current, it is true, will drop to a minimum value at ω_{ar}, but this is not discernible to a voltmeter across the input terminals of the antiresonant circuit. Now if some very high resistance is placed in series with the constant-voltage source, converting it to a constant-current source, the voltage across the antiresonant circuit will rise to a maximum at ω_{ar} because of the constant current flowing through the maximum impedance R_{ar} at ω_{ar}. Therefore, we may conclude that a current source is the ideal way to drive an antiresonant circuit.

Consider Fig. 8-7*a*, which illustrates an antiresonant circuit being driven by a current source having an internal impedance R_g. Let us try to convert this circuit to some equivalent series circuit which we already know how to analyze.

The coil and its series resistor may be resolved into a parallel resistor and inductor, as shown in Fig. 8-7*b*. The two resistors form an equivalent resistor

1) $$R_{eq} = \frac{R_g R_L(Q_L{}^2 + 1)}{R_g + R_L(Q_L{}^2 + 1)}$$

where $Q_L = \omega_L/R_L$. It is also expedient to relocate the position of the capacitor to that shown in Fig. 8-7*c*. Now the inductor and resistor of Fig. 8-7*c* are reconverted to an equivalent series combination, as shown in Fig. 8-7*d*. Using the series-parallel conversions of Chap. 7 again, we have

2) $$R_S = \frac{R_{eq}}{Q_P{}^2 + 1}$$

where

3) $$Q_P = \frac{R_{eq}}{\omega L(1 + 1/Q_L{}^2)}$$

[3] Maximum impedance and unity power factor is not truly simultaneous at ω_{ar}, but the difference is negligible if $Q_T > 5$.

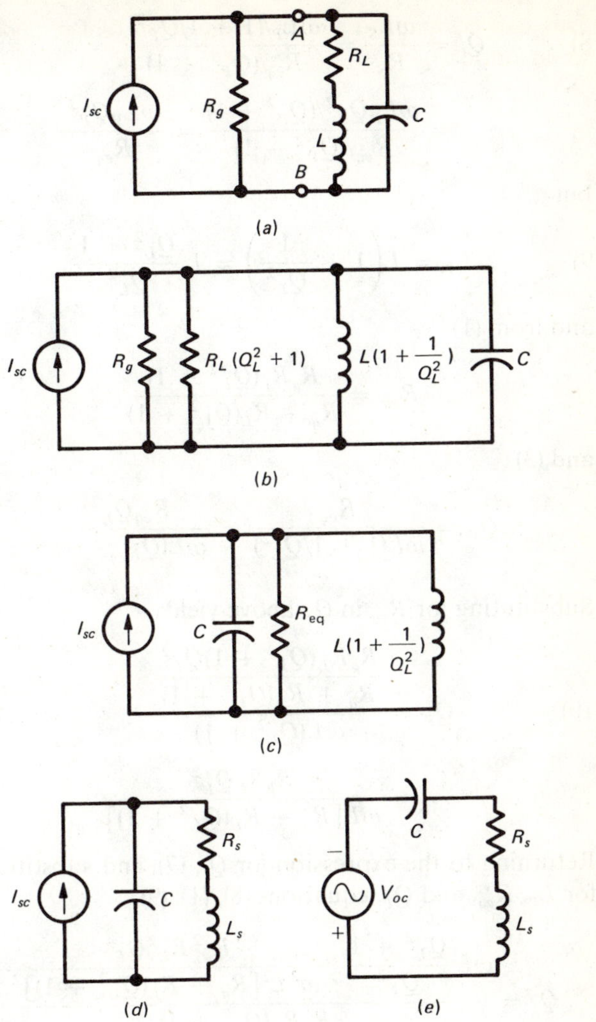

FIGURE 8-7

and

4) $$L_S = \frac{L_P}{1 + 1/Q_P{}^2}$$

With R_g absorbed into R_S, we can convert Fig. 8-7*d* into a voltage-source equivalent circuit which drives an equivalent series-resonant circuit, as shown in Fig. 8-7*e*. The selectivity of a series-resonant circuit is given by

5) $$\frac{\omega_2 - \omega_1}{\omega_r} = \frac{1}{Q_t}$$

where

6) $$Q_t = \omega L_S/R_S$$

and $\omega_2 - \omega_1$ is the bandwidth between the half-power points, and ω_r is the series-resonant frequency for Fig. 8-7*e*. We must evaluate Q_t as follows:

7) $$Q_t = \frac{\omega L_S}{R_S}$$

8) $$Q_t = \frac{\omega L_S}{R_S} = \frac{\omega L_P/(1 + 1/Q_P^2)}{R_{eq}/(Q_P^2 + 1)}$$

$$= \frac{\omega L_P Q_P^2/(Q_P^2 + 1)}{R_{eq}/(Q_P^2 + 1)} = \frac{\omega L_P Q_P^2}{R_{eq}}$$

but

9) $$L_P = L\left(1 + \frac{1}{Q_L^2}\right) = L\frac{Q_L^2 + 1}{Q_L^2}$$

and from (1)

$$R_{eq} = \frac{R_g R_L(Q_L^2 + 1)}{R_g + R_L(Q_L^2 + 1)}$$

and (3)

$$Q_P = \frac{R_{eq}}{\omega L(1 + 1/Q_L^2)} = \frac{R_{eq} Q_L^2}{\omega L(Q_L^2 + 1)}$$

Substituting for R_{eq} in Q_P above yields

10) $$Q_P = \frac{\dfrac{R_g R_L(Q_L^2 + 1)Q_L^2}{R_g + R_L(Q_L^2 + 1)}}{\omega L(Q_L^2 + 1)}$$

$$= \frac{R_g R_L Q_L^2}{\omega L[R_g + R_L(Q_L^2 + 1)]}$$

Returning to the expression for Q_t, (7), and substituting for L_P, R_{eq}, and Q_P, equations (8), (1), (9),

$$Q_t = \frac{\omega L \dfrac{Q_L^2 + 1}{Q_L^2} \cdot \dfrac{R_g^2 R_L^2 Q_L^4}{\omega^2 L^2[R_g + R_L(Q_L^2 + 1)]^2}}{\dfrac{R_g R_L(Q_L^2 + 1)}{R_g + R_L(Q_L^2 + 1)}}$$

11) $$= \frac{R_g R_L Q_L^2}{\omega L[R_g + R_L(Q_L^2 + 1)]}$$

12) Since $R_L/\omega L = 1/Q_L$,

13) $$Q_t = \frac{R_g Q_L}{R_g + R_L(Q_L^2 + 1)} \quad \text{and} \quad R_L = \frac{\omega L}{Q_L}$$

14) $$Q_t = \frac{R_g Q_L}{R_g + (\omega L/Q_L)(Q_L^2 + 1)}$$

$$= \frac{R_g Q_L^2}{Q_L R_g + \omega L(Q_L^2 + 1)}$$

Therefore,

15) $$\frac{\omega_2 - \omega_1}{\omega_r} = \frac{1}{Q_t} = \frac{Q_L R_g + \omega L(Q_L^2 + 1)}{R_g Q_L^2}$$

$$= \frac{1}{Q_L} + \frac{\omega L(Q_L^2 + 1)}{R_g Q_L^2}$$

and since we are considering ω at ω_{ar},

$$\frac{\omega_2 - \omega_1}{\omega_r} = \frac{1}{Q_t} = \frac{1}{Q_L} + \frac{\omega_{ar} L}{R_g}\left(\frac{Q_L^2 + 1}{Q_L^2}\right) \quad (8\text{-}25a)$$

If $Q_L = 10$, $Q_L^2 + 1 \approx Q_L^2$, and since from Eq. (8-21*b*), $\omega_{ar} L Q_L \approx R_{ar}$, we may obtain

$$\frac{\omega_2 - \omega_1}{\omega_r} = \frac{1}{Q_t} \approx \frac{1}{Q_L} + \frac{\omega_{ar} L}{R_g} \quad (8\text{-}25b)$$

$$\frac{\omega_2 - \omega_1}{\omega_r} = \frac{1}{Q_t} \approx \frac{1}{Q_L} + \frac{1}{Q_L}\frac{R_{ar}}{R_g} \quad (8\text{-}25c)$$

Equations (8-25*a*), (8-25*b*), and (8-25*c*) indicate that the greater R_g is, the larger the circuit Q will be.

Now carefully work through problem PS 8-13.

8-10 Universal resonance curve for the antiresonant circuit

The universal resonance curve in Fig. 8-4 may also be applied to the antiresonant circuit of Fig. 8-6 if Q_L is high. The validity of this statement may be proved in the following manner.

The general expression for the input impedance in Fig. 8-6 is

1) $$\mathbf{Z}_{in} = \frac{(R + jX_L)(-jX_C)}{R_L + j(X_L - X_C)}$$

If Q_L is high, say 10 or more, $X_L \gg R$ and Eq. (1) simplifies to

2) $$\mathbf{Z}_{in} = \frac{jX_L(-jX_C)}{R + j(X_L - X_C)}$$

$$= \frac{L/C}{R + j(\omega L - 1/\omega C)}$$

The R term in the denominator could not be discarded, since $X_L - X_C$ becomes very small in the vicinity of resonance. At $\omega = \omega_{ar} \cong \omega_r$, the reactive component in the denominator of (2) equals zero, and we again derive $Z_{ar} = L/R_r C$, where R_r is the coil resistance at ω_r. Then

3) $$\frac{\mathbf{Z}_{in}}{\mathbf{Z}_{ar}} = \frac{L/C}{R + j(\omega L - 1/\omega C)} \cdot \frac{1}{L/R_r C}$$

$$= \frac{R_r}{R + j(\omega L - 1/\omega C)} \quad (8\text{-}26)$$

For a series-resonant circuit we have

$$\frac{\mathbf{I}}{\mathbf{I}_r} = \frac{\mathbf{E}/\mathbf{Z}_{in}}{\mathbf{E}/\mathbf{Z}_r} = \frac{\mathbf{Z}_r}{\mathbf{Z}_{in}} = \frac{R_r}{R + j(\omega L - 1/\omega C)}$$

which is identical in form with Eq. (8-26). Therefore, the universal resonance curve may be applied to an antiresonant circuit with the following considerations:

1. The Q is high, say 10 or more.
2. The vertical scale (ordinate values) indicates Z_{in}/R_{ar}.

3. The Q is constant over the frequency range of interest.
4. The phase angles for the antiresonant circuit are exactly the opposite of those shown for series resonance. For example, at frequencies below ω_{ar}, the reactance of the capacitive branch is high, and most of the current flows through the coil. The impedance is predominantly inductive, and therefore Z_{in}/Z_{ar} will have an angle between zero and 90°. On the other hand, a series-resonant circuit will exhibit capacitive reactance below ω_r, and therefore $\mathbf{I}/\mathbf{I}_r$ is between zero and minus 90°. Since the phase angles in the universal resonance curve are shown for the series circuit, the reader must remember to invert the angles when dealing with antiresonant circuits.
5. The universal resonance curve in itself does not include the effect of the source resistance.

Now study problem PS 8-14 in detail.

PROBLEMS WITH SOLUTIONS

PS 8-1 Determine the resonant frequency (f_r) of the circuit shown in Fig. PS 8-1. Also determine the circuit Q at f_r, and the half-power frequencies.

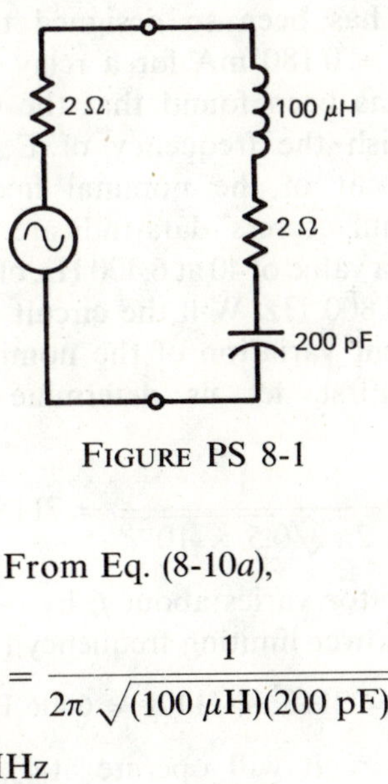

FIGURE PS 8-1

SOLUTION From Eq. (8-10*a*),

$$f_r = \frac{1}{2\pi\sqrt{LC}} = \frac{1}{2\pi\sqrt{(100\ \mu\text{H})(200\ \text{pF})}}$$

$$= 1.1254\ \text{MHz}$$

$$\omega_r = 2\pi f_r = 7.071 \times 10^6\ \text{rad/sec}$$

$$Q = \frac{\omega_r L}{R_t} = \frac{707.1\ \Omega}{4\ \Omega} = 176.8$$

From Eq. (8-10*e*),

$$\text{B} = f_2 - f_1 = \frac{f_r}{Q} = \frac{1.125\ \text{MHz}}{176.8} = 6.366\ \text{kHz}$$

From Eqs. (8-10*i*) and 8-10*j*), since Q > 10,

$$f_2 \cong f_r + \frac{\text{B}}{2} = 1.1286\ \text{MHz}$$

$$f_1 \cong f_r - \frac{\text{B}}{2} = 1.1222\ \text{MHz}$$

PS 8-2 Determine in the circuit of PS 8-2 the following: f_r, the exact f_1 and f_2 values, the exact B, Q, and the approximate B.

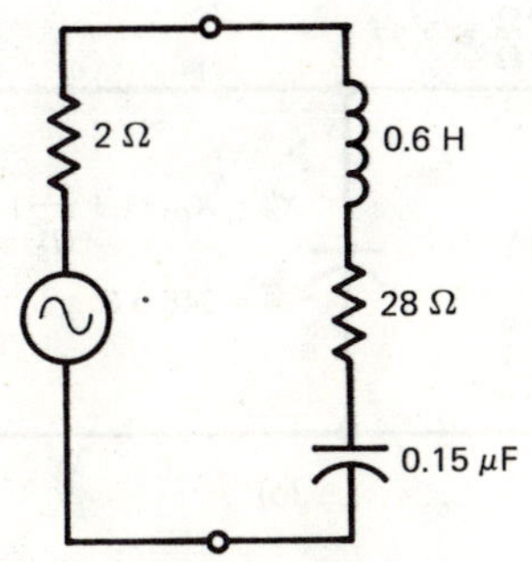

FIGURE PS 8-2

SOLUTION From Eq. (8-9*d*),

$$f_r = \frac{1}{2\pi\sqrt{LC}} = \frac{1}{2\pi\sqrt{(0.6\ \text{H})(0.15\ \mu\text{F})}} = 530.52\ \text{Hz}$$

From Eq. (8-9*c*),

$$\omega_1 = \frac{-30(0.15 \times 10^{-6})}{2(0.6)(0.15 \times 10^{-6})} + \frac{\sqrt{(30)^2(0.15 \times 10^{-6})^2 + 4(0.6)(0.15 \times 10^{-6})}}{2(0.6)(0.15 \times 10^{-6})}$$

$$= 3308.43\ \text{rad/sec}$$

or

$$f_1 = 526.55\ \text{Hz}$$

Similarly from Eq. (8-9*b*),

$$\omega_2 = 3358.43\ \text{rad/sec}$$

$$f_2 = 534.51\ \text{Hz}$$

Therefore

$$\text{B} = f_2 - f_1 = 7.96\ \text{Hz}$$

We may obtain Q from

$$Q = \frac{\omega_r L}{R_t} = 66.7$$

or alternatively by

$$Q = \frac{f_r}{f_2 - f_1} = 66.7$$

PS 8-3 Determine L in Fig. PS 8-3*a* so that R_L looks back into a series-resonant circuit, and also determine R_L for maximum power transfer.

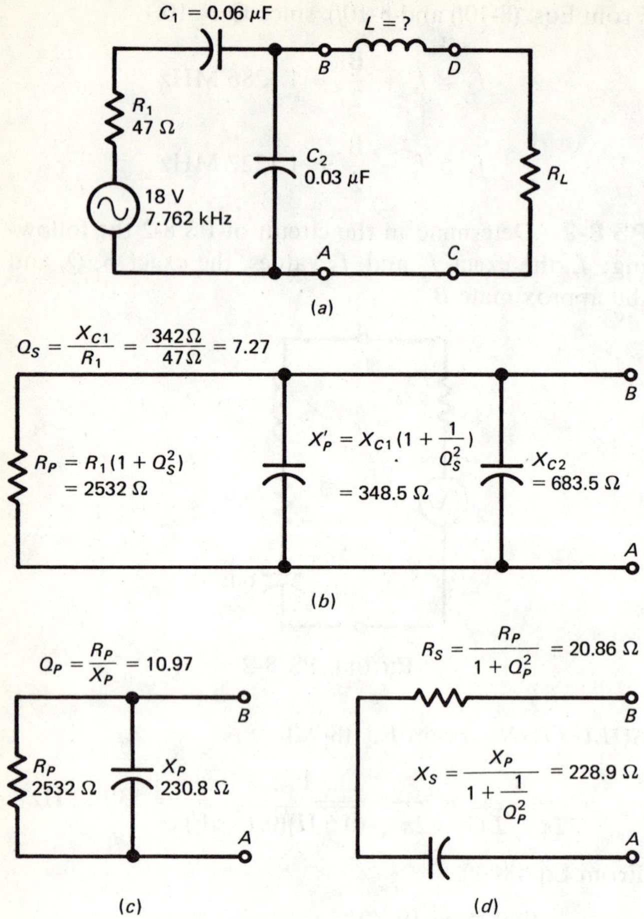

FIGURE PS 8-3

SOLUTION Replacing the voltage source by its internal impedance (R_1) and looking back to the left of the *AB* terminals in Fig. PS 8-3*a*, we see the series R_1C_1 combination in parallel with C_2. Applying the series-parallel conversions yields Fig. PS 8-3*b*. The reactance X_P' and X_{C_2} have an equivalent parallel reactance X_P, as shown in Fig. PS 8-3*c*. Converting Fig. PS 8-3*c* to its series equivalent yields PS 8-3*d*, from which we conclude that for series resonance

$$X_L = X_S$$

$$2\pi f L = 228.9\ \Omega$$

Solving for L yields

$$L = 4.69 \text{ mH}$$

For maximum power transfer

$$R_L = R_S = 20.86\ \Omega$$

PS 8-4 The radio command circuit of Fig. PS 8-4*a* operates in the following manner. As the input E_{in} approaches the vicinity of resonance, I increases. Since I also flows through the input resistance of the transistor, the resonant current rise will be amplified and in subsequent stages may be rectified and used to actuate a relay. The resistance seen looking into the transistor (r_{in}) terminals is about 600 ohms. Now R_b is in the order of 100 kilohms, which is so much larger than r_{in} that any shunting effect due to R_b may be neglected. The ac equivalent circuit is that of Fig. PS 8-4*b*.

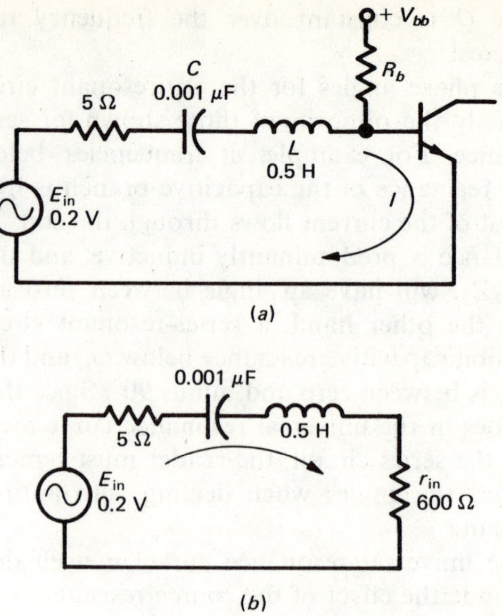

FIGURE PS 8-4

The system has been so designed that it takes a minimum of $I = 0.180$ mA for a relay to be actuated elsewhere. It has been found that the tone generator used to establish the frequency of E_{in} will vary by minus 10 percent of the nominal frequency during operation. Manufacturers' data indicate that the coil Q (Q_L, that is) has a value of 40 at 6,400 Hz, of 80 at 7,100 Hz, and of 100 at 7,800 Hz. Will the circuit operate with a minus 10 percent variation of the nominal frequency?

SOLUTION First, let us determine the resonant frequency.

$$f_r = \frac{1}{2\pi\sqrt{0.5 \times 10^{-9}}} = 7118 \text{ Hz}$$

The tone generator varies about f_r by -10 percent, or 711.8 Hz. The lower limiting frequency (f_a) is then

$$f_a = 7118 - 711.8 = 6406 \text{ Hz}$$

To see if the circuit will operate at f_a, we plug into Eq. (8-13). However, Q and Q_r in Eq. (8-13) are circuit Q's and not just coil Q's. In general, with R_g equal to $R_s = 5$ ohms plus $r_{\text{in}} = 600$ ohms, $R_g = 605$ ohms.

$$Q = \frac{\omega L}{R_g + R_L}$$

or

$$\frac{1}{Q} = \frac{R_g + R_L}{\omega L} = \frac{R_g}{\omega L} + \frac{1}{Q_L}$$

At f_a,

$$\frac{1}{Q_a} = \frac{605}{2\pi \times 6406 \times 0.5} + \frac{1}{40} = 0.0301 + 0.025 = 0.0551$$

Therefore $Q_a = 18.16$. Similarly,

$$\frac{1}{Q_r} = \frac{R_g}{\omega_r L} + \frac{1}{Q_{Lr}} = \frac{605}{2\pi \times 7.118 \times 10^3 \times 0.5} + \frac{1}{80}$$

$$= 0.0271 + 0.0125 = 0.03956$$

Therefore $Q_r = 25.28$. We can now plug into Eq. (8-13)

$$\frac{\mathbf{I}}{\mathbf{I}_r} = \frac{1}{\dfrac{6406}{7118} \times \dfrac{25.28}{18.16} + j25.28\left(\dfrac{6406}{7118} - \dfrac{7118}{6406}\right)}$$

$$= \frac{1}{1.253 - j5.337} = \frac{1}{5.482\underline{/-76.79}}$$

$$= 0.1824\underline{/+76.79^\circ}$$

Since

$$I_r = \frac{E_{in}}{R_g + R_{Lr}} = \frac{0.2}{R_g + (\omega_r L/Q_{Lr})} = \frac{0.2}{884.5} = 0.2261 \text{ mA}$$

we obtain

$$I = (0.1824\underline{/76.79^\circ})(0.2261\underline{/0^\circ} \text{ mA}) = 0.0412\underline{/76.79^\circ} \text{ mA}$$

Since 180 μA is needed but only 41.2 μA is available at minus 10 percent shift in command frequency, the relay will not actuate. For greater accuracy we could have plotted Q_L versus frequency from the given data and interpolated for more accurate Q_L values at f_a.

PS 8-5 With or without the aid of the universal resonance curve plot, the current ratio $|\mathbf{Y}/\mathbf{Y}_r|$ as a function of frequency for $R = 400$ ohms in the circuit of Fig. PS 8-5*a*.

SOLUTION The universal resonance curve plots $\mathbf{I}/\mathbf{I}_r$. This is actually identical with a normalized admittance plot, since

1) $$\frac{\mathbf{I}}{\mathbf{I}_r} = \frac{\mathbf{YE}}{\mathbf{Y}_r\mathbf{E}} = \frac{\mathbf{Y}}{\mathbf{Y}_r}$$

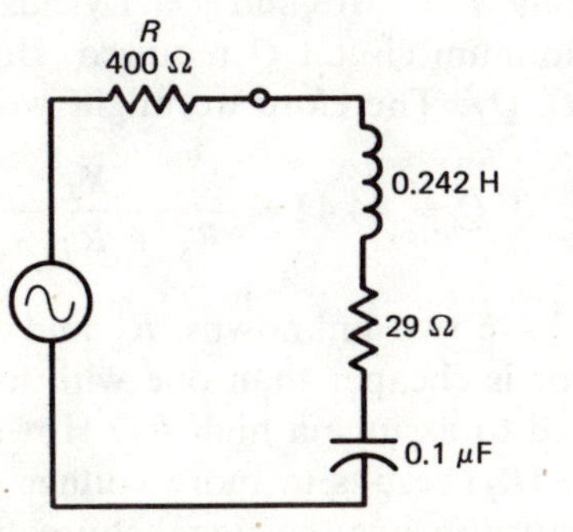

(*a*)

FIGURE PS 8-5

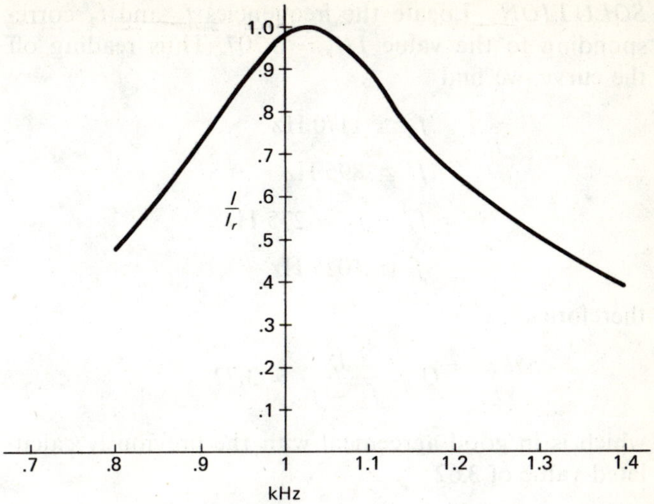

(*b*)

FIGURE PS 8-5 (*Continued*)

Now let us locate f_r from

2) $$f_r = \frac{1}{2\pi\sqrt{(0.242 \text{ H})(0.1\ \mu\text{F})}} = 1023 \text{ Hz}$$

The circuit Q at f_r with $R = 400$ ohms is

3) $$Q_r = \frac{\omega_r L}{R} = \frac{2\pi(1023)(0.242)}{429} = 3.63$$

Evidently we have a low Q circuit, and since the universal resonance curve of Fig. 8-4 does not include the curve for $Q < 10$, it is not of much value. Instead, we will assume some ω values, calculate the corresponding k values from Eq. (8-14*d*), and then with Q_r as determined in step 3, plug into Eq. (8-14*e*) to determine $\mathbf{I}/\mathbf{I}_r = \mathbf{Y}/\mathbf{Y}_r$. Thus we have as some sample values the following table:

f (Hz)	$k = \dfrac{f - f_r}{f_r}$	$\mathbf{Y}/\mathbf{Y}_r$
800	−0.2181	$0.485\underline{/60.9^\circ}$
900	−0.1203	$0.731\underline{/43.0^\circ}$
1000	−0.02257	$0.988\underline{/9.4^\circ}$
1023	0.0	$1.0\ \underline{/0^\circ}$
1030	0.0076	$0.999\underline{/-2.79^\circ}$
1500	0.4662	$0.332\underline{/-70.6^\circ}$

PS 8-6 Estimate the circuit Q from the resonance curve of Fig. PS 8-5*b* and check it against the solution in problem PS 8-5.

SOLUTION Locate the frequencies f_1 and f_2 corresponding to the value $I/I_r = 0.707$. Thus reading off the curve, we find

$$f_2 \cong 1170 \text{ Hz}$$

$$f_1 \cong 895 \text{ Hz}$$

$$f_2 - f_1 = 275 \text{ Hz}$$

$$f_r \cong 1025 \text{ Hz}$$

therefore

$$Q = \frac{f_r}{f_2 - f_1} = 3.72$$

which is in good agreement with the previously calculated value of 3.62.

PS 8-7 A sinusoidal voltage of 12 ohms internal resistance source is to deliver a pure 6-volt, 1000-Hz tone signal to a 50-ohm load (R_o). Unfortunately the generated tone signal is contaminated with some second harmonic content which is 18 percent of the fundamental. It is proposed that a series-resonant circuit be inserted between the load and signal source in such a manner that the resultant output only contain not more than 1.5 percent second harmonic. Devise a circuit to accomplish this goal.

SOLUTION We will insert the series-resonant circuit as shown in Fig. PS 8-7 so that it offers minimum impedance when tuned to the fundamental while it discriminates against the 2-kHz signal.

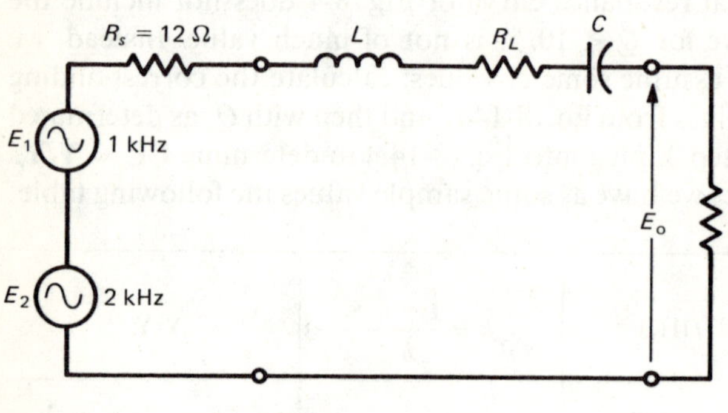

FIGURE PS 8-7

The signal source can be represented as shown in Fig. PS 8-7. Since the output amplitude of the 1-kHz signal must be 6 volts, we have

1) $$I_r = \frac{6 \text{ V}}{50\ \Omega} = 0.12 \text{ A}$$

at 1 kHz. At 2 kHz

2) $$E_o = 1.5\% \times 6 \text{ V} = 0.09 \text{ V}$$

Therefore at 2 kHz

3) $$I = \frac{0.09 \text{ V}}{50\ \Omega} = 0.0018 \text{ A}$$

We are not concerned with phase angles, as only the magnitudes are of interest. Thus we at least know

4) $$\frac{I}{I_r} = \frac{0.0018 \text{ A}}{0.12 \text{ A}} = 0.015$$

However, if we enter the universal resonance curve at this value, we find that the corresponding Q_rk values are off the curve. That is unfortunate, since we could have read off Q_rk, evaluated k from

5) $$k = \frac{f - f_r}{f_r} = \frac{2 \text{ kHz} - 1 \text{ kHz}}{1 \text{ kHz}} = 1$$

and then have solved for Q. Eyeballing the curve, it would seem that for $I/I_r = 0.015$, Q_rk lies anywhere between 5 and 200, hardly a satisfactory estimate. To find the necessary circuit Q, we can either try successive Q approximations in the universal resonance curve equation

6) $$\mathbf{A} = \frac{\mathbf{I}}{\mathbf{I}_r} = \frac{1}{1 + k + jQk\left(\dfrac{2+k}{1+k}\right)}$$

and zero in on an $I/I_r = A = 0.015$, or we can solve it for Q. Suppose we take the latter route. Then

7) $$A = \frac{1}{\sqrt{(1+k)^2 + \left[Qk\left(\dfrac{2+k}{1+k}\right)\right]^2}}$$

Squaring both sides and solving for Q yields

8) $$Q = \left[\frac{\dfrac{1}{A^2} - (1+k)^2}{\left[k^2\left(\dfrac{2+k}{1+k}\right)\right]^2}\right]^{1/2}$$

Substituting $A = 0.015$ and $k = 1$ yields $Q = 44.42$. This is the minimum circuit Q required. However, we need the coil Q, Q_L. Therefore we might write

9) $$Q = 44.42 = \frac{X_L}{R_S + R_o + R_L}$$

Now we have two unknowns, R_L and X_L. Since a low Q inductor is cheaper than one with high Q, we might be tempted to assume a high R_L. However, a high coil resistance (R_L) results in more voltage drop across R_L, which in turn means the signal source must have more amplitude to develop the necessary output voltage at f_r. Therefore, let us choose R_L so that the required source voltage at f_r is, say, twice E_o or 12 volts. Therefore we may write

10) $$2E_o = I_r(R_S + R_o + R_L)$$
$$12 \text{ V} = 0.12 \text{ A } (12 + 50 + R_L)\ \Omega$$
$$R_L = 38\ \Omega$$

Substituting the result of (10) into (9) and solving for X_L, we have

11) $$X_L = Q(R_S + R_o + R_L)$$
$$= 44.42\,(12 + 50 + 38) = 4442\ \Omega$$

The required coil Q is then

12) $$Q_L = 4442\ \Omega/38\ \Omega = 117$$

which is not impractical. The coil inductance is determined from

13) $$L = \frac{X_L}{2\pi f_r} = \frac{4442}{2\pi(1000)} = 0.707 \text{ H}$$

Since

14) $$f_r = \frac{1}{2\pi\sqrt{LC}}$$

we may obtain

15) $$C = \frac{1}{(2\pi f_r)^2 L} = 0.03583\ \mu\text{F}$$

which completes the design of the circuit. In actual practice a slightly smaller value of C would be used with a small adjustable (trimmer) capacitor shunted across it to tune the circuit, since the very exact L and C values called for are not readily available. The selectivity of the circuit and Q of the inductor could be trimmed by starting with a $Q_L > 117$ and shunting a variable resistor across the coil (or capacitor for that matter).

PS 8-8 In Fig. PS 8-8, determine (*a*) the resonant frequency (f_r), (*b*) the circuit Q at f_r (Q_{t_r}), (*c*) the maximum voltage across the coil, (*d*) the bandwidth, (*e*) the upper and lower half-power frequencies, (*f*) the frequencies at which the current is down to one-half the resonant current, (*g*) the value of resistance that would have to be inserted in series with the rest of the circuit to double the existing bandwidth.

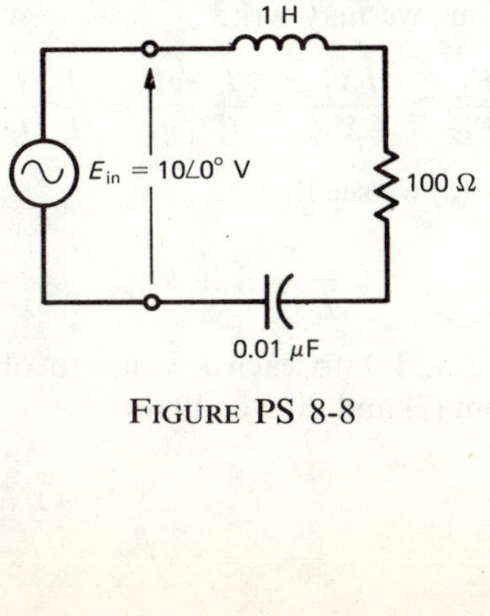

FIGURE PS 8-8

SOLUTION

1) $$f_r = \frac{\omega_r}{2\pi} = \frac{1}{2\pi\sqrt{LC}} = \frac{0.159}{\sqrt{1(10^{-8})}}$$
$$= \frac{0.159}{10^{-4}} = 1{,}591 \text{ Hz}$$

2) $$Q_{t_r} = \frac{\omega_r L}{R_t} = \frac{2\pi f_r L}{R_t} = \frac{6.28(1.59)10^3(1)}{10^2} = 100$$
$$\omega_r = 10^4 \text{ rad/sec}$$

3) Since Q_{t_r} is greater than 10, the coil voltage will be considered maximum at ω_r, in which case

$$E_{L\max} = Q_{t_r} E_{in} = 100(10) = 1{,}000 \text{ V}$$

4) The bandwidth is given by

$$B = \frac{\omega_r}{Q_{t_r}} = \frac{10^4}{10^2} = 100 \text{ rad/sec, or } 15.92 \text{ Hz}$$

5) Since Q_{t_r} is greater than 10, the half-power frequencies may be considered to be equally displaced about ω_r. Therefore,

$$f_2 = f_r + \frac{B}{2} = 1{,}591 + \frac{15.92}{2} = 1{,}599 \text{ Hz}$$

$$f_1 = f_r - \frac{B}{2} = 1{,}583 \text{ Hz}$$

6) To find the frequencies at which the current is one-half the resonant current, we shall use the universal resonance curve of Fig. 8-4. In Fig. 8-4, we see that the current is 50 percent below the resonant current at $Q_r k \cong \pm 0.85$. Since Q_r in Fig. PS 8-8 is total circuit Q or Q_{t_r}, we have

$$k = \pm\frac{0.85}{100}$$

but

$$k = \frac{\omega - \omega_r}{\omega_r}$$

Therefore

$$\omega = \omega_r(1 + k) = 10^4(1 + 0.0085) \text{ and } 10^4(1 - 0.0085)$$
$$= 10{,}085 \text{ rad/sec and } 9{,}915 \text{ rad/sec}$$

which respectively correspond to 1,605 Hz and 1,578 Hz.

7) The present B is 15.9 Hz. The double of this amount is 31.8 Hz. Since $B = \omega_r/Q_{t_r}$, or $Q_{t_r} = \omega_r/B$, and we are setting B equal to 31.8 Hz, the new Q_{t_r} must be $Q_{t_r} = {}^{1{,}590}/_{31.8} = 50$, which is to be expected, since Q and B are inversely proportional. To reduce the Q by a factor of 2, we must double the resistance. Therefore, an additional 100 ohms must be inserted into the circuit.

PS 8-9 The circuit of Fig. PS 8-9*a* is used as part of a radio control system. The principle of operation is that by switching the signal source (E_{in}) between frequencies f_{r1} and f_{r2} to which the resonant branch circuits L_1C_1 and L_2C_2 are tuned, we can, by additional circuitry not shown, detect the resonant voltage rise in E_1 and E_2 across L_1 and L_2. This voltage rise may ultimately be used to actuate relays which control lights, motors, or whatever. Our objective is to sketch the response curves of E_1 and E_2 versus frequency, so that we could visualize the separation between the two control channels.

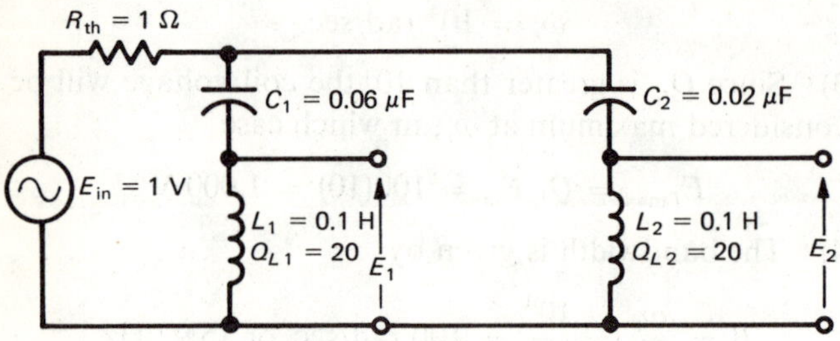

(*a*)

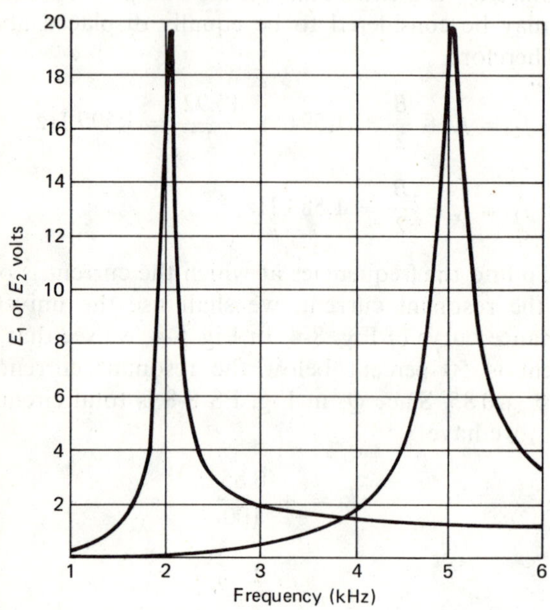

(*b*)

FIGURE PS 8-9

SOLUTION The resonant frequency of the L_1C_1 branch is

1) $$f_{r1} = \frac{1}{2\pi\sqrt{L_1C_1}} = 2055 \text{ Hz}$$

and that of the L_2C_2 branch is

2) $$f_{r2} = \frac{1}{2\pi\sqrt{L_2C_2}} = 5033 \text{ Hz}$$

A point to consider is whether or not the tuned circuits interact due to R_{th}, which is common to both branches. We can check this by determining the minimum impedance in each branch which, of course, occurs at f_r. Thus for the L_1C_1 branch

3) $$R_1 = \frac{2\pi f_{r1} L_1}{Q_{L1}} = 64.56\ \Omega$$

and for the L_2C_2 branch

4) $$R_2 = \frac{2\pi f_{r2} L_2}{Q_{L2}} = 158.1\ \Omega$$

Since R_1 and R_2 are so much greater than the source impedance R_{th}, we can neglect R_{th} and safely assume that the resonant circuits are driven from a constant voltage source and hence do not interact.

Another point to consider is that since $Q > 10$, the frequencies at which E_1 and E_2 are at their respective maximum values are f_{r1} and f_{r2}.

We may now proceed to sketch the frequency-response curves either with the aid of Eq. (8-14*e*) or the universal resonance curve. If the universal resonance curve is used, let us round off f_{r1} to 2 kHz and f_{r2} to 5 kHz, since we will never read the curve or plot points that accurately. Then we can choose convenient Q_rk values (say $Q_rk = -1$) for the L_1C_1 branch and then for the L_2C_2 branch. For any chosen Q_rk, we next calculate k.

5) $$k = \frac{(\text{chosen } Q_rk)}{20}$$

$$= \frac{-1}{20} = -0.05$$

Next we determine

6) $$f = f_r(1 + k)$$

$$= 2 \text{ kHz}(1 - .05) = 1.9 \text{ kHz}$$

for the L_1C_1 branch and where $f_r = f_{r1}$ in table A, and f_{r2} in table B. This yields f. Then we find I/I_r, which is simply the ordinate value as read off the curve for each abscissa (Q_rk) value. In this example, $I/I_r \cong 0.43$. Now we are seeking E_L/E_{Lr} rather than I/I_r. These quantities can readily by related by first noting that since $Q_L = 20$, X_L is in general much larger than the coil resistances R_1 or R_2. Thus we may write

7) $$\frac{E_L}{E_{Lr}} \cong \frac{IX_L}{I_rX_{Lr}} = \frac{I}{I_r}\frac{\omega L}{\omega_r L} = \frac{I}{I_r}\frac{f}{f_r}$$

but from Eq. (6) we see that

8) $$\frac{f}{f_r} = 1 + k$$

Therefore we add 1 to each k value to obtain $1 + k$, and then from (7) and (8) we obtain

9) $$\frac{E_L}{E_{Lr}} = \frac{I}{I_r}\frac{f}{f_r} = \frac{I}{I_r}(1 + k)$$

$$= 0.43(1 - 0.05) = 0.409$$

10) $$E_r \cong Q_r E_{in} = 20(1\text{ V}) = 20\text{ V}$$

we may substitute (10) into (9) to obtain

11) $$E_L = 20\text{ VX (the result of (9)}$$

$$= 20\text{ V }(0.409) \cong 8.2$$

Equation (11) is then plotted as a function of (6) to ultimately yield Fig. PS 8-9*b*.

PS 8-10 What are the parallel- and series-resonant frequencies of the circuit in Fig. PS 8-10?

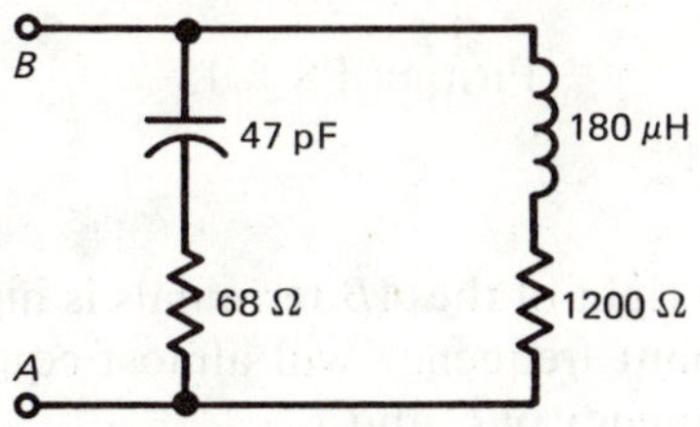

FIGURE PS 8-10

SOLUTION The exact parallel or antiresonant frequency by Eq. (8-17) is

1) $$f_{ar} = \frac{1}{2\pi\sqrt{LC}}\sqrt{\frac{L - CR_L^2}{L - CR_C^2}} = \frac{1}{2\pi\sqrt{(180\ \mu\text{H})(47\text{ pF})}} \cdot \sqrt{\frac{180\ \mu\text{H} - 47\text{ pF}(1200)^2}{180\ \mu\text{H} - 47\text{ pF}(68)^2}}$$

$$= 1.730\text{ MHz}(0.7904) = 1.368\text{ MHz}$$

The series-resonant frequency around the tank circuit (loop formed by $L_1 R_{L1} R_C$ and C) is

2) $$f_r = \frac{1}{2\pi\sqrt{LC}} = 1.730\text{ MHz}$$

As you can see, the difference between the parallel and series resonant frequencies is significant. This implies that the Q around the tank circuit is low.

PS 8-11 Using the series-parallel conversions of Chap. 7, determine Q_{AB}, the Q seen looking into the AB terminals of Fig. PS 8-10, and the input impedance Z_{ar}.
SOLUTION For the inductive branch, at $f_{ar} = 1.368$ MHz

1) $$Q_s = \frac{X_L}{R_L} = \frac{2\pi f_{ar}(180\ \mu\text{H})}{1200\ \Omega} = \frac{1547}{1200} = 1.289$$

Converting to the parallel equivalent,

2) $$R_{LP} = R_S(1 + Q_S^2) = 3195\ \Omega$$

(where $R_S = R_L = 1200\ \Omega$)

3) $$X_{LP} = X_S\left(1 + \frac{1}{Q_S^2}\right) = 2478\ \Omega$$

(where $X_S = X_L = 1547\ \Omega$)

For the capacitive branch

4) $$Q_S = \frac{X_{CS}}{R_C} = \frac{1}{2\pi f_{ar}(47\text{ pF})(68\ \Omega)} = 36.4$$

Converting to the parallel equivalent,

5) $$R_{CP} = R_S(1 + Q_S^2) = 90{,}180\ \Omega$$

(where $R_S = R_C = 68\ \Omega$)

6) $$X_{CP} = X_{CS}\left(1 + \frac{1}{Q_S^2}\right) = 2477\ \Omega$$

(where $X_{CS} = 2475\ \Omega$)

Therefore, looking into the AB terminals we see, in parallel, R_{LP}, R_{CP}, X_{LP}, and X_{CP}. Simplifying,[4]

7) $$R_P = R_{LP} \| R_{CP} = 3085\ \Omega$$

8) $$X_P = jX_{LP} \| (-jX_{CP}) = \frac{(j2478)(-j2477)}{j(2478 - 2477)}$$

$$\cong -j6.138 \times 10^6\ \Omega$$

Since for all practical purposes $R_P \ll X_P$, $Z_{ar} = R_{ar} \cong R_P = 3085$ ohms. The equivalent $Q_P = Q_{AB}$ is

9) $$Q_P = \frac{R_P}{X_{LP}} = \frac{3085}{2478} = 1.24$$

PS 8-12 Determine the exact f_{ar} that E_g should be tuned to for I_g to be in phase with E_g. Also determine E_{AB}, I_g, I_C, and I_L at f_{ar}.
SOLUTION Setting $R_C = 0$ in Eq. (8-17) we have

1) $$\omega_{ar} = \frac{1}{\sqrt{(0.8\text{ H})(0.02\ \mu\text{F})}}\sqrt{\frac{0.8\text{ H} - (0.02\ \mu\text{F})(36)^2}{0.8\text{ H} - 0}}$$

$$= 7905.7\sqrt{0.9997}$$

$$= 7905.6\text{ rad/sec or } f_{ar} = 1258.2\text{ Hz}$$

Note that if we tried to estimate f_{ar} by assuming

2) $$f_{ar} \cong f_r = \frac{1}{2\pi\sqrt{LC}} = 1258.2\text{ Hz}$$

we get, at least for most practical purposes, the same result. This indicates a high Q circuit which also means $X_L \cong X_C$. Thus

3) $$X_C \cong X_L = \omega_{ar} L = 6324\ \Omega$$

[4]This value of X_P is a gross approximation since it depends upon the difference between two large numbers of limited accuracy.

The tank circuit Q is then ($R_C = 0$)

4) $$Q_T = \frac{X_L}{R_L} = \frac{6324}{36} = 175.7$$

Since Q_T is greater than 10 we can safely approximate R_{ar} by

5) $$R_{ar} \cong Q_T X_L = 1.111 \text{ M}\Omega$$

Therefore

6) $$\mathbf{E}_{AB} = \frac{R_{ar}}{R_g + R_{ar}} E_g\underline{/0^\circ}$$

$$= 0.7027\, E_g = 12.65\underline{/0^\circ} \text{ V}$$

7) $$I_g = \frac{\mathbf{E}_{AB}}{R_{ar}} = 11.39\ \mu\text{A}$$

Since $Q_T > 10$ we can, by Eq. (8-24*a*) and (8-24*b*), assume that the circulating tank current is

$$|I_C| = |I_L| = Q_T I_g = 2.001 \text{ mA}$$

Alternatively,

$$\mathbf{I}_C = \frac{\mathbf{E}_{AB}}{X_C\underline{/-90^\circ}} = 2.000\underline{/90^\circ}$$

$$\mathbf{I}_L = \frac{\mathbf{E}_{AB}}{X_L\underline{/90^\circ}} = 2.000\underline{/-90^\circ}$$

Note that when the phase relationships of I_C and I_L are considered, one of these currents flows in a direction opposite to that assumed in Fig. PS 8-12. Thus if we

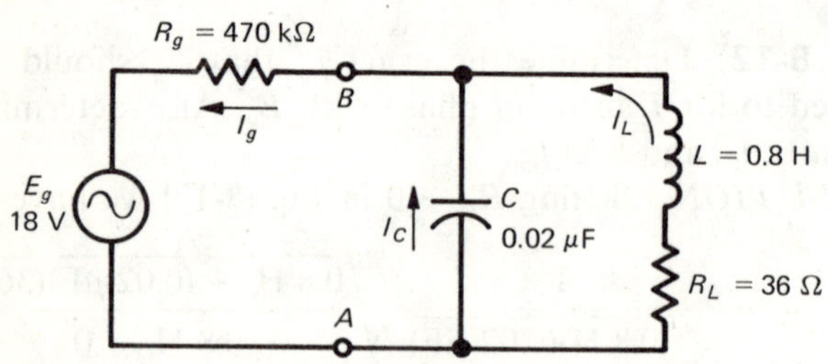

FIGURE PS 8-12

reverse, say, I_C and neglect any small I_g that trickles out of the tank circuit, it would seem that I_L merges into I_C, forming a circulating tank circuit current approximately Q_T times greater than the line current. The larger Q_T is, the more identical $|I_L|$ and $|I_C|$ will be. This is the reasoning behind the circulating current concept in the parallel-resonant circuit.

PS 8-13 In the circuit of Fig. PS 8-13*a*, determine $\omega_{ar}, f_{ar}, R_{ar}, E_o, I_g, I_C, I_L$ at ω_{ar}, Q_t, the per unit bandwidth, and the upper and lower half-power frequencies.

SOLUTION As a first step, thevenize R_1, R_2, and E_g to obtain Fig. PS 8-13*b*. Since the Q of the antiresonant

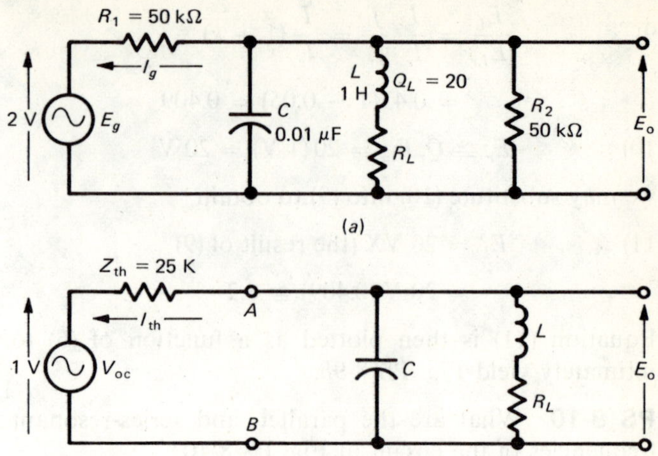

FIGURE PS 8-13

circuit to the right of the AB terminals is high, $Q_L = 20$, the antiresonant frequency will almost equal the series resonant frequency of C and L,

$$\omega_{ar} \approx \frac{1}{\sqrt{LC}} = \frac{1}{\sqrt{10^{-8}(1)}} = 10^4 \text{ rad/sec}$$

therefore,

$$f_{ar} = \frac{\omega_{ar}}{2\pi} = 1{,}592 \text{ Hz}$$

The apparent resistance of the antiresonant circuit at ω_{ar} is

$$R_{ar} \approx \omega_{ar} L Q_L = 10^4(1)(20) = 200 \text{ k}\Omega$$

Since the tuned circuit appears as a pure resistance, R_{ar} at ω_{ar}, E_o may be determined by the voltage-divider relationship

$$E_o = \frac{V_{oc}(R_{ar})}{Z_{\text{th}} + R_{ar}} = \frac{1(200)}{225} = 0.889 \text{ V}$$

The generator current in Fig. PS 8-13*a* will be

$$I_g = \frac{E_g - E_o}{R_1} = \frac{2.00 - 0.889 \text{ V}}{50 \text{ k}\Omega} = 22.2\ \mu\text{A}$$

The capacitor current is

$$I_C = \frac{E_o}{X_C} = 0.889(10^4)(10^{-8}) = 0.0889 \text{ mA}$$

or, by Eq. (8-24*b*),

$$I_C = I_{\text{th}} Q_T = I_{\text{th}} Q_L$$

where I_{th} in Fig. PS 8-13*b* equals

$$I_{\text{th}} = \frac{V_{oc} - E_o}{Z_{\text{th}}} = \frac{1.00 - 0.889 \text{ V}}{25 \text{ k}\Omega} = 4.44\ \mu\text{A}$$

Therefore, $I_C = 4.44\ \mu A\ (20) = 88.8\ \mu A$, which checks closely. Notice that I_g and I_{th} are not the same. Thévenin's theorem must be used with caution if conditions inside the equivalent circuit rather than the load are to be determined. The current in the inductive branch will essentially equal the capacitor current since there is no resistance in the capacitive branch and Q_L is large. The per unit bandwidth is, by Eq. (8-25c),

$$\frac{\omega_2 - \omega_1}{\omega_r} = \frac{1}{Q_t} = \frac{1}{Q_L} + \frac{1}{Q_L}\frac{R_{ar}}{R_g} = \frac{1}{20} + \frac{1}{20}\left(\frac{200}{25}\right)$$

where $R_g = Z_{th}$,

$$\frac{\omega_2 - \omega_1}{\omega_r} = \frac{1}{Q_t} = 0.05 + 0.05(8) = 0.45$$

The overall circuit Q is $Q_t = 1/0.45 = 2.2$. The bandwidth is $\Delta\omega = \omega_2 - \omega_1 = 0.45(\omega_r) = 4{,}500$ rad/sec, or 716 Hz. Since the circuit Q is only 2.2, the upper and lower half-power frequencies will not be symmetrically located about f_{ar}. As a first approximation, however, we may write

$$f_2 = f_{ar} + \frac{\Delta f}{2} = 1{,}592 + \frac{716}{2} = 1{,}950 \text{ Hz},$$

and

$$f_1 = f_{ar} - \frac{\Delta f}{2} = 1{,}592 - \frac{716}{2} = 1{,}234 \text{ Hz}$$

PS 8-14 The circuit of Fig. PS 8-14a is to be used in a radio command system. Resistors R_a and R_b are used to isolate the number 1 and number 2 channels from each other and thereby prevent interaction through the common source impedance R_g. This is why a low R_g is desirable. Resistors R_a and R_b, which are relatively large, also serve to convert the voltage source E_g to more of a current source, which improves the selectivity of the antiresonant circuits.

When the frequency of E_g is adjusted to that of the channel 1 antiresonant frequency, E_1 will be a maximum. Voltage E_1 may be subsequently amplified and rectified and used to drive a control relay. Channel 2 will behave in a similar manner.

With the aid of the universal resonance curve, roughly sketch on linear graph paper the values of E_1 and E_2 versus frequency. Assume the following parameters:

$$E_g = 20 \text{ V} \qquad C_1 = 0.06\ \mu\text{F}$$
$$R_g = 100\ \Omega \qquad L_1 = 0.1 \text{ H}$$
$$R_a = 100 \text{ k}\Omega \qquad C_2 = 0.01\ \mu\text{F}$$
$$R_b = 100 \text{ k}\Omega \qquad L_2 = 0.1 \text{ H}$$
$$Q_1 = Q_2 = 20$$

where Q_1 and Q_2 are the coil Q values.

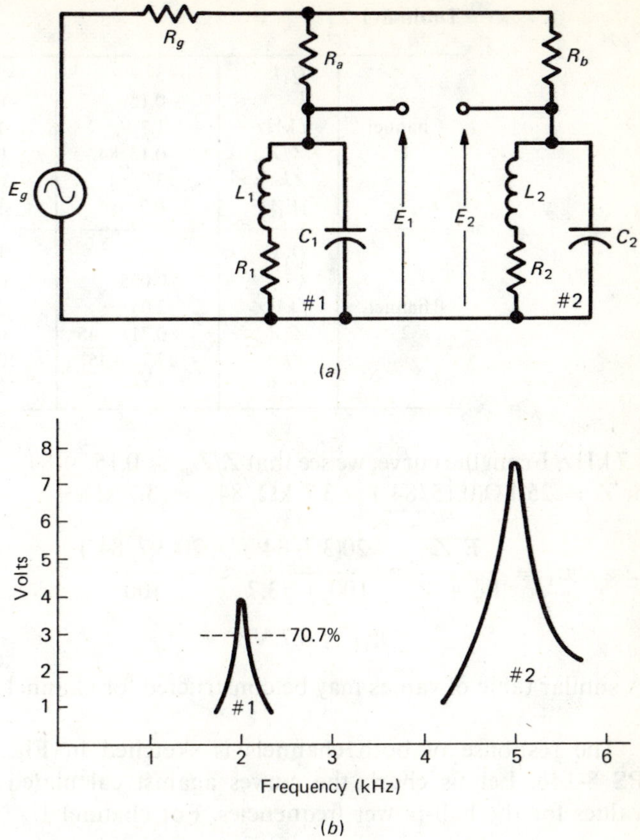

FIGURE PS 8-14

SOLUTION Since the coil Q's are high and graphic methods are used, we may write

$$f_{r_1} = \frac{1}{2\pi\sqrt{L_1 C_1}} = 2.055 \text{ kHz} \cong 2 \text{ kHz}$$

$$f_{r_2} = \frac{1}{2\pi\sqrt{L_2 C_2}} = 5.033 \text{ kHz} \cong 5 \text{ kHz}$$

$$R_{ar_1} = 2\pi f_{r_1} L_1 Q_1 = 6.28(2)10^3(0.1)(20) \cong 25 \text{ k}\Omega$$

$$R_{ar_2} = 2\pi f_{r_2} L_2 Q_2 = 6.28(5)10^3(0.1)20 \cong 63 \text{ k}\Omega$$

At f_{r_1}

$$E_1 = \frac{R_{ar_1} E_g}{R_{ar_1} + R_a} = \frac{25 \text{ k}\Omega}{125 \text{ k}\Omega} \times 20 \text{ V} \cong 4.0 \text{ V}$$

At f_{r_2}

$$E_2 = \frac{R_{ar_2}}{R_{ar_2} + R_b} E_g = \frac{63 \text{ k}\Omega}{163 \text{ k}\Omega} \times 20 \text{ V} \cong 7.7 \text{ V}$$

Table 8-1 may be compiled for channel 1 as follows.

An arbitrary value of $Q_1 k$ is chosen from the universal resonance curve, say $Q_1 k = -3$. With $Q_1 = 20$, we find $k = -3/20 = -0.15$. Since $k = (f - f_r)/f_r$, we can solve for f_1 which yields $f = f_r(1 + k) = 2$ kHz$(1 - 0.15) =$

Table 8-1

Channel 1	Q_1k	-3	-2	-1	-0.5	0
	k	-0.15	-0.1	-0.05	-0.025	0
	f kHz	1.7	1.8	1.9	1.95	2
	Z/Z_{ar}	$0.15\angle 84^\circ$	$0.23\angle 80^\circ$	$0.43\angle 65^\circ$	$0.71\angle 45^\circ$	1
	Zk	$3.7\angle 84^\circ$	$6.3\angle 80^\circ$	$10.7\angle 65^\circ$	$17.7\angle 45^\circ$	25
	$\lvert E_1\rvert$	0.7	1.3	2.1	3.16	4
Channel 2	Q_1k	0.5	1	2	3	0
	k	0.025	0.05	0.1	0.15	0
	f kHz	2.05	2.1	2.2	2.3	2
	Z/Z_{ar}	$0.71\angle -45^\circ$	$0.43\angle -65^\circ$	$0.25\angle -73^\circ$	$0.19\angle -75^\circ$	1
	Zk	$17.7\angle -45^\circ$	$10.7\angle -65^\circ$	$6.3\angle -73^\circ$	$4.75\angle -75^\circ$	25
	$\lvert E_1\rvert$	3.16	2.1	1.3	0.91	4

1.7 kHz. From the curve, we see that $\mathbf{Z}/\mathbf{Z}_{ar} \cong 0.15\angle +84^\circ$, or $\mathbf{Z} = 25\ \text{k}\Omega(0.15\angle 84^\circ) = 3.7\ \text{k}\Omega\angle 84^\circ \approx j3.7\ \text{k}\Omega$.

$$E_1 = \frac{E_g\mathbf{Z}}{R_a + \mathbf{Z}} \approx \frac{20(3.7\angle 84^\circ)}{100 + j3.7} \approx \frac{20(3.7\angle 84^\circ)}{100}$$

$$|E_1| \cong 0.7\ \text{V}$$

A similar table of values may be constructed for channel 2.

The response of both channels is sketched in Fig. PS 8-14*b*. Let us check the curves against calculated values for the half-power frequencies. For channel 1,

$$\frac{f_2 - f_1}{f_{r_1}} = \frac{1}{Q_t} = \frac{1}{Q_L} + \frac{1}{Q_L}\frac{R_{ar}}{R_a}$$

$$= 0.05 + 0.05\left(\frac{25}{100}\right) = 0.063$$

therefore,

$$Q_t = 1/0.063 \cong 16$$

and

$$f_2 - f_1 = 0.063(2\ \text{kHz}) = 126\ \text{Hz}$$

therefore, $f_2 \approx 2{,}063$ Hz and $f_1 \approx 1{,}940$ Hz, which roughly checks with the 70.7 percent level for channel 1.

PROBLEMS WITH ANSWERS

PA 8-1 Determine the series-resonant frequency and bandwidth of the circuit in Fig. PA 8-1.

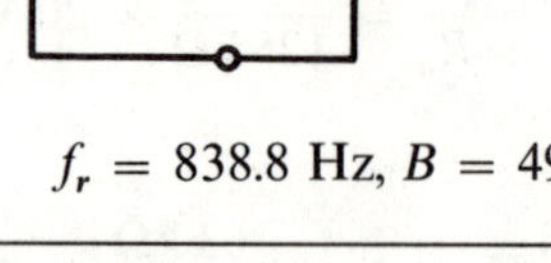

ANSWER $f_r = 838.8$ Hz, $B = 49.5$ Hz

PA 8-2 Determine the current at f_r in the circuit of Fig. PA 8-1 if $E_s = 15\angle 0^\circ$ volts.

ANSWER $I_r = 0.02679$ A

PA 8-3 What is the power dissipated in the 60-ohm resistor at f_r in the circuit of Fig. PA 8-1?

ANSWER 0.0431 W

PA 8-4 The series-resonant frequency of a tuned circuit is 27 MHz. What circuit Q is required for a 3-MHz bandwidth?

ANSWER $Q \cong 9$

PA 8-5 Assume that the resonant frequency of the previous problem is increased but the same bandwidth is maintained. What happens to Q?

ANSWER Q must increase.

PA 8-6 What kind of load, $\mathbf{Z}_L$, would you connect across the output terminals of Fig. PA 8-6 for maximum power transfer?

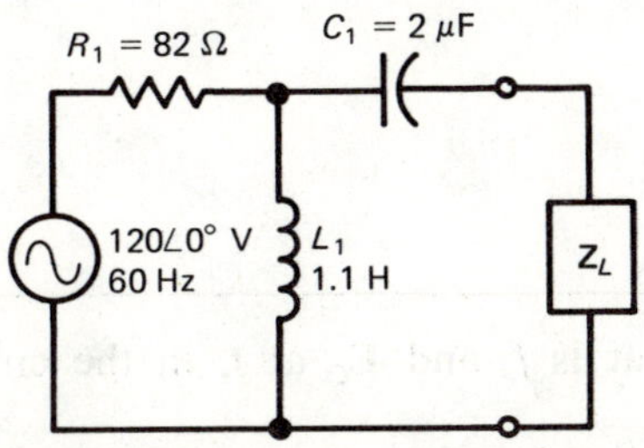

ANSWER $\mathbf{Z}_L = 78.9 + j1310\ \Omega$

PA 8-7 What is the resonant frequency f_r of the circuit in Fig. PA 8-7? What is the current I_r at f_r? At approximately what frequencies is the current I equal to half of I_r?

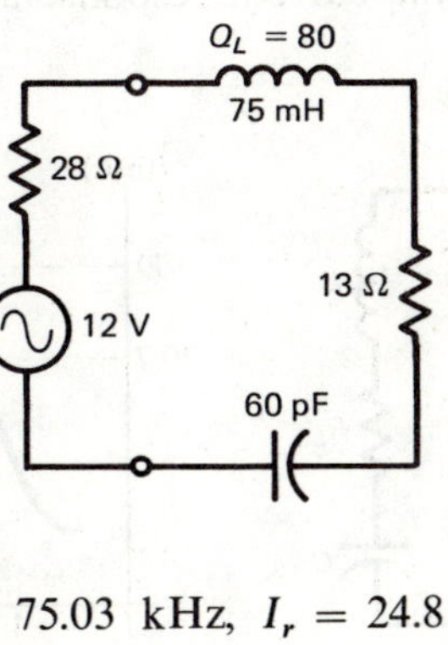

ANSWER $f_r = 75.03$ kHz, $I_r = 24.8$ mA, $f_2 \cong 75.9$ kHz, $f_1 \cong 74.1$ kHz.

PA 8-8 The basic circuit of a Q meter is shown in Fig. PA 8-8. Generator E_g, with source impedance R_g, pumps rf current, measured by the rf milliammeter, through R_1 and through the series R_LLC circuit which shunts R_1. Capacitor C is adjusted until the R_LLC branch is resonant at the generator frequency. A very high-impedance VTVM measures the voltage across the capacitor, which at resonance is Q times the voltage applied to the tuned circuit. This voltage is the same as that across R_1, since the two branches are in parallel. Resistor R_1 is deliberately chosen at about 0.04 ohm, so that the source impedance that the series-resonant circuit looks into will be at least 0.04 ohm. Resistor R_g is about 100 ohms. With R_1 so low, the loading effect of the tuned circuit at resonance will be negligible except for the very highest Q values, that is, lowest R_L values. Therefore the generator amplitude may be adjusted to pump a known current through R_1, which produces a calculable driving voltage for the tuned circuit. This voltage will not vary significantly as C is adjusted.

If the milliammeter is adjusted to 50 mA and the VTVM reads 0.1 volt when C is tuned to resonance, what is the coil Q?

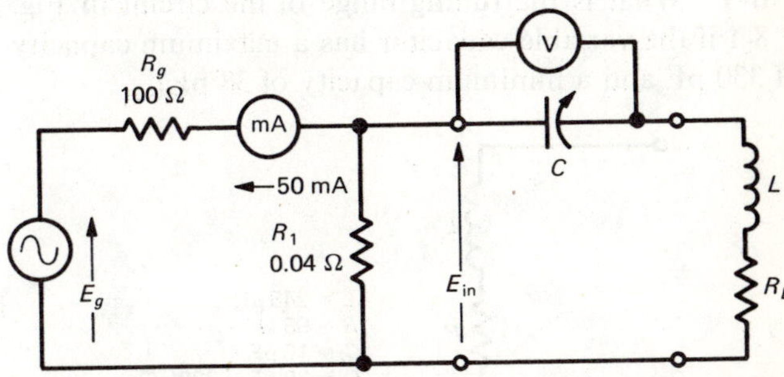

ANSWER $Q_L = 50$

PA 8-9 Given in Fig. PA 8-9 that $I_{sc} = 1$ mA, $R_g = 100$ kΩ, $C = 159$ pF, L = 159 μH, and $R_L = 10$ Ω, determine f_{ar}, E_o, I_C, I_L, the power dissipated in the tank circuit, and the circuit Q at f_{ar}. Also determine the bandwidth.

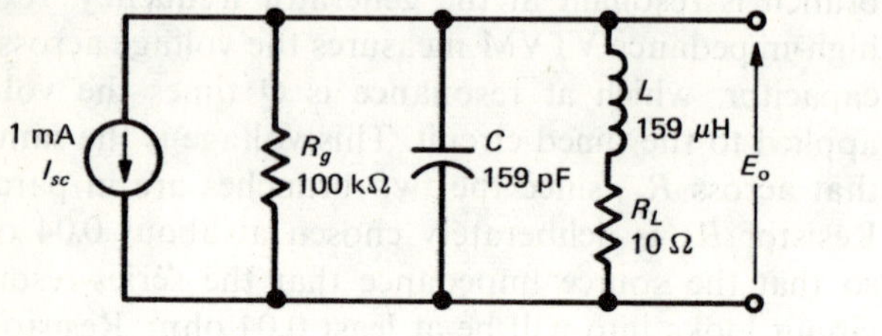

ANSWER $f_{ar} = 1.0009$ MHz, $I_L = 5\underline{/-90°}$ mA, $I_C = 5\underline{/90°}$ mA, $P = 25$ mW, $Q_t = 50$, $B = 20.02$ kHz.

PA 8-10 Design the tuned circuit of Fig. PA 8-10 so that at 500 kHz, $E_o = (0.6\underline{/0°})E_{in}$. Assume $Q_L = 50$ at 500 kHz. What is the resultant selectivity?

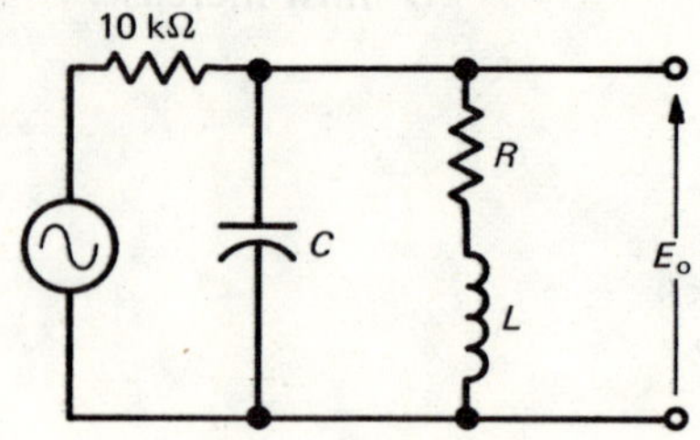

ANSWER $L = 95.5$ μH, $C = 1061$ pF, $B = 25$ kHz

PROBLEMS WITHOUT ANSWERS

P 8-1 What is the tuning range of the circuit in Fig. P 8-1 if the variable capacitor has a maximum capacity of 330 pF and a minimum capacity of 38 pF?

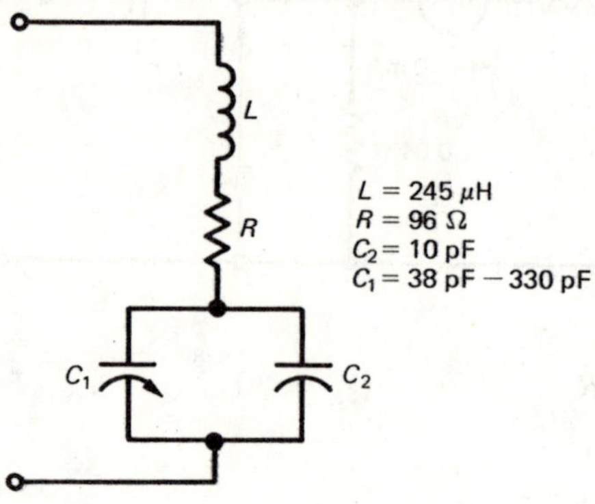

P 8-2 A series-resonant circuit exhibits a $Q = 150$ at $f_r = 1$ MHz. What is the bandwidth?

P 8-3 What is f_r and E_C at f_r in the circuit of Fig. P 8-3?

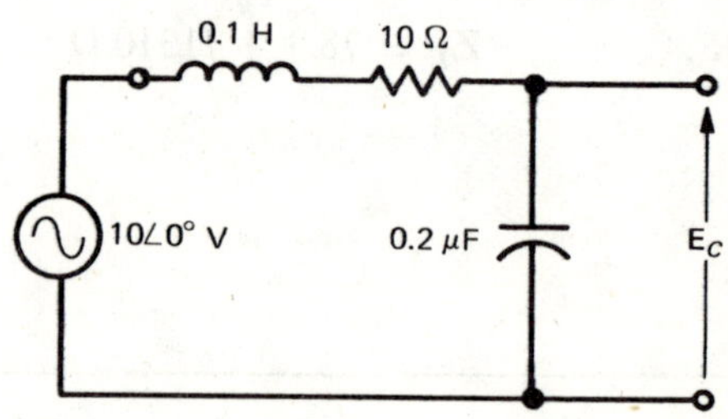

P 8-4 Determine L, Q_L, and C in the circuit of Fig. P 8-4*a* so that the current response appears as shown in Fig. P 8-4*b*.

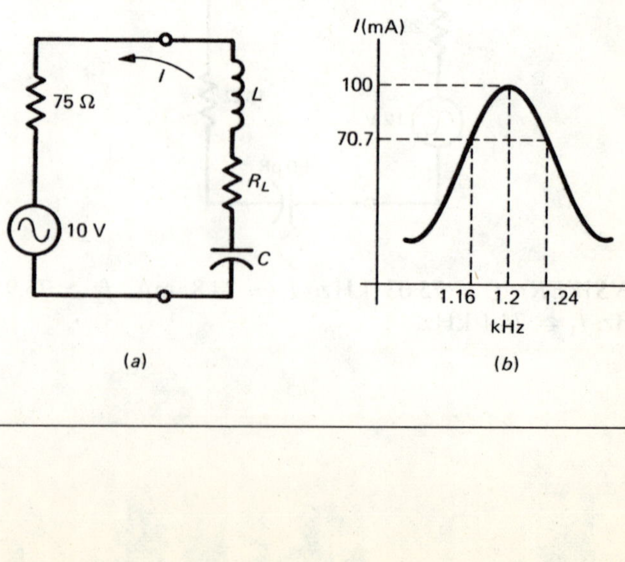

(*a*) (*b*)

P 8-5 Sketch the voltage transfer function $\frac{E_o}{E_{in}}$ versus frequency for the circuit of Fig. P 8-5.

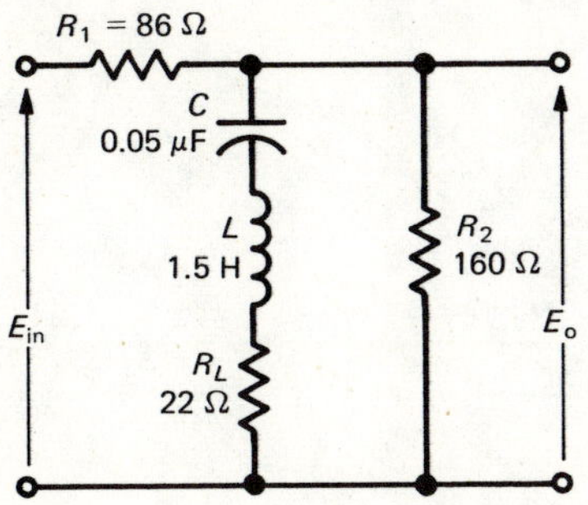

P 8-6 Determine f_{ar} if $R_L = R_C = \sqrt{\frac{L}{C}}$ in Fig. P 8-6.

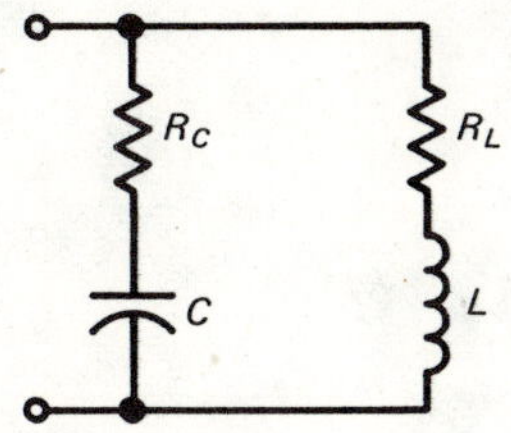

P 8-7 How might the circuit of Fig. P 8-7 be modified so that the line current I_g is always in phase with the line voltage E_g despite the poor frequency regulation of the system? How does I_g vary with the frequency shift in E_g?

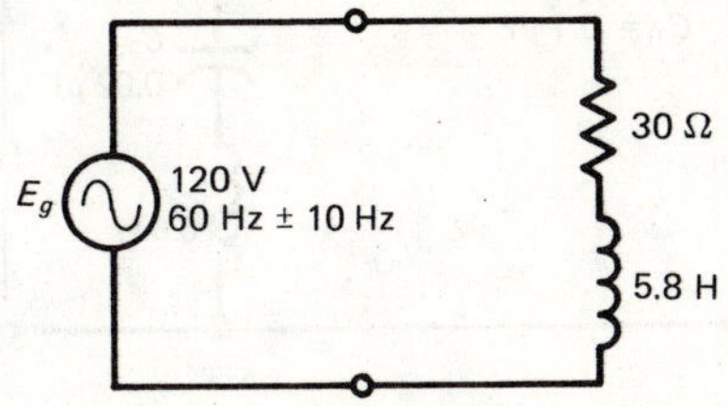

P 8-8 Determine f_{ar}, I_g at f_{ar}, and the B of the circuit in Fig. P 8-8.

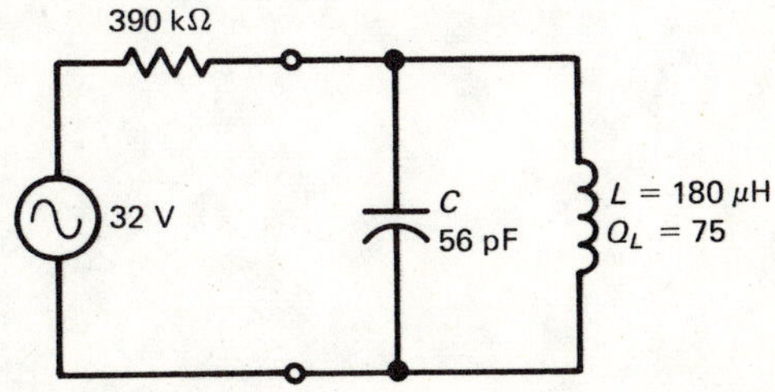

P 8-9 Design an antiresonant circuit so that it resonates at 1 MHz and exhibits an R_{ar} = 60 kΩ. Also the magnitude of tank circuit impedance is equal to 30 kΩ at $f_{ar} \pm 10$ kHz. Assume the resonant circuit will be driven from a 270 kΩ source impedance.

P 8-10 The triangles labeled A in Fig. P 8-10 represent ideal amplifiers with infinite input and zero output impedance. Each tuned circuit is identical in every way. What must the amplifier gain (A) equal if $E_g = E_1 = E_2 = E_o$ at f_{ar}?

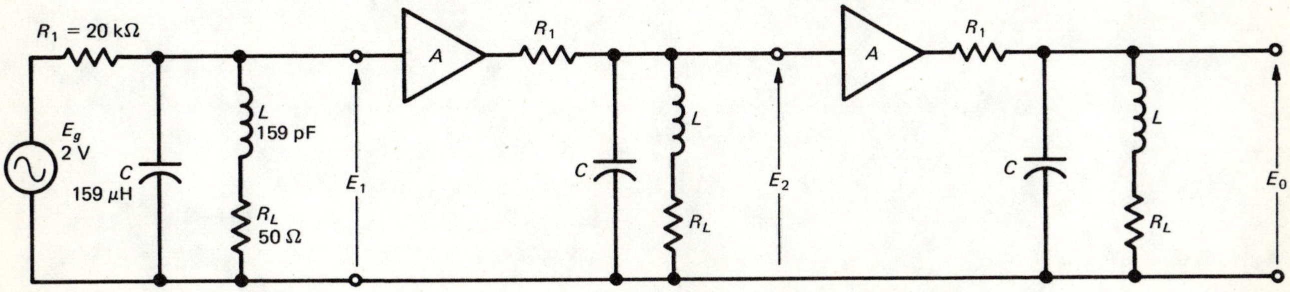

P 8-11 Plot E_o/E_{in} versus frequency for the circuit of Fig. P 8-11. Explain the resultant curve.

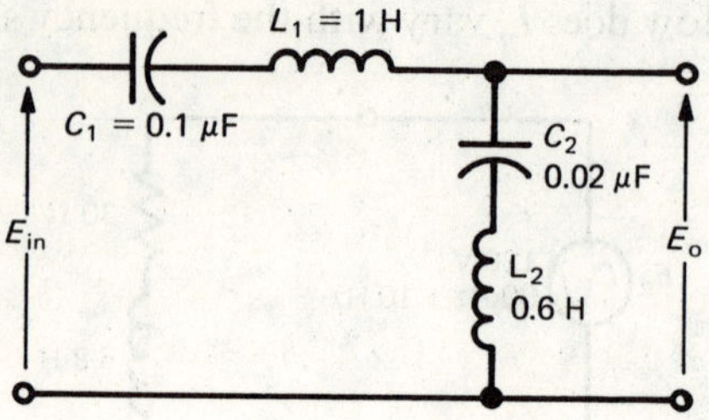

9 four-terminal (two-port) networks

In Chap. 15 of the preceding text in this series[1] we learned some techniques of developing equivalent circuits from data presented in graphic or numerical form. A major portion of this chapter was devoted to the analysis of four-terminal or two-port networks and in particular the R, G, and h parameters were introduced. The presentation, however, was restricted to purely resistive circuits at that time. In this chapter we will build upon that foundation and investigate four-terminal networks with ac excitation. The same theoretical concepts hold true. The only difference is that the effects of L and C in addition to R must be considered. If the concept of a four-terminal network is new or hazy to you, it is *strongly* recommended that you review Chap. 15 in the preceding text.

9-1 The z parameters

The z parameters are a more general form of the R parameters as the impedance, z, also includes the effects of L and C. Therefore the z parameters will be dependent upon frequency.

You will recall that given a four-terminal network as shown in Fig. 9-1*a*, with its indicated dc voltage and

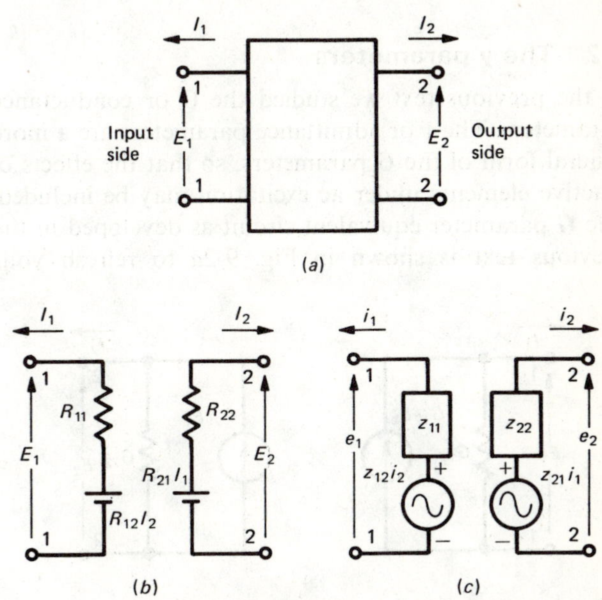

FIGURE 9-1

current reference directions, it is possible to derive the R parameter network of Fig. 9-1*b*. The more general but analogous case is the z parameter network of Fig. 9-1*c*. As you can see, the only change involved is the substitution of z for R so that ac as well as dc excitation can be accommodated. The z parameters (also in a

[1]Phillip Cutler, "DC Circuit Analysis with Illustrative Problems," McGraw-Hill Book Company, New York, 1974.

manner analogous to the R parameters) are defined by

$$z_{11} = \frac{e_1}{i_1}\bigg|_{i_2=0} \qquad z_{12} = \frac{e_1}{i_2}\bigg|_{i_1=0}$$
$$z_{21} = \frac{e_2}{i_1}\bigg|_{i_2=0} \qquad z_{22} = \frac{e_2}{i_2}\bigg|_{i_1=0} \tag{9-1}$$

Although the e's and i's in Eq. (9-1) are taken as instantaneous values they could just as well all be peak, pk-pk, or effective values.

Inspection of Fig. 9-1c indicates that the descriptive z parameter equations which express the terminal conditions are

1) $$e_1 = z_{11}i_1 + z_{12}i_2 \tag{9-2a}$$

2) $$e_2 = z_{21}i_1 + z_{22}i_2 \tag{9-2b}$$

Now work through problems PS 9-1, PS 9-2, and PS 9-3. If you find difficulty with these problems, it is because you need to review the necessary prerequisite material in Chap. 15 of the previous text.

The z or impedance parameters are sometimes called the open-circuit parameters because their measurement involves opening a pair of terminals which forces a terminal current to zero.

9-2 The y parameters

In the previous text we studied the G or conductance parameters. The y or admittance parameters are a more general form of the G parameters, so that the effects of reactive elements under ac excitation may be included. The G parameter equivalent circuit as developed in the previous text is shown in Fig. 9-2a to refresh your memory. If, by analogy, we substitute y's for the G's, we have the y parameter equivalent circuit of Fig. 9-2b.

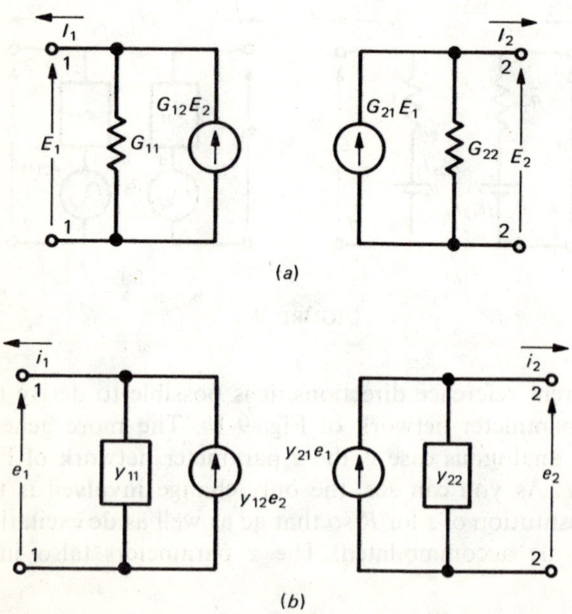

FIGURE 9-2

The y parameters, also in a manner analogous to the G parameters, are defined by the following measurements:

$$y_{11} = \frac{i_1}{e_1}\bigg|_{e_2=0} \qquad y_{12} = \frac{i_1}{e_2}\bigg|_{e_1=0}$$
$$y_{21} = \frac{i_2}{e_1}\bigg|_{e_2=0} \qquad y_{22} = \frac{i_2}{e_2}\bigg|_{e_1=0} \tag{9-3}$$

The y or admittance parameters are also known as the short-circuit parameters, since their measurement involves shorting a pair of terminals which forces a terminal voltage to zero.

The y parameter equivalent circuit is based upon the nodal equations

$$i_1 = y_{11}e_1 + y_{12}e_2 \tag{9-4a}$$
$$i_2 = y_{21}e_1 + y_{22}e_2 \tag{9-4b}$$

Now work through problems PS 9-4, PS 9-5, and PS 9-6.

9-3 Parameter conversions

Occasionally situations arise in which we are given or can readily determine one set of parameters but wish to use a different set. The problem then is one of being able to convert from the z to the y parameters and vice versa. These parameter conversions may be developed as follows:

The z parameter equations are

1) $$e_1 = z_{11}i_1 + z_{12}i_2$$

2) $$e_2 = z_{21}i_1 + z_{22}i_2$$

Solving for i_1 and i_2 by determinants yields

3) $$i_1 = \frac{\begin{vmatrix} e_1 & z_{12} \\ e_2 & z_{22} \end{vmatrix}}{D^z} = \frac{z_{22}e_1 - z_{12}e_2}{D^z} \tag{9-5a}$$

4) $$i_2 = \frac{\begin{vmatrix} z_{11} & e_1 \\ z_{21} & e_2 \end{vmatrix}}{D^z} = \frac{-z_{21}e_1 + z_{11}e_2}{D^z} \tag{9-5b}$$

where

5) $$D^z = z_{11}z_{22} - z_{12}z_{21} \tag{9-5c}$$

The y parameter equations are

6) $$i_1 = y_{11}e_1 + y_{12}e_2$$

7) $$i_2 = y_{21}e_1 + y_{22}e_2$$

Solving (6) and (7) for e_1 and e_2 yields

$$8)\qquad e_1 = \frac{\begin{vmatrix} i_1 & y_{12} \\ i_2 & y_{22} \end{vmatrix}}{D^y} = \frac{y_{22}i_1 - y_{12}i_2}{D^y} \qquad (9\text{-}6a)$$

$$9)\qquad e_2 = \frac{\begin{vmatrix} y_{11} & i_1 \\ y_{21} & i_2 \end{vmatrix}}{D^y} = \frac{-y_{21}i_1 + y_{11}i_2}{D^y} \qquad (9\text{-}6b)$$

where

$$10)\qquad D^y = y_{11}y_{22} - y_{12}y_{21} \qquad (9\text{-}6c)$$

Now if we compare Eqs. (3) and (4) with (6) and (7) we see that they are in the same form with regard to what the dependent and independent variables are. Hence we may compare the coefficients of the variables to obtain

$$11a)\ y_{11} = \frac{z_{22}}{D^z} \qquad 11b)\ y_{12} = \frac{-z_{12}}{D^z}$$

$$11c)\ y_{21} = \frac{-z_{21}}{D^z} \qquad 11d)\ y_{22} = \frac{z_{11}}{D^z} \qquad (9\text{-}7)$$

Thus we have expressed the y's in terms of the z's. To express the z's in terms of the y's, we compare the coefficients of (8) and (9) with (1) and (2) to obtain

$$12a)\ z_{11} = \frac{y_{22}}{D^y} \qquad 12b)\ z_{12} = \frac{-y_{12}}{D^y}$$

$$12c)\ z_{21} = \frac{-y_{21}}{D^y} \qquad 12d)\ z_{22} = \frac{y_{11}}{D^y} \qquad (9\text{-}8)$$

Now work through problems PS 9-7 and PS 9-8.

9-4 The pi-T conversions

In Chap. 15 of the preceding text you were introduced to the pi and T networks as well as the means for converting from one form to the other. These networks and the conversion equations are shown here in Fig. 9-3 with impedances rather than resistances. Aside from the fact that the pi and T networks are four-terminal networks, we mention them again because they frequently are useful in simplifying networks or parameter conversions. Their utility in parameter conversions arises from the fact that we can easily see the z parameters in a T network and the y parameters in a pi network. Thus if we should happen to have a pi network but want its z parameters, it might be useful to use the pi to T conversions first and then, virtually by inspection, to read off the z parameters.

Now work through problem PS 9-9.

There are many other four-terminal network equivalent circuits with their respective parameters. Some of these are the hybrid parameters, the *ABCD* parameters, the scattering (*S*) parameters, etc. You will learn more about these in future lessons.

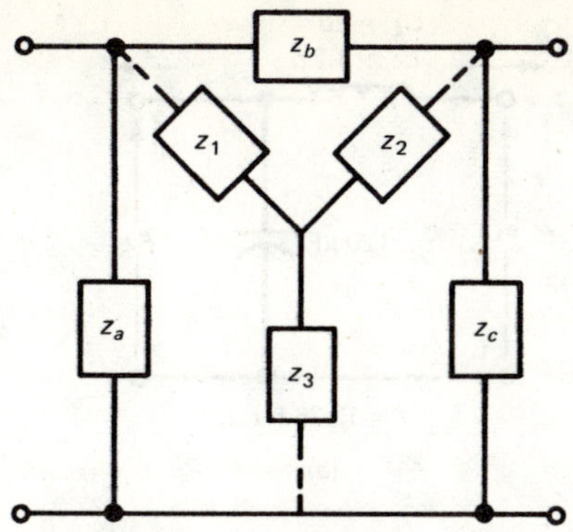

$$z_1 = \frac{z_a z_b}{z_a + z_b + z_c};\ z_2 = \frac{z_b z_c}{z_a + z_b + z_c};\ z_3 = \frac{z_a z_c}{z_a + z_b + z_c}$$

$$z_a = \frac{z_1z_2 + z_1z_3 + z_2z_3}{z_2}$$

$$z_b = \frac{z_1z_2 + z_1z_3 + z_2z_3}{z_3}$$

$$z_c = \frac{z_1z_2 + z_1z_3 + z_2z_3}{z_1}$$

FIGURE 9-3

PROBLEMS WITH SOLUTIONS

PS 9-1 The z parameters obtained from measurements made upon a four-terminal network at a particular frequency are $z_{11} = 12 - j6$ ohms, $z_{12} = 8$ ohms, $z_{21} = 8$ ohms, and $z_{22} = 18 + j15$ ohms. Sketch the z parameter equivalent circuit applicable at this frequency.

SOLUTION See Fig. PS 9-1, which is constructed by analogy with Fig. 9-1*c*.

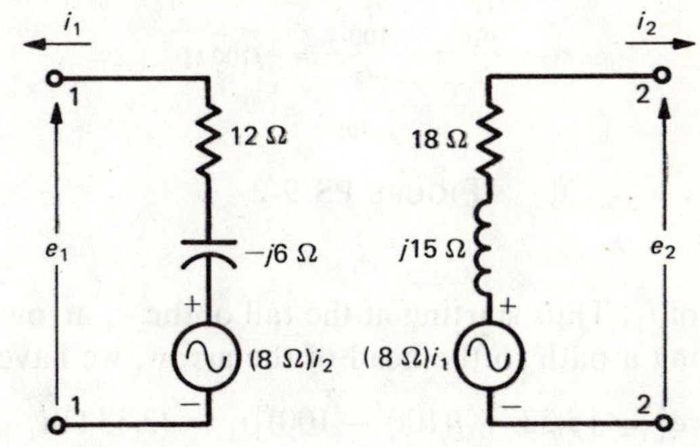

FIGURE PS 9-1

PS 9-2 What are the z parameters of the network shown in Fig. PS 9-2*a* at a frequency of 13.26 Hz?

SOLUTION At 13.26 Hz, $X_C = 100$ ohms, $X_L = 100$ ohms, and the coil resistance is $R_L = X_L/Q_L = 100$ ohms/30 $= 3.33$ ohms. To determine z_{11} and z_{21} we open the output side so that $i_2 = 0$ as shown in Fig. PS 9-2*b* and proceed to write an expression for e_1 in

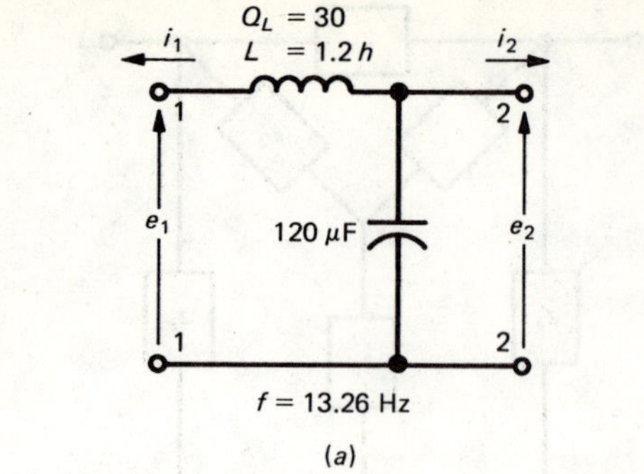

(a)

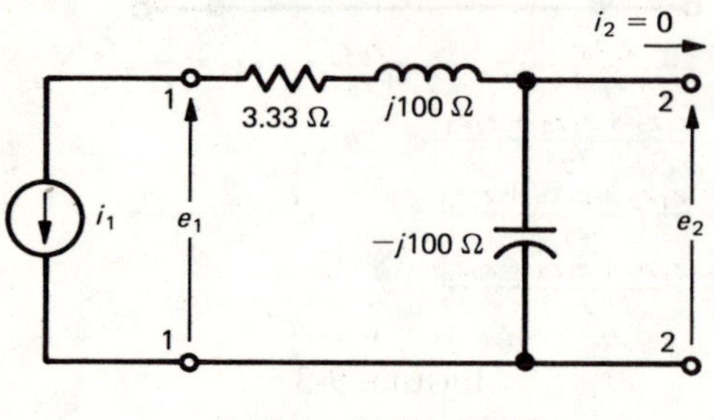

$$z_{11} = \frac{e_1}{i_1} = \frac{[3.33\ \Omega + j(100 - 100)\ \Omega]i_1}{i_1} = 3.33\ \Omega$$

$$z_{21} = \frac{e_2}{i_1} = \frac{(-j100\ \Omega)i_1}{i_1} = -j100\ \Omega$$

(b)

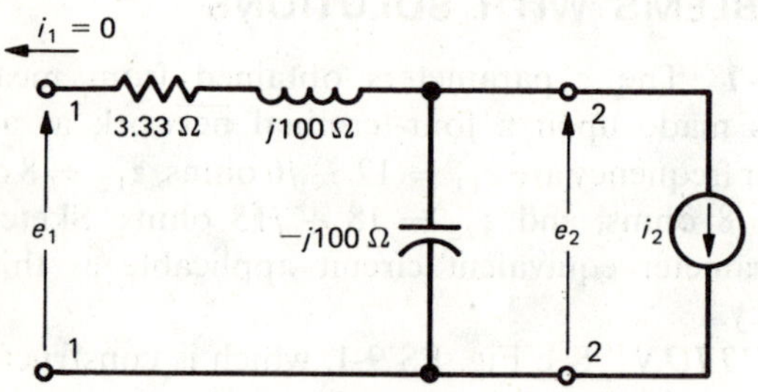

$$z_{22} = \frac{e_2}{i_2} = \frac{-j100 i_2}{i_2} = -j100\ \Omega$$

$$z_{12} = \frac{e_1}{i_2} = \frac{-j100 i_2}{i_2} = -j100\ \Omega$$

(c)

FIGURE PS 9-2

terms of i_1. Thus starting at the tail of the e_1 arrow and following a path to the head of the arrow, we have

1) $\quad e_1 = [3.33 + j(100 - 100)]i_1 = (3.33\ \Omega)i_1$

Since $z_{11} = e_1/i_1$ (remember, this is with $i_2 = 0$), we can divide (1) by i_1 to obtain

2) $$z_{11} = \frac{e_1}{i_1} = \frac{(3.33\ \Omega)i_1}{i_1} = 3.33\ \Omega$$

Similar reasoning may be applied to the evaluation of z_{21} as shown in Fig. PS 9-2*b*. The parameters z_{22} and z_{12} are similarly evaluated in Fig. PS 9-2*c* except that i_1 is now forced to zero by opening the input terminals.

PS 9-3 What is the input impedance of the network in Fig. PS 9-3*a* when the output is terminated in a 24-ohm resistive load?

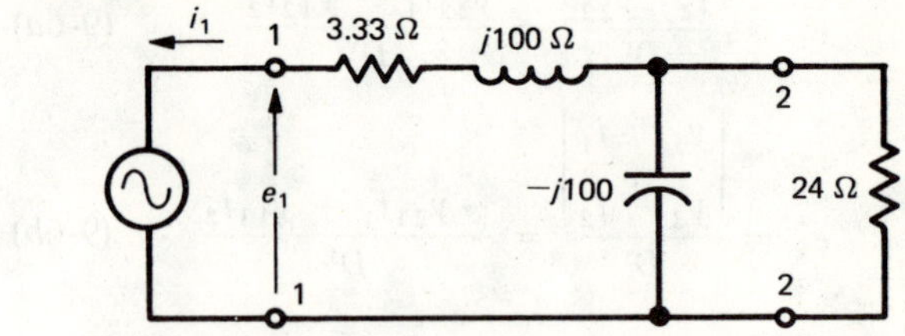

(a)

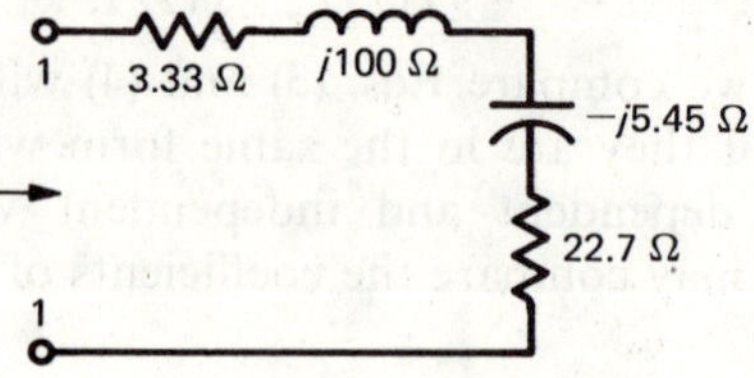

(b)

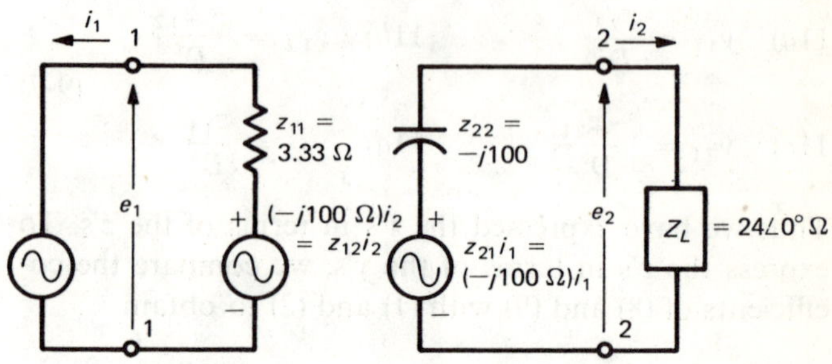

(c)

FIGURE PS 9-3

SOLUTION Several methods could be used. One approach is to convert the capacitive reactance (−*j*100 ohms) and 24-ohm load to its series equivalent by means of the parallel-to-series conversions. Thus

1) $$Q_P = \frac{R_P}{X_P} = \frac{24}{100} = 0.24$$

2) $$X_S = \frac{X_P}{1 + (1/Q_P)^2} = \frac{100}{1 + (1/0.24)^2} = 5.45\ \Omega$$

3) $$R_S = \frac{R_P}{1 + Q_P{}^2} = \frac{24}{1 + (0.24)^2} = 22.7\ \Omega$$

Therefore e_1 effectively sees the circuit of Fig. PS 9-3*b*, from which

4) $$z_{\text{in}} = 26 + j94.6\ \Omega$$

An alternative approach is to use the previously derived *z* parameters of problem PS 9-2, which was based on the same four-terminal network. We could use them to construct the equivalent circuit of Fig. PS 9-3*c* and then solve for

5) $$z_{\text{in}} = \frac{e_1}{i_1}$$

by writing the following equations:

Around the input loop

$$6)\qquad e_1 = z_{11}i_1 + z_{12}i_2$$

Around the output loop

$$7)\qquad 0 = z_{21}i_1 + (z_{22} + Z_L)i_2$$

Solving for i_1 by determinants,

$$8)\quad i_1 = \frac{\begin{vmatrix} e_1 & z_{12} \\ 0 & (z_{22} + Z_L) \end{vmatrix}}{\begin{vmatrix} z_{11} & z_{12} \\ z_{21} & (z_{22} + Z_L) \end{vmatrix}} = \frac{(z_{22} + Z_L)e_1}{z_{11}(Z_{22} + Z_L) - z_{12}z_{21}}$$

Therefore

$$9)\qquad z_{in} = \frac{e_1}{i_1} = \frac{z_{11}(z_{22} + Z_L) - z_{12}z_{21}}{z_{22} + Z_L}$$

Substituting values,

$$10)\quad z_{in} = \frac{3.33\underline{/0°}\,(24 - j100) - (-j100)(-j100)}{24 - j100}$$

$$= 98.1\underline{/74.6°}\ \Omega$$

$$z_{in} = 26 + j94.6\ \Omega$$

which checks our previous solution.

Note how easily the problem was solved by the first method. We hope this illustrates why it is advisable to have several analytical tools at our disposal.

PS 9-4 What are the y parameters of the pi network shown in Fig. PS 9-4a?

SOLUTION By shorting the output side so that $e_2 = 0$ and applying a voltage e_1 across the input terminals as shown in Fig. PS 9-4b, we are prepared to determine y_{11} and y_{21}. Clearly i_1 is the sum of the currents through C_1 and L, which respectively are $e_1/(-jX_{C1})$ and $e_1/(R + jX_L)$. Thus we have the first equation for y_{11} in this figure. To evaluate y_{21}, we simply express i_2 as a function of e_1, as shown in the second equation in Fig. PS 9-4b.

To determine y_{22} and y_{12}, we short the input side and drive the output side with a voltage source as shown in

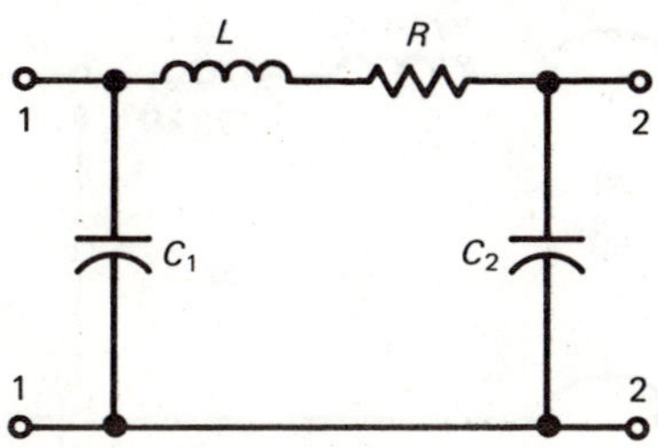

(a)

FIGURE PS 9-4

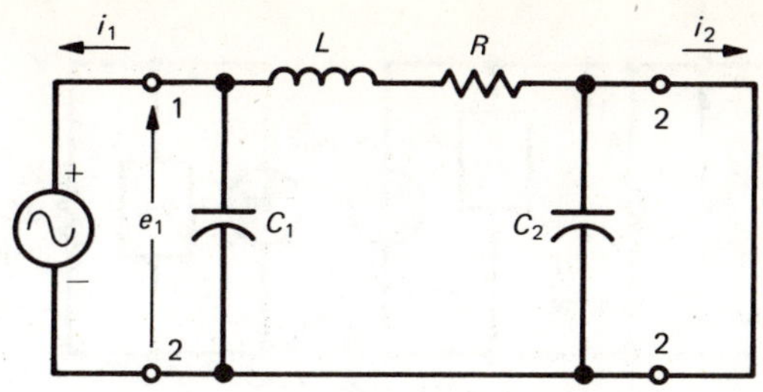

$$y_{11} = \frac{i_1}{e_1} = \frac{\dfrac{e_1}{-jX_{C1}} + \dfrac{e_1}{R + jX_L}}{e_1} = \frac{1}{-jX_{C1}} + \frac{1}{R + jX_L}$$

$$y_{21} = \frac{i_2}{e_1} = -\frac{\dfrac{e_1}{R + jX_L}}{e_1} = -\frac{1}{R + jX_L}$$

(b)

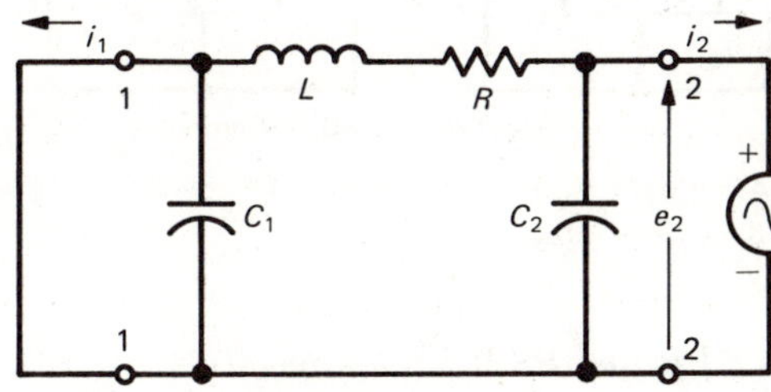

$$y_{22} = \frac{i_2}{e_2} = \frac{\dfrac{e_2}{-jX_{C2}} + \dfrac{e_2}{R + jX_L}}{e_2} = \frac{1}{-jX_{C2}} + \frac{1}{R + jX_L}$$

$$y_{12} = \frac{i_1}{e_2} = -\frac{\dfrac{e_2}{R + jX_L}}{e_2} = -\frac{1}{R + jX_L}$$

(c)

FIGURE PS 9-4 (*Continued*)

Fig. PS 9-4c. An analysis similar to that in Fig. PS 9-4b yields y_{22} and y_{12}.

PS 9-5 Using the information obtained in the previous problem, determine the voltage gain $A_V = e_2/e_g$ for the circuit shown in Fig. PS 9-5a.

SOLUTION The general approach we will use to solve this problem is to determine the y parameters of the four-terminal network by using the general y parameters of this network developed in the previous problem. We can then construct a y parameter equivalent circuit, hang the source and load on the input and output

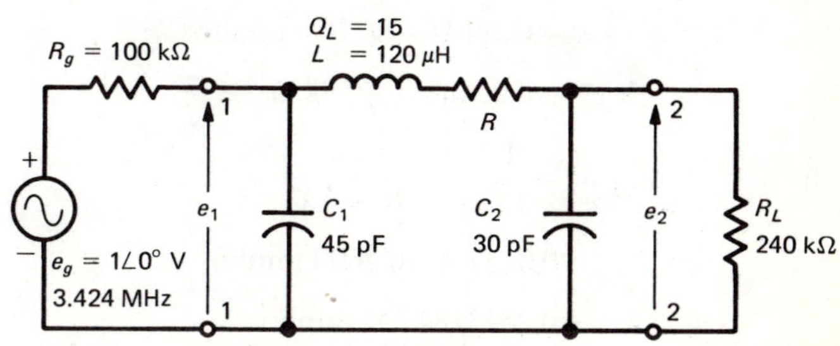

(a)

FIGURE PS 9-5

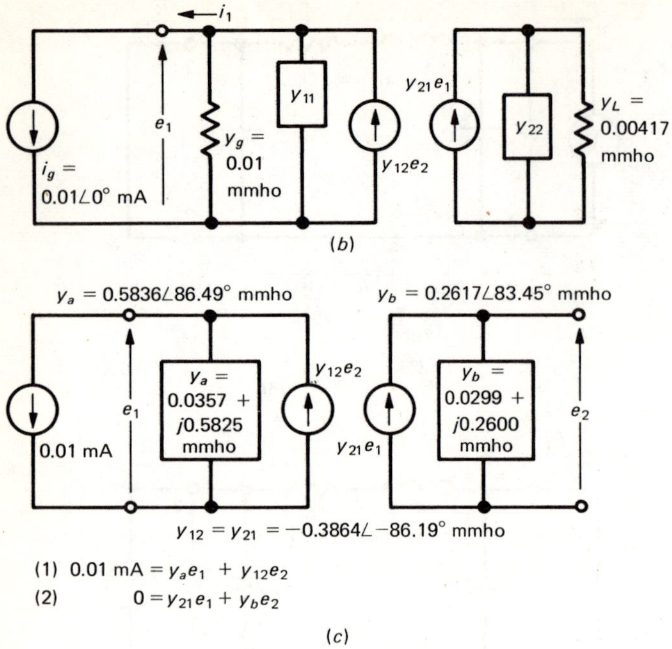

FIGURE PS 9-5 (*Continued*)

respectively, and then solve for e_2. However, we will change the e_g, R_g voltage source to a current source so that R_g appears in parallel with y_{11}, in which case we need only add their admittances.

We will need the following data based upon the driving frequency of 3.424 MHz:

$$f = 3.424 \text{ MHz} \qquad R = X_L/Q_L = 172.1\ \Omega$$

$$\omega = 21.51 \times 10^6 \text{ rad/sec} \qquad X_{C1} = 1033\ \Omega$$

$$X_L = 2582\ \Omega \qquad X_{C2} = 1549\ \Omega$$

From the previous problem

1) $$y_{11} = \frac{1}{-jX_{C1}} + \frac{1}{R + jX_L} = 0.0257 + j0.5825 \text{ mmhos} = 0.5830\underline{/87.47^\circ} \text{ mmhos}$$

2) $$y_{12} = y_{21} = -\frac{1}{R + jX_L} = -0.3864\underline{/-86.19^\circ} \text{ mmhos} = -0.02568 + j0.3855$$

3) $$y_{22} = \frac{1}{-jX_{C2}} + \frac{1}{R + jX_L} = 0.0257 + j0.2600 \text{ mmhos} = 0.2613\underline{/84.35^\circ} \text{ mmhos}$$

Converting the voltage source to an equivalent current source and expressing R_L as an admittance, we have Fig. PS 9-5*b*. Combining y_g with y_{11} and y_L with y_{22}, we obtain Fig. PS 9-5*c*, from which the following nodal equations are derived:

1) $$0.01 \text{ mA} = y_a e_1 + y_{12} e_2$$

2) $$0 = y_{21} e_1 + y_b e_2$$

Solving the above simultaneous equations for e_2 and noting that $y_{12} = y_{21}$,

$$e_2 = \frac{\begin{vmatrix} y_a & 0.01 \text{ mA} \\ y_{21} & 0 \end{vmatrix}}{\begin{vmatrix} y_a & y_{12} \\ y_{21} & y_b \end{vmatrix}} = -\frac{y_{12}(0.01 \text{ mA})}{y_a y_b - y_{12}^{\;2}}$$

$$= -\frac{(-0.3864\underline{/-86.19^\circ} \text{ mmho})(0.01 \text{ mA})}{(0.5836\underline{/86.49^\circ})(0.2617\underline{/83.45^\circ}) \text{ mmho}^2 - (0.3864\underline{/-86.19^\circ} \text{ mmho})^2}$$

$$= -\frac{(-0.3864\underline{/-86.19^\circ} \text{ mmho})(0.01 \text{ mA})}{0.04654\underline{/92.95^\circ} \text{ mmho}^2}$$

$$\cong 0.0831\underline{/-179.1^\circ} \text{ V}$$

$$A_v = \frac{\mathbf{e}_2}{\mathbf{e}_g} = \frac{0.0831\underline{/-179.1^\circ}}{1\underline{/0^\circ}} \cong 0.0831\underline{/-180^\circ}$$

PS 9-6 Make a rough check of your solution in the previous problem using an alternative approach.

SOLUTION One approach is to take e_g, R_g, and C_1 in Fig. PS 9-5*a* and apply Thévenin's theorem. Therefore, we obtain Fig. PS 9-6*a*. Since $R_g \gg X_{C1}$, we may to a first approximation assume that the impedance presented to e_g in Fig. PS 9-6*a* is essentially R_g. Hence, the current is $i \cong 1\underline{/0^\circ}$ V/100 kΩ = $0.01\underline{/0^\circ}$ mA, and it is essentially in phase with the applied voltage. In flowing

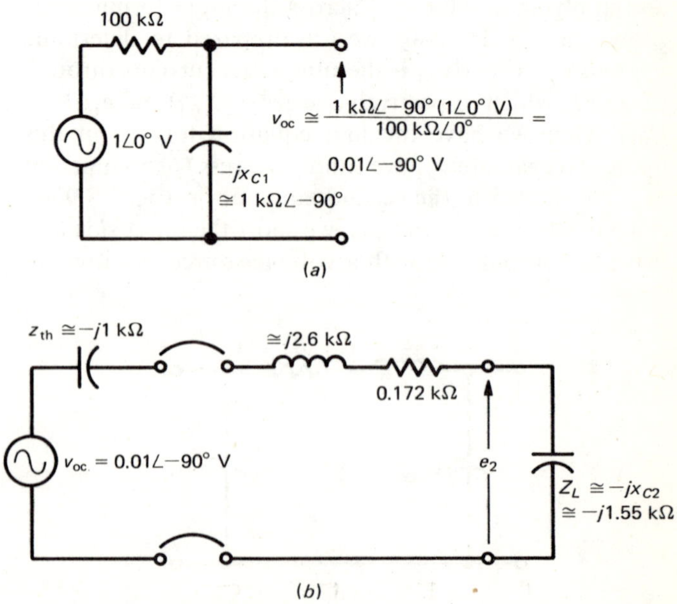

FIGURE PS 9-6

through C_1, a lagging voltage $v_{oc} = -jX_{C1}\ i \cong (1\ \text{k}\Omega \underline{/-90^\circ})(0.01\underline{/0^\circ}\ \text{mA}) = 0.01\underline{/-90^\circ}$ volts. Since $|X_{C1}| \ll R_g$, the Thévenin's equivalent impedance $z_{th} \cong -jX_{C1}$.

On the output side, we might note that $|X_{C2}| \ll R_L$, so that we can assume e_2 is essentially developed across C_2 only. With these assumptions we can hang it all together as shown in Fig. PS 9-6*b*.

A glance at Fig. PS 9-6*b* reveals that the net capacitive reactance equals the inductive reactance, so that the circuit is series-resonant. In that case the current in Fig. PS 9-6*b* is $i = 0.01\underline{/-90^\circ}\ \text{V}/0.172\ \text{k}\Omega = 0.0581\underline{/-90^\circ}$ mA. This 90° lagging current flowing through C_2 develops a voltage which lags 90° behind the current so that e_2 actually exhibits a 180° phase shift relative to e_g. Specifically $e_2 = -jX_{C2}(i) = (1.55\ \text{k}\Omega\underline{/-90^\circ})(0.0581\underline{/-90^\circ}\ \text{mA}) = 0.0901\underline{/-180^\circ}$ volts which is in substantial agreement with the solution previously obtained.

PS 9-7 What are the *z* parameters of the network in Fig. PS 9-7?

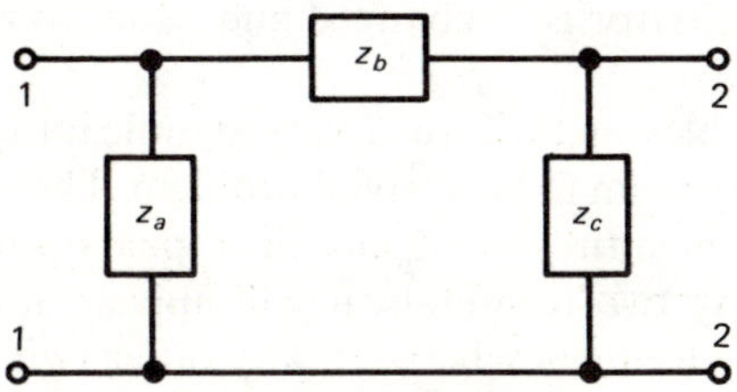

FIGURE PS 9-7

SOLUTION One approach is to imagine the input or the output terminals shorted as if the *y* parameters were actually being measured. In that case we would obtain

1) $$y_{11} = \frac{1}{z_a} + \frac{1}{z_b} = y_a + y_b$$

2) $$y_{12} = y_{21} = -\frac{1}{z_b} = -y_b$$

3) $$y_{22} = \frac{1}{z_b} + \frac{1}{z_c} = y_b + y_c$$

4) $$D^y = y_{11}y_{22} - y_{12}y_{21}$$
$$= (y_a + y_b)(y_b + y_c) - y_b{}^2$$
$$= y_a y_b + y_a y_c + y_b y_c$$

And from Eq. (9-8),

5) $$z_{11} = \frac{y_{22}}{D^y} = \frac{y_b + y_c}{y_a y_b + y_a y_c + y_b y_c}$$

6) $$z_{12} = z_{21} = -\frac{y_{12}}{D^y} = \frac{-y_b}{D^y}$$

7) $$z_{22} = \frac{y_{11}}{D^y} = \frac{y_a + y_b}{D^y}$$

PS 9-8 Verify the results of the previous problem by actually setting up the network of Fig. PS 9-7 for a direct evaluation of the *z* parameters. In other words, imagine the required open circuits at the input and output terminals. In that case we would obtain, after some analysis,

1) $$z_{11} = z_a \| (z_b + z_c) = \frac{z_a(z_b + z_c)}{z_a + z_b + z_c}$$

2) $$z_{12} = \frac{z_a z_c}{z_a + z_b + z_c} = z_{21}$$

3) $$z_{22} = z_c \| (z_a + z_b) = \frac{z_c(z_a + z_b)}{z_a + z_b + z_c}$$

To verify the results of the previous problem, we must show that the *z*'s obtained here are actually identical with the *z*'s obtained in PS 9-7. Thus we must, for example, show that z_{11} of Eq. (1) in this problem is actually identical with z_{11} of Eq. (5) in the previous problem. If we express (1) in this problem in terms of admittances, we should have

4) $$z_{11} = \frac{\dfrac{1}{y_a}\left(\dfrac{1}{y_b} + \dfrac{1}{y_c}\right)}{\dfrac{1}{y_a} + \dfrac{1}{y_b} + \dfrac{1}{y_c}}$$

Multiplying numerator and denominator by $y_a y_b y_c$ yields

5) $$\frac{y_b y_c\left(\dfrac{1}{y_b} + \dfrac{1}{y_c}\right)}{y_b y_c + y_a y_c + y_a y_b} = \frac{y_b + y_c}{y_b y_c + y_a y_c + y_a y_b}$$

which happily is identical with z_{11} as given in (6) of the previous problem. The remainder of these conversions is left as an exercise for the student who needs practice in algebra.

PS 9-9 Suppose that in Fig. 9-9*a* the pi network is operated at a frequency such that $X_L = X_{C1} + X_{C2}$, and that there are no resistive losses in the circuit. Determine the elements of the corresponding T network shown in Fig. PS 9-9*b*.

SOLUTION Using the pi to T conversions of Fig. 9-3 we have, for example,

$$z_1 = \frac{-jX_{C1}(jX_L)}{j(X_L - X_{C1} - X_{C2})} = \frac{X_{C1}X_L}{j0} = |\infty|$$

Clearly z_1, z_2, and z_3 will be infinite due to the zero denominator.

Intuitively we might suspect that this is due to an antiresonant condition looking into the 1-1 or the 2-2 terminal, but on second thought we might be a little disturbed as to why the circuit should look antiresonant at 1-1 terminals and the 2-2 terminals at the same frequency since there is no stipulation that $C_1 = C_2$.

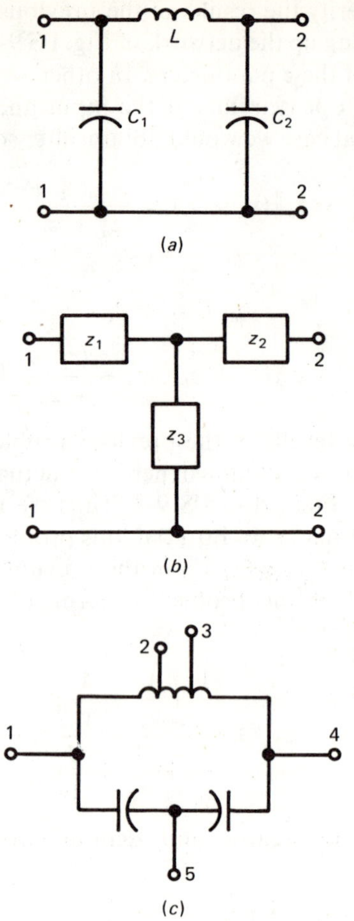

FIGURE PS 9-9

To explore this further let us express the input impedance looking into the 1-1 terminals. Thus

$$z_{\text{in}(1\text{-}1)} = -jX_{C1} \| [j(X_L - X_{C2})]$$
$$= \frac{-jX_{C1}[j(X_L - X_{C2})]}{j(X_L - X_{C1} - X_{C2})}$$

Since the denominator is zero, $z_{\text{in}(1\text{-}1)}$ is infinite at that f_{ar} which causes the zero denominator.

Looking into the 2-2 terminals,

$$z_{\text{in}(2\text{-}2)} = \frac{-jX_{C2}[j(X_L - X_{C2})]}{j(X_L - X_{C1} - X_{C2})}$$

which again indicates antiresonance at the 2-2 terminals at the same f_{ar} since the denominator is the same and equal to zero.

In fact, looking into the upper 1-2 terminals, we see

$$z_{\text{in}(1\text{-}2)} = \frac{jX_L(-jX_{C1} - jX_{C2})}{j(X_L - X_{C1} - X_{C2})}$$

The denominator is unchanged and equal to zero at the same f_{ar}.

This enables us to draw a very significant conclusion in what may seem to be a trivial problem. The conclusion is that if an antiresonant circuit appears antiresonant between any two terminals, it will appear antiresonant at the same frequency between any other two terminals. For example, in Fig. PS 9-9*c* we may reason that if the circuit looks antiresonant between any two terminals, say 1 and 2, it will also look antiresonant between any other two terminals, say 4 and 5, or 3 and 5, etc.

PROBLEMS WITH ANSWERS

PA 9-1 Determine the *z*, *y*, and *h* parameters of the network in Fig. PA 9-1. Review the *h* parameters in Chap. 15 of the preceding text.

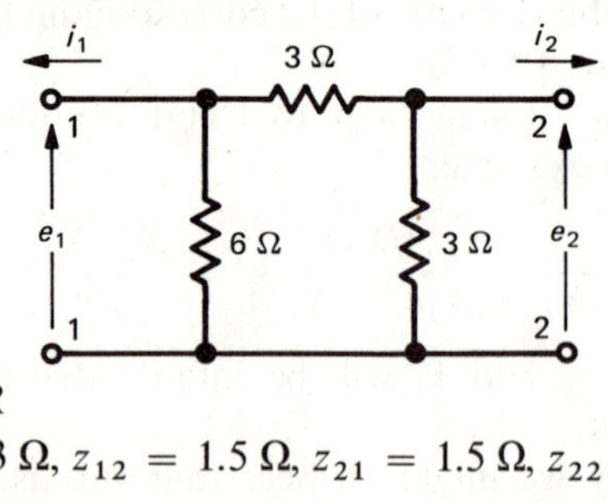

ANSWER

$$z_{11} = 3\ \Omega,\ z_{12} = 1.5\ \Omega,\ z_{21} = 1.5\ \Omega,\ z_{22} = 2.25\ \Omega$$
$$y_{11} = 1/2 \text{ mho},\ y_{12} = -1/3 \text{ mho},$$
$$y_{21} = -1/3 \text{ mho},\ y_{22} = 1/1.5 \text{ mho}$$
$$h_{11} = 2\ \Omega,\ h_{12} = 2/3,\ h_{21} = -2/3,\ h_{22} = 1/2.25 \text{ mho}$$

PA 9-2 Determine e_2 in the circuit of Fig. PA 9-2.

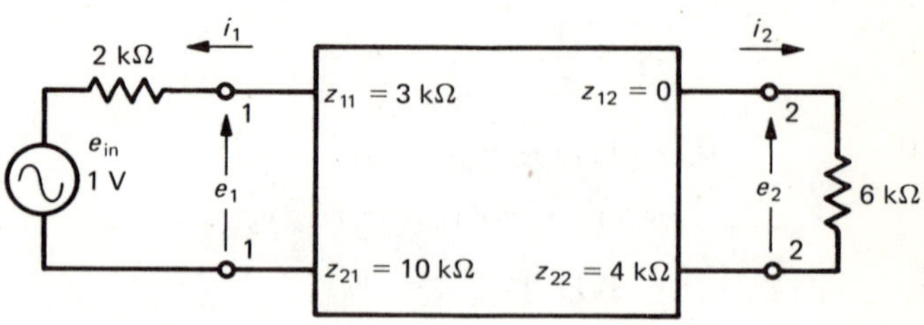

ANSWER $e_2 = 12$ kV

PA 9-3 Determine i_2' in the circuit of Fig. PA 9-3.

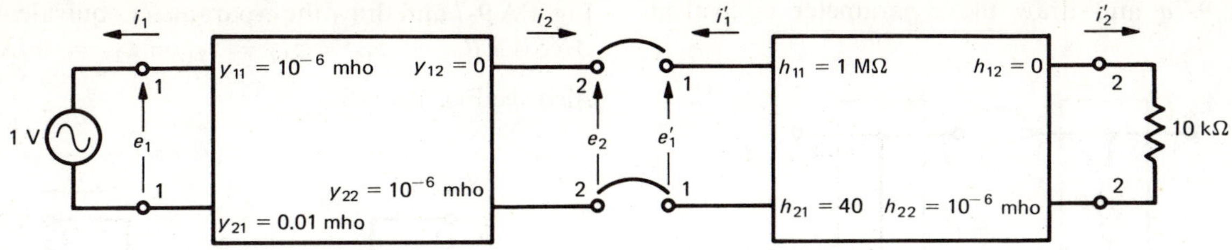

ANSWER $i_2' = -0.198$ A

PA 9-4 What are the z and y parameters of the network in Fig. PA 9-4 at an $\omega = 1/\sqrt{LC}$? Assume $Q_L = \infty$.

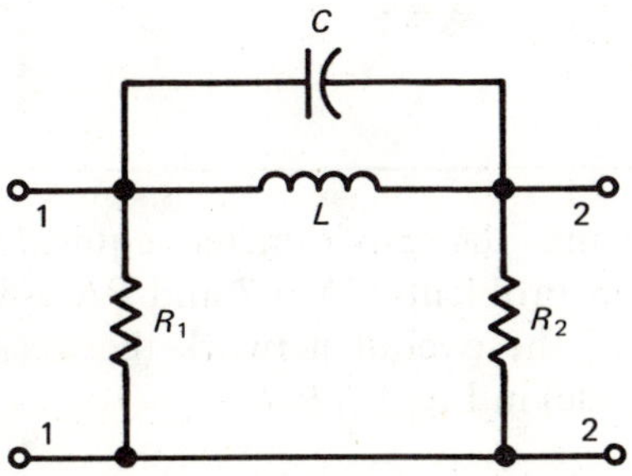

ANSWER

$$z_{11} = R_1,\ z_{22} = R_2,\ z_{12} = z_{21} = 0,$$

$$y_{11} = 1/R_1,\ y_{22} = 1/R_2,\ y_{12} = y_{21} = 0$$

PA 9-5 What are the z and y parameters of the network in Fig. PA 9-5 at an $\omega = 1/\sqrt{C(L_1 + L_2)}$? Assume the network is lossless.

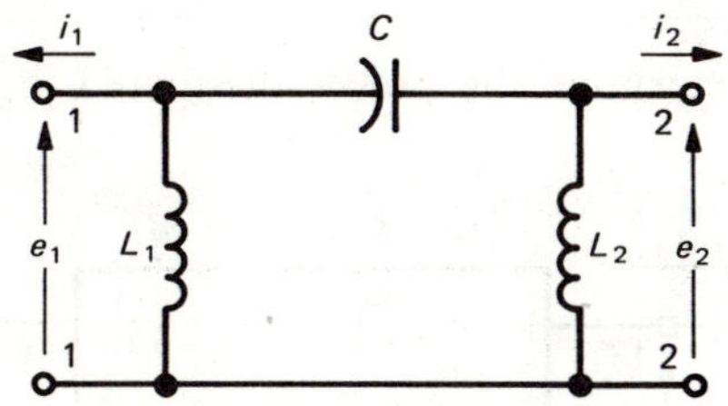

ANSWER

$$z_{11} = z_{12} = z_{21} = z_{22} = \infty,$$

$$y_{11} = 1/j(X_{L1} - X_C),\ y_{12} = y_{21} = 1/jX_C,$$

$$y_{22} = 1/j(X_{L2} - X_C)$$

PA 9-6 What are the z parameters of the network in Fig. PA 9-6?

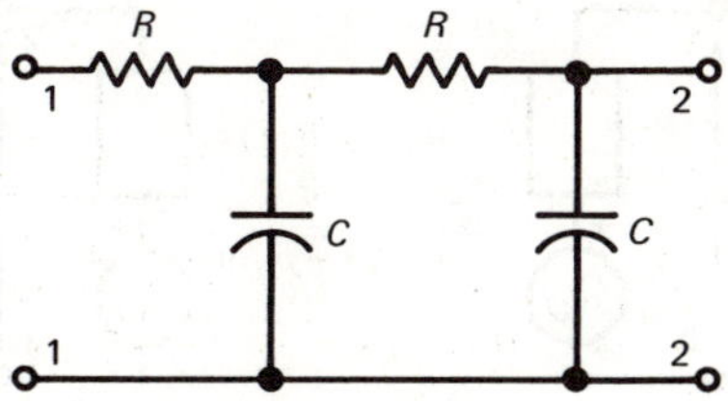

ANSWER

$$z_{11} = R + \frac{R - X_C^2 - jX_C}{R - j2X_C} \qquad z_{12} = z_{21} = \frac{-X_C^2}{R - j2X_C}$$

$$z_{22} = \frac{R - X_C^2 - jX_C}{R - j2X_C}$$

PA 9-7 Find the z parameters of network A only in Fig. PA 9-7a and draw the z-parameter equivalent circuit.

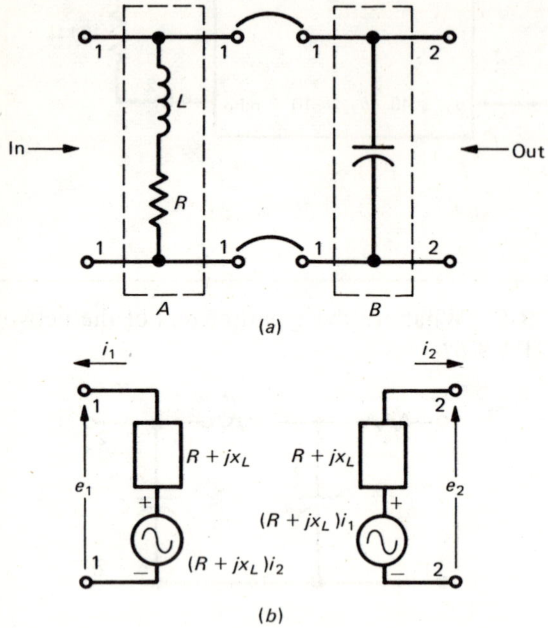

ANSWER $z_{11} = z_{12} = z_{21} = z_{22} = R + jX_L$.

Also see Fig. PA 9-7b.

PA 9-8 Find the z parameters of network B only in Fig. PA 9-7 and draw the z-parameter equivalent circuit.

ANSWER $z_{11} = z_{12} = z_{21} = z_{22} = -jX_C$.

Also see Fig. PA 9-8.

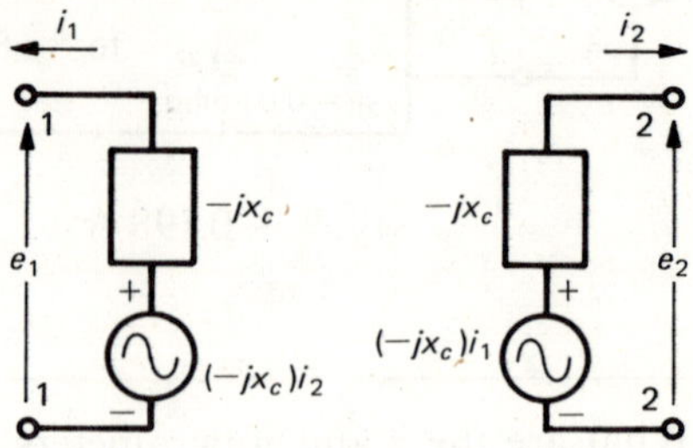

PA 9-9 Using the z-parameter equivalent circuits determined in problems PA 9-7 and PA 9-8, find the z parameters of the overall network (between the input and output side) in Fig. PA 9-7.

ANSWER

$$z_{11} = z_{12} = z_{21} = z_{22} = (R + jX_L)\|(-jX_C)$$

PROBLEMS WITHOUT ANSWERS

P 9-1 Determine the z and y parameters of the network in Fig. P 9-1.

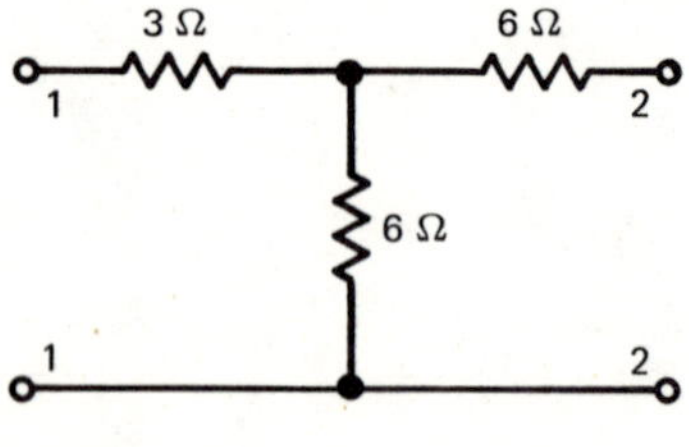

P 9-2 Determine the power dissipated in R_L in Fig. P 9-2.

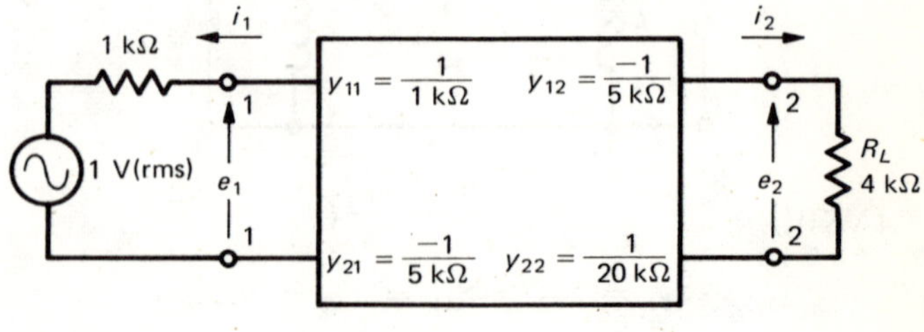

P 9-3 Find the z parameters of the network in Fig. P 9-3.

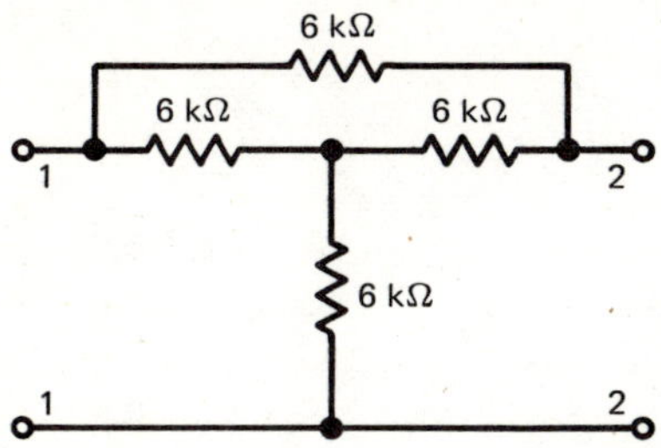

P 9-4 Determine e_2 in Fig. P 9-4.

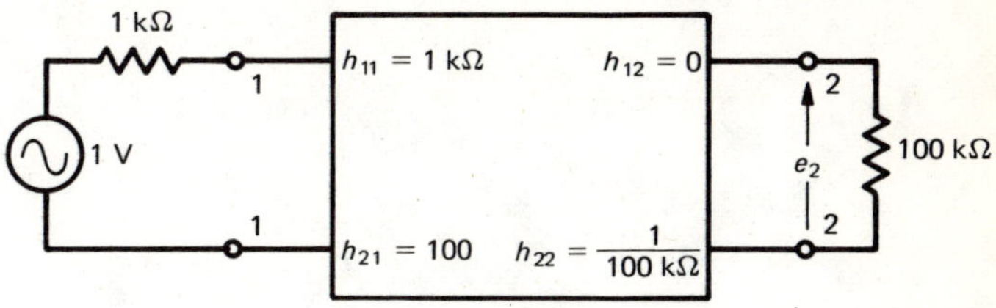

P 9-5 Determine the Thévenin's equivalent circuit, v_{oc} and z_{th}, seen looking into the output terminals of the network in Fig. P 9-5.

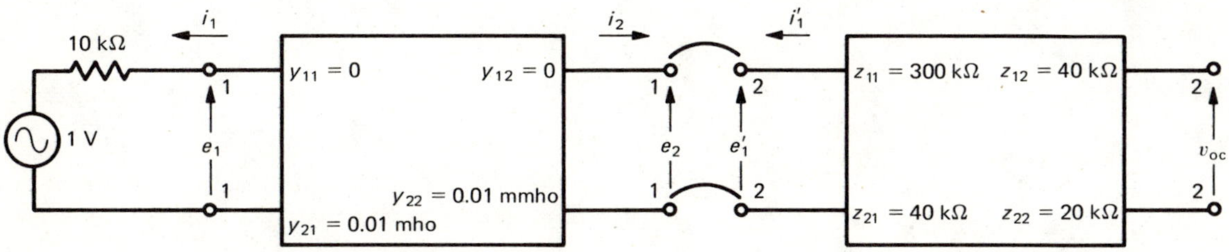

P 9-6 What are the y parameters of the network in Fig. P 9-6 if $R_L = 76$ ohms, $X_L = 12.3$ kΩ, $X_C = 12.3$ kΩ, and $R_C = 240$ kΩ?

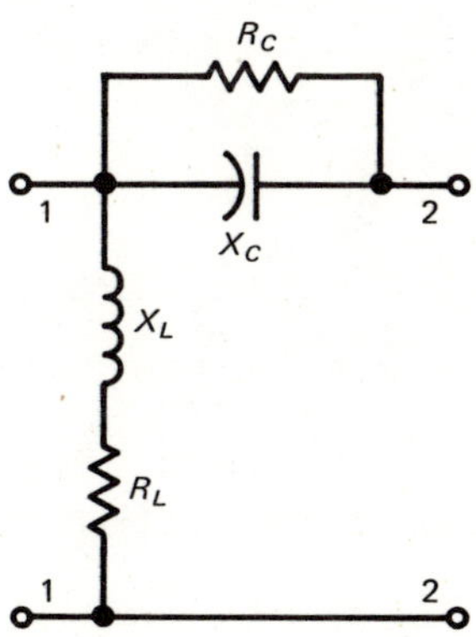

10 the transformer

10-1 Transformer action

When we studied inductors, it was pointed out that the self-induced voltage across an inductor was given by

1) $$e_L = L\frac{di}{dt}$$

where L is the self-inductance and di/dt is the time rate of change of current. A deeper interpretation of Eq. (1) is to be found in Faraday's law which has the form

2) $$e_L = N\frac{d\phi}{dt} \qquad (10\text{-}1a)$$

where N is the number of turns in a coil that is being linked by a magnetic flux changing at a rate $d\phi/dt$. We need not explore Eq. (2) in any great depth for our purposes. Suffice it to say that Eq. (2) implies that whenever a flux change occurs relative to a coil (or conductor), a voltage is induced in that coil.

From the theory of magnetic circuits it may be shown that

3) $$\phi = \frac{\text{mmf}}{\mathcal{R}} = \frac{Ni}{\mathcal{R}} \qquad (10\text{-}1b)$$

where the magnetomotive force (mmf) equals the ampere-turns (Ni), and the reluctance $\mathcal{R}$ is determined by the geometry of the coil and the magnetic properties of the core material. If (3) is substituted into (2), there results

4) $$e_L = \left(\frac{N^2}{\mathcal{R}}\right)\frac{di}{dt}$$

If we compare (4) with (1), we see that

5) $$L = \frac{N^2}{\mathcal{R}} \qquad (10\text{-}1c)$$

Thus we have established the connection between Eqs. (2) and (1), which enables us to conclude that where there exists magnetic flux, there exists a related inductance.

Now (2) simply states that when a changing flux links a coil (inductor), a voltage is induced in that coil. It really says nothing about where the flux comes from. If the flux under consideration is due to the current within the inductor under consideration, then the induced voltage is called a self-induced voltage. However, if the flux variation is due to the current change in a second coil whose magnetic field links the first coil, then the voltage in the first coil is simply called the induced voltage. In essence, this is the operating principle of the transformer.

10-2 The transformer

A transformer in its simplest form consists of two inductors which are magnetically coupled as shown in

Fig. 10-1. Since in this circuit the only coupling between the two loops is through their magnetic fields and since it is only changing currents which induce voltages, it follows that dc voltage (current) components in one loop will not affect the other loop. This means that there is no dc transmission through a transformer which is purely inductively coupled. Only ac or other time-varying waveforms can pass through.[1]

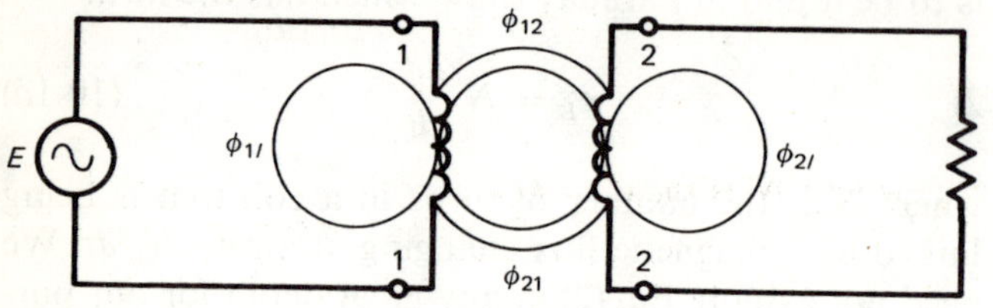

FIGURE 10-1

It is necessary to define some basic parameters. Consider Fig. 10-1. The current flowing through the 1-1 winding sets up a magnetic field (flux) about this winding. A part of this flux, ϕ_{1l}, is the leakage flux which links only the 1-1 winding. The other part, ϕ_{21}, links both the 1-1 and 2-2 windings and therefore is a component of mutual flux. Similarly the current in the 2-2 winding will result in ϕ_{2l}, the leakage flux about the 2-2 winding which does not link the 1-1 winding, and ϕ_{12}, which is a flux component linking both windings. The net mutual flux (ϕ_M) is equal to $\phi_{12} + \phi_{21}$.

Wherever there is magnetic flux, there is inductance. The inductance may be that of a straight wire or of a many-turn coil, but always magnetic flux and inductance go together. We then say that the leakage flux ϕ_{1l} is due to a leakage inductance L_{1l}, which is common only to loop 1. Similarly, ϕ_{2l} is due to the leakage inductance L_{2l}, common only to loop 2. The mutual flux ϕ_M is set up by a mutual inductance M, which is common to both loops. The position and function of these inductances can perhaps be visualized by imagining the transformer inductances in the form of a *T*, as shown in Fig. 10-2.

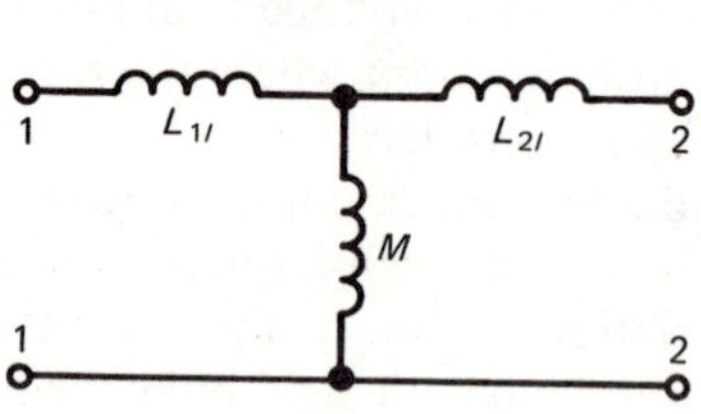

FIGURE 10-2

Just as the self-inductance of a coil depends upon the number of turns, core material, wire size, and so forth, so the mutual inductance M between two coils depends upon their geometric configuration (proximity and orientation), permeability, and the individual inductance of each coil.[2] The tightness of coupling between the two windings is defined by the coefficient of coupling. A formula relating the coefficient of coupling k to the mutual and self-inductances is

$$k = \frac{M}{\sqrt{L_1 L_2}} \tag{10-2}$$

When the core material is other than air, M, L_1, and L_2 may not be constant. Instead, these parameters depend upon the coil currents.

In most communication circuits, the value of k is about 0.5 or less. In power work, where high efficiency is important, k is about 0.9 or more. The maximum value of k is 1.

Now we know that the voltage induced in one winding depends upon a current change in another winding. The exact relationship is

$$e_{21} = \pm M \frac{di_1}{dt} \approx \pm M \frac{\Delta i_1}{\Delta t} \tag{10-3}$$

where e_{21} means the voltage induced in coil 2 by a current change in coil 1. M is the mutual inductance between these coils, and di_1/dt is the instantaneous rate of change of current with respect to time in coil 1. The meaning of the plus and minus signs will soon be made clear. For the moment just ignore them.

Suppose two coils were so coupled that $M = 1$ henry. If the current in one coil changes at a rate of 1 amp/sec, the voltage induced in the other is 1 volt. Increasing the frequency of the current variations raises the induced voltage. For example, were the current to change at a rate of 1 amp/msec, the induced voltage would be 1000 volts.

When the current is a sine wave, the induced voltage will be a cosine wave. When the sine wave goes through zero, it is *changing* most rapidly. The induced voltage will therefore be a maximum at this time, or 90° out of phase with the current causing it. Similarly, we can write.

$$e_{12} = \pm M \frac{di_2}{dt} \approx \pm M \frac{\Delta i_2}{\Delta t} \tag{10-4}$$

where e_{12} is the voltage induced in coil 1 by a current change in coil 2.

10-3 The dot notation

If we could write loop equations for an inductively coupled circuit, the loop currents could then be evaluated. Once the currents are known, all other network characteristics can readily be determined. To write the loop equations, we must know something about the

[1] Nonsinusoidal waveforms may be terribly distorted.

[2] Permeability is a magnetic property of the medium the inductor is immersed in, or the core material in the case of ferromagnetic cores.

phasing of the windings. This information is necessary to determine whether the voltage induced in loop 1 by a current in loop 2 aids or opposes the assumed direction of current in loop 1. Similarly, we must determine the effect of $M\ di_1/dt$ upon the assumed direction of secondary current.

In order to determine the polarity of the induced voltage in one winding due to a current change in another, a simple experiment is performed. The circuit of Fig. 10-3 is set up, and the switch is closed. The voltmeter

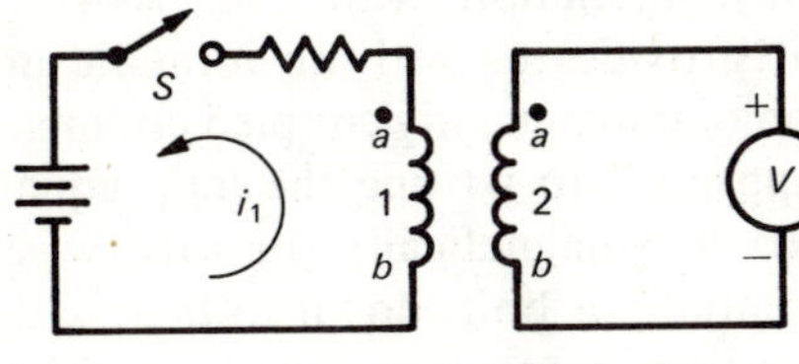

FIGURE 10-3

will kick either up-scale or backwards and then return to zero. The current i_1, in building up through coil 1, must have made the b side of coil 1 negative and the a side positive. If the meter kicks up-scale, we know that making the a side of coil 1 positive drives the a side of coil 2 positive.

Similarly, making the a side of coil 1 negative, which could be due to a decrease in i_1, causes the a side of coil 2 to become negative with respect to the b side. A convenient way of denoting the winding sense is to place a dot on the end of the coils which follow each other in phase.

If in Fig. 10-3 it had been determined that closing S caused the meter to read backwards, then the dot on the a side of coil 2 would instead be placed on the b side of coil 2. The dots have nothing to do with the circuit or assumed current directions. The only thing that determines the dots is the manner in which the coil is wound.

The location of the dots must be known before loop equations can be written. This section presents a step-by-step method of writing loop equations for inductively coupled circuits.

Assume that we have the circuit of Fig. 10-4a to analyze. The location of the dots has already been determined as shown. To write the loop equations, use the following procedure:

1. Replace each coil with two generators showing no polarity, as in Fig. 10-4b. One generator in each coil represents the counter emf due to the self-inductance of the winding, and the other represents the induced voltage due to a current in the other winding.
2. Arbitrarily choose a direction of current flow in loop 1, say, counterclockwise. This determines the polarity of the voltage across L_1 as i_1 builds up to its steady-state value. The counter emf develops across L_1 so that it tends to keep the assumed current from increasing. This counter emf ($L_1\ di_1/dt$) will then have the polarity shown in Fig. 10-4c.
3. Observe in Fig. 10-4c that the voltage developed

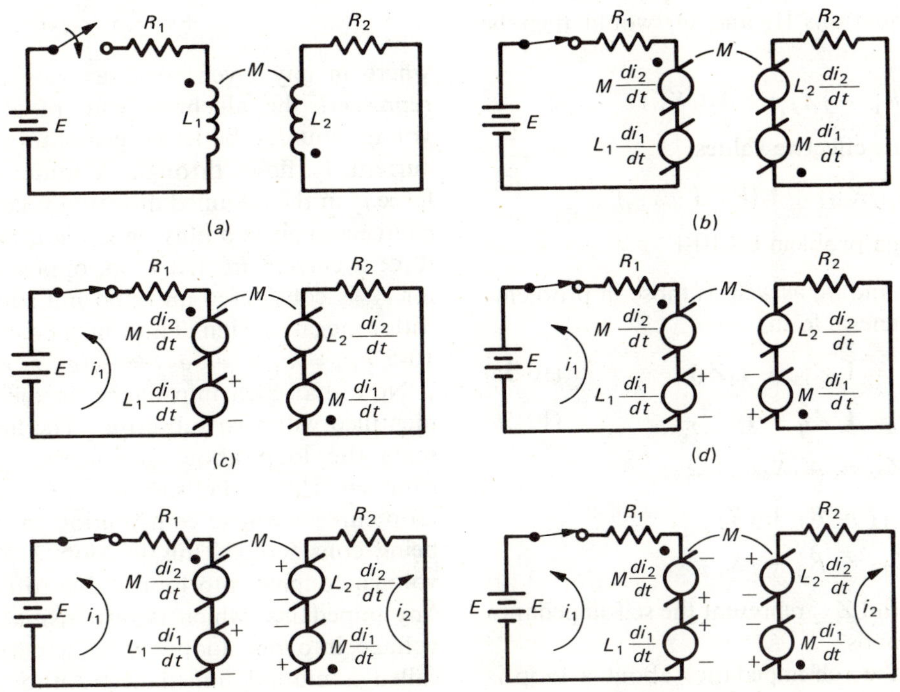

FIGURE 10-4

across L_1 because of its self-inductance is such that the dot end of coil 1 is driven positive. Since the dot voltages follow in phase, the dot end of the $M\,di_1/dt$ generator in loop 2 must be driven positive. Up to this point, everything can be represented by Fig. 10-4*d*.

4. Now arbitrarily assume a current direction for loop 2, and, as in loop 1, note how the $L_2\,di_2/dt$ generator is polarized. Let us assume a clockwise current direction in loop 2, as in Fig. 10-4*e*, which polarizes L_2 as shown.
5. In Fig. 10-4*e*, we see that the self-induced voltage across L_2 ($L_2\,di_2/dt$) drives the dot end of coil 2 negative. The dot end of the induced generator in coil 1 ($M\,di_2/dt$) must then also go negative. At this point, our analysis can be summed up by Fig. 10-4*f*. We are now in a position to write our loop equations.

1) $$E = i_1R_1 + L_1\frac{di_1}{dt} - M\frac{di_2}{dt}$$

2) $$0 = -M\frac{di_1}{dt} + i_2R_2 + L_2\frac{di_2}{dt}$$

Simultaneous Eqs. (1) and (2) can be solved by the methods of calculus for either i_1 or i_2 as functions of time. Our intention at this point is only to set up the loop equations. If, however, the battery E were replaced with a sinusoidal generator, we could solve the problem with our present tools. For a sine wave of current, the $L\,di/dt$ generators become $\mathbf{I}(jX_L)$ and the $M\,di/dt$ generators become $\mathbf{I}(jX_M)$. Equations (1) and (2) would then be written as

3) $$\mathbf{E} = \mathbf{I}_1(R_1 + jX_{L_1}) - \mathbf{I}_2(jX_M)$$

where **E** and **I** are the effective values,

4) $$0 = -\mathbf{I}_1(jX_M) + \mathbf{I}_2(R_2 + jX_{L_2})$$

Now work through problem PS 10-1.

Notice that Eqs. (3) and (4) as well as those in problems PS 10-1 are in the general form

5) $$\mathbf{E} = \mathbf{I}_1Z_{11} + \mathbf{I}_2Z_M \qquad (10\text{-}5a)$$

6) $$0 = \mathbf{I}_1Z_M + \mathbf{I}_2Z_{22} \qquad (10\text{-}5b)$$

where
$$Z_M = \pm jX_M$$
$$Z_{11} = R_1 + jX_{L1}$$
$$Z_{22} = R_2 + jX_{L2}$$

The quantities Z_{11} and Z_{22} represent the self-impedance of their respective loops.

In other words, the self-impedance about a loop is simply the total impedance presented to that particular loop current.

The quantity Z_M is called the mutual impedance between loops 1 and 2 and, as you can see from (5) and (6), it allows for the effect of one loop current to influence the other. In other words, Z_M is the parameter which allows for coupling between the two loops. Thus, if Z_M were equal to zero, the $\mathbf{I}_2Z_M$ term in (5) would be zero and $\mathbf{I}_1$ would be totally independent of $\mathbf{I}_2$. Similar reasoning may be applied to (6).

Since $Z_M = \pm jX_M$, we must be sure we know how to determine the sign of X_M. The previous procedure as described in conjunction with Fig. 10-4 is perfectly valid, and it provides us with an intimate insight into the operation of magnetically coupled circuits. However, a quicker approach to writing the loop equations may be developed for magnetically (or otherwise) coupled circuits. Essentially it boils down to this:

No matter how many loops we have, how we have chosen the loop currents, or how they are connected, the *general* form of loop equations can be written in the form

$$\begin{aligned}
&1)\ e_{11} = Z_{11}i_1 + Z_{12}i_2 + Z_{13}i_3 + \cdots + Z_{1N}i_N \quad \text{(loop 1)}\\
&2)\ e_{22} = Z_{21}i_1 + Z_{22}i_2 + Z_{23}i_3 + \cdots + Z_{2N}i_N \quad \text{(loop 2)}\\
&3)\ e_{33} = Z_{31}i_1 + Z_{32}i_2 + Z_{33}i_3 + \cdots + Z_{3N}i_N \quad \text{(loop 3)}\\
&4)\ e_{NN} = Z_{N1}i_1 + Z_{N2}i_2 + Z_{N3}i_3 + \cdots + Z_{NN}i_N \quad \text{(loop } N\text{)}
\end{aligned} \qquad (10\text{-}6)$$

where in any given row, say row 2 for example, e_{22} represents the algebraic sum of all the independent voltage sources (batteries, generators, etc.) that the loop current, i_2, flows through. A source which is trying to force i_2 in the assumed direction is considered an aiding source and given a plus sign. A source which is trying to force a current in that loop, opposite to that assumed for i_2, is considered an opposing source and is written with a negative sign. Thus, in the circuit shown in Fig. 10-5, $e_{11} = e_1 - e_2$, $e_{22} = -(e_2 + e_3)$, $e_{33} = 0$.

Note that each impedance on the right-hand side is identified with two subscripts. The first subscript represents the loop about which the voltages are being summed. The second subscript represents the particular loop current whose contribution to the loop voltage is being considered. Thus in loop 2, Z_{21} represents the voltage coupled into loop 2 due to the flow of i_1 only. Any impedance which is responsible for introducing a voltage into one loop due to a current in another is called a mutual impedance. Although in the *general* form of Eq. (10-5) all the mutual (and self) impedance terms are written with plus signs, the actual mutual

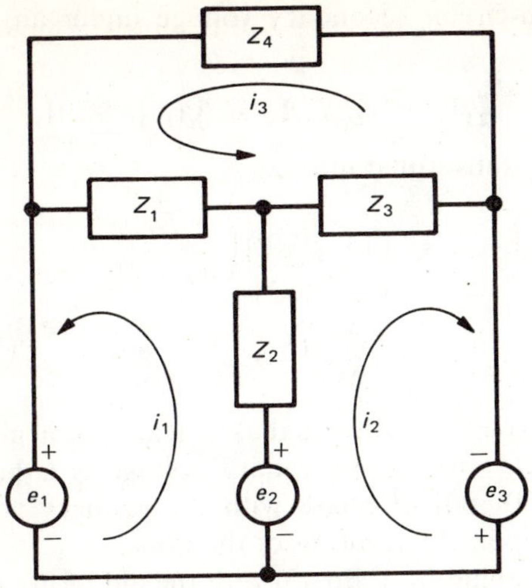

$e_{11} = e_1 - e_2;\ e_{22} = -e_2 - e_3$
$e_{33} = 0$
$z_{11} = z_1 + z_2;\ z_{22} = z_2 + z_3;\ z_{33} = z_1 + z_3 + z_4$
$z_{12} = z_{21} = +z_2;\ z_{13} = z_{31} = -z_1;\ z_{23} = z_{32} = +z_3$

FIGURE 10-5

impedance in a particular circuit may have a plus or minus sign associated with it.[3] An easy rule to remember is that whenever the loop current (i_2) and the voltage-inducing current (i_1 in Z_{21}) flow through the mutual impedance in the *same* direction, the mutual impedance is positive. If they flow through it in opposite directions, the mutual impedance is negative. Thus in Fig. 10-5, for example, Z_{21}, the impedance common (mutual) to both loops 2 and 1 is $Z_{21} = +Z_2$, since i_1 and i_2 flow through Z_2 in the same direction.[4] The mutual impedance $Z_{13} = -Z_1$, $Z_{32} = +Z_3$.

It might also be noted that in circuits involving bilateral elements, as in this text, a mutual impedance $Z_{xy} = Z_{yx}$. Thus $Z_{13} = -Z_1 = Z_{31}$.

Whenever both subscripts of an impedance term in Eq. (10-6) are the same, say Z_{22}, it means we are concerned only with the net impedance seen by i_2 as it goes around loop 2. Thus Z_{22}, the self-impedance of loop 2, is given by $Z_{22} = Z_2 + Z_3$. The self-impedance terms are always positive unless a negative impedance element is in some manner synthesized or introduced into the circuit. Generally this requires electronic devices such as tubes, transistors, tunnel diodes, four-layer devices, etc., which are capable of providing power gain and need not concern us in this text.

[3]The polarity associated with a mutual impedance term, say Z_{xy}, is determined by whether or not the voltage coupled into loop x due to the current in loop y looks like an aiding or opposing voltage source to i_x. If it aids, $Z_{xy} = -Z_M$, and it opposes $Z_{xy} = +Z_M$.

[4]Do not be confused by the fact that i_1 flows in a counterclockwise direction while i_2 flows in a clockwise direction. Although these are opposite directions, the point is that they both flow through Z_2 in the same (upward) direction.

A useful point to remember is that if all the loop currents are chosen in the same direction, say counterclockwise, all the mutual impedance terms will be negative. The self-impedance terms remain positive.

Now work through problem PS 10-2.

In the case of inductively coupled circuits, the sign of a mutual impedance term, say $Z_{12} = \pm j\omega M$, is easily determined by the following rule. If the two loop currents, say I_1 and I_2, both flow into or out of the dotted ends of the transformer, the sign of $j\omega M$ is positive. However, if one current is directed into a dot and the other out, then $j\omega M$ is negative. For example, in Fig. 10-6*a*, $Z_{xy} = +j\omega M$, and in Fig. 10-6*b* and *c*, $Z_{xy} = -j\omega M$.

Now work through PS 10-3.

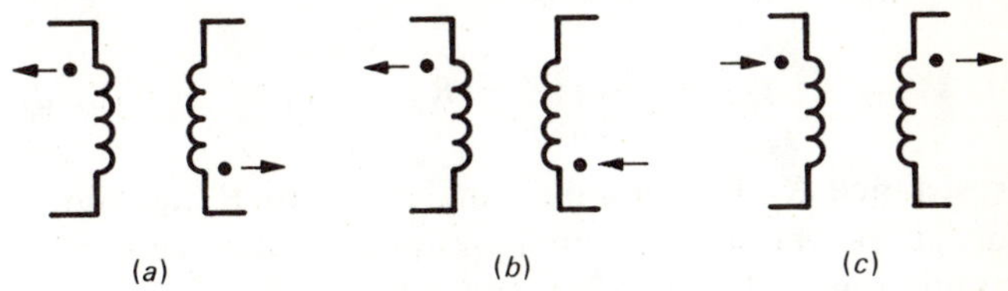

FIGURE 10-6

10-4 The transformer as a four-terminal network

The simplest type of transformer may be considered to be a four-terminal network. If we neglect any stray capacitances across or between windings, the circuit of Fig. 10-7 may be used to represent a transformer.

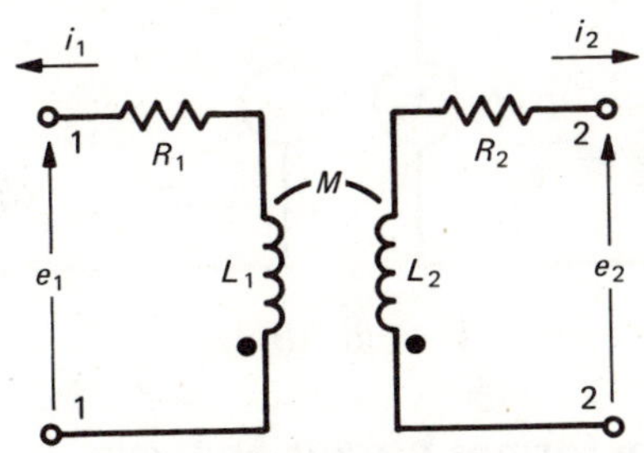

FIGURE 10-7

If we measure the z parameters of the transformer with sinusoidal excitation, we obtain

$$1)\qquad z_{11} = \left.\frac{e_1}{i_2}\right|_{i_2=0} = R_1 + jX_{L1} \qquad (10\text{-}7a)$$

The resistance R_1 accounts for all the power losses seen looking into the 1-1 terminals (also called input or primary winding) caused by winding resistance, skin effect, and core losses if a core is used. The inductance L_1 is the self-inductance of the primary winding and in iron-core transformers is sometimes called the magnetizing inductance, which ideally is as large as possible.

2) $$z_{21} = \frac{e_2}{i_1}\bigg|_{i_2=0} = j\omega M \qquad (10\text{-}7b)$$

The mutual impedance z_{21} with $i_2 = 0$ determines the open-circuit voltage developed across the 2-2 terminals (also called output or secondary winding) due to a current in the input side. The sign of $j\omega M$ is positive because of the way we placed the dots in Fig. 10-7. Note that the setup for determining z_{21} may actually be used to measure the mutual inductance of a transformer.

3) $$z_{12} = \frac{e_1}{i_2}\bigg|_{i_1=0} = j\omega M \qquad (10\text{-}7c)$$

The mutual impedance $z_{12} = z_{21}$, and it is a measure of the open-circuit voltage induced in the primary due to a current in the secondary.

4) $$z_{22} = \frac{e_2}{i_2}\bigg|_{i_1=0} = R_2 + jX_{L2} \qquad (10\text{-}7d)$$

Resistance R_2 in a manner analogous to R_1 accounts for all power losses seen looking into the secondary winding with the primary winding open. Inductance L_2 is simply the self-inductance of the secondary winding.

The general z parameter equivalent circuit of the transformer will then appear as shown in Fig. 10-8. Since no dots are included in Fig. 10-8, we use $\pm$ signs for the mutual impedance.

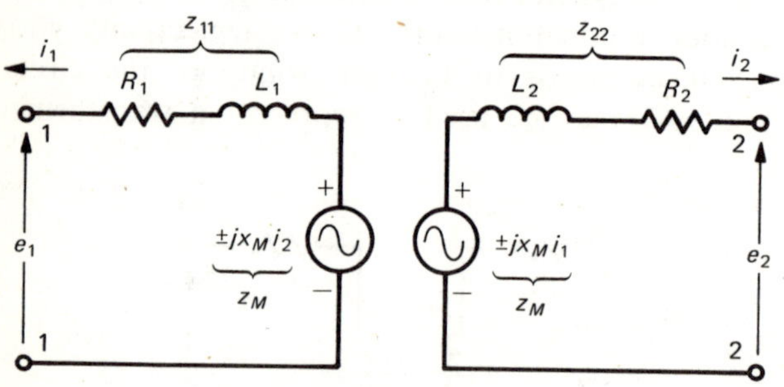

FIGURE 10-8

10-5 Transformer circuit analysis

The equivalent circuit of Fig. 10-8 enables us to draw some useful conclusions regarding transformer operation. First of all we see that if the secondary winding is not loaded ($i_2 = 0$), the primary current i_1, is determined by e_1 and z_{11} only, since the $z_{12}i_2 = z_M i_2$ voltage source is equal to zero when $i_2 = 0$. In a high-quality iron-core-type transformer, $X_{L1} \gg R_1$, so that the primary current i_1 is essentially limited by the magnetizing inductance and it lags the primary voltage e_1 by approximately 90°. Thus, under the conditions $X_{L1} \gg R_1$, and $i_2 = 0$, we have (effective values assumed)

1) $$\mathbf{I}_1 \cong \frac{\mathbf{E}_1}{jX_{L1}} = \frac{E_1/0^\circ}{X_{L1}/90^\circ} = \left|\frac{E_1}{X_{L1}}\right| /-90^\circ \qquad (10\text{-}8)$$

The open-circuit secondary voltage under any conditions is

2) $$\mathbf{E}_{2(\text{oc})} = z_{21}\mathbf{I}_1 = \pm jX_M\mathbf{I}_1 = \pm(X_M/90^\circ)\mathbf{I}_1 \qquad (10\text{-}9)$$

and with (1) substituted into (2),

3) $$\mathbf{E}_{2(\text{oc})} = \pm(X_M/90^\circ)\left(\frac{E_1}{X_{L1}/90^\circ}\right) = \pm\frac{X_M}{X_{L1}}(E_1/0^\circ) \qquad (10\text{-}10)$$

Equation (3) clearly indicates that with zero or negligible secondary current, the secondary voltage is either in-phase or 180° out of phase with the primary voltage depending upon the location of the dots.

The interpretation of Eq. (3) with the aid of Eq. (10-2) may be extended further as follows

$$\frac{\mathbf{E}_{2(\text{oc})}}{\mathbf{E}_1} = \pm\frac{X_M}{X_{L1}} = \pm\frac{\omega M}{\omega L_1} = \pm\frac{M}{L_1} = \pm\frac{k\sqrt{L_1L_2}}{L_1} = \pm k\sqrt{\frac{L_2}{L_1}} \qquad (10\text{-}11)$$

Now it turns out [see Eq. (10-1c)] that the self-inductance of a coil is proportional to the square of the number of turns in the winding. Thus equation (4) may be written as

5) $$\frac{\mathbf{E}_{2(\text{oc})}}{\mathbf{E}_1} = \pm k\sqrt{\frac{N_2^2}{N_1^2}} = \pm k\left(\frac{N_2}{N_1}\right) \qquad (10\text{-}12)$$

The quantity N_2/N_1 (or N_S/N_P) is often referred to as the turns ratio, and (5) indicates that in *any* transformer in which $X_{L1} \gg R_1$, the open-circuit secondary voltage is proportional to k, the coefficient of coupling, and the turns ratio. In transformers using ferromagnetic cores, the flux is essentially confined to the core with little leakage, and hence $k \cong 1$. Thus in transformers of this type

6) $$\frac{\mathbf{E}_{2(\text{oc})}}{\mathbf{E}_1} \cong \pm\frac{N_2}{N_1} \qquad (10\text{-}13)$$

Now let us investigate what happens to the primary current if a load is connected across the secondary so that $\mathbf{I}_2 \neq 0$. Let us assume that z_M is positive in Fig. 10-8 and that some load Z_L is connected across the 2-2 terminals as shown in Fig. 10-9a. Clearly the $z_M\mathbf{I}_1 = jX_M\mathbf{I}_1$ open-circuit voltage source will actually force a current in the secondary winding which is opposite to $\mathbf{I}_2$. If $\mathbf{I}_2$ reverses, the dependent $z_M\mathbf{I}_2 = jX_M\mathbf{I}_2$ voltage source in the primary side must also reverse polarity, so that we have the situation shown in Fig. 10-9a. Note that the flow of secondary current has coupled (induced) a voltage into the primary side of such a polarity that it series aids the driving source $\mathbf{E}_1$ in forcing $\mathbf{I}_1$. This means $\mathbf{I}_1$ will tend to increase. This seems intuitively

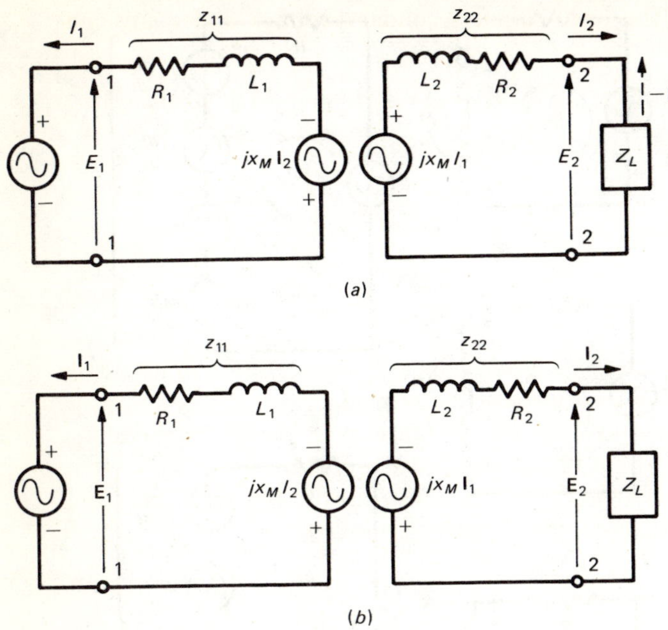

FIGURE 10-9

reasonable, since current flow in the secondary side means power dissipation, at least in resistive components. This power has to come from the driving source, which means $\mathbf{I}_1$ must increase since $\mathbf{E}_1$ is fixed.

Does the same hold true if z_M is negative? In this case the polarity of the $z_M\mathbf{I}_1$ voltage source in Fig. 10-9*a* would be reversed and $\mathbf{I}_2$ would actually flow in the indicated reference direction as shown in Fig. 10-9*b*. Thus I_2 is a positive quantity. However, with z_M negative, the $z_M\mathbf{I}_2$ voltage source in the primary side in effect reverses polarity, so that it still series aids the driving voltage. We may therefore conclude that loading the secondary apparently increases the primary current independently of the sign of z_M, which only affects the direction of $\mathbf{I}_2$.

Up to this point our analysis has been rather informal. For an exact determination of the currents in the general circuit of Fig. 10-10 we might write, in general form, for any two-loop network

1) $$\mathbf{E}_1 = Z_{11}\mathbf{I}_1 + \mathbf{Z}_{12}\mathbf{I}_2$$

2) $$0 = Z_{12}\mathbf{I}_1 + Z_{22}\mathbf{I}_2$$

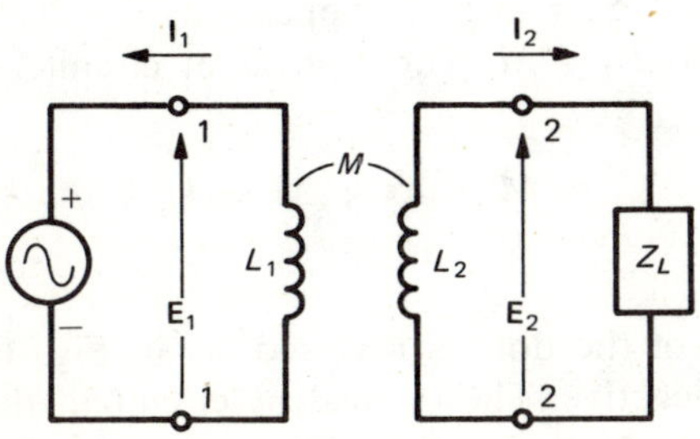

FIGURE 10-10

Since $Z_{12} = Z_{21} = Z_M$, we may instead write

3) $$\mathbf{E}_1 = Z_{11}\mathbf{I}_1 + Z_M\mathbf{I}_2 \qquad (10\text{-}14a)$$

4) $$0 = Z_M\mathbf{I}_1 + Z_{22}\mathbf{I}_2 \qquad (10\text{-}14b)$$

No dots are shown in Fig. 10-10, so that Z_M may actually be a positive or negative quantity. The impedances Z_{11} and Z_{22} are the self-impedances around loops 1 and 2 respectively. Thus

5) $$Z_{11} = z_{11} = R_1 + jX_{L1} \qquad (10\text{-}15a)$$

6) $$Z_{22} = z_{22} + Z_L = R_2 + jX_{L2} + Z_L \qquad (10\text{-}15b)$$

7) $$Z_M = z_M = \pm jX_M \qquad (10\text{-}15c)$$

Solving (3) and (4) by determinants for $\mathbf{I}_1$ yields

8) $$\mathbf{I}_1 = \frac{\mathbf{E}_1 Z_{22}}{Z_{11}Z_{22} - Z_M^2} = \frac{\mathbf{E}_1}{Z_{11} - \dfrac{Z_M^2}{Z_{22}}} \qquad (10\text{-}16)$$

Now the net input impedance Z_{in} that $\mathbf{E}_1$ forces $\mathbf{I}_1$ through must be the denominator of Eq. (8). Thus

9) $$Z_{\text{in}} = Z_{\text{in}(1\text{-}1)} = \frac{\mathbf{E}_1}{\mathbf{I}_1} = Z_{11} - \frac{Z_M^2}{Z_{22}} \qquad (10\text{-}17a)$$

Note that (9) is made up of two terms: Z_{11}, which is the self-impedance of the primary, and $-(Z_M^2/Z_{22})$, which is called the reflected impedance since it reflects the effects of what is in loop 2 into loop 1.

Now $Z_M = \pm jX_M$ and therefore $Z_M^2 = +j^2X_M^2 = -X_M^2$ irrespective of the sign of Z_M. Hence, if we substitute this result into Eq. (9), we obtain

10) $$Z_{\text{in}} = Z_{\text{in}(1\text{-}1)} = Z_{11} - \frac{(-X_M^2)}{Z_{22}}$$

$$= Z_{11} + \frac{X_M^2}{Z_{22}} \qquad (10\text{-}17b)$$

Be sure you understand that (10) is valid for any sign of Z_M and hence is independent of the dot locations.

Now work through problems PS 10-4, PS 10-5, and PS 10-6.

10-6 Thevenizing the transformer

To further facilitate our analysis of inductively coupled circuits, it is most useful to develop the Thévenin's equivalent circuit that the load Z_L in Fig. 10-10 looks back into. This is a simple matter, since we can readily see that with $\mathbf{I}_2 = 0$ in Fig. 10-9*a* or 10-9*b*,

1) $$\mathbf{E}_{2(\text{oc})} = Z_M\mathbf{I}_1 = \pm jX_M\mathbf{I}_1 \qquad (10\text{-}18)$$

Since

2) $$\mathbf{I}_1 = \frac{\mathbf{E}_1}{Z_{\text{in}(1\text{-}1)}} = \frac{\mathbf{E}_1}{Z_{\text{in}}}$$

By substituting (2) into (1), we obtain

3) $$\mathbf{E}_{2(\mathrm{oc})} = \frac{Z_M \mathbf{E}_1}{Z_{\mathrm{in}}} = \frac{\pm jX_M \mathbf{E}_1}{Z_{11} + \dfrac{X_M{}^2}{Z_{22}}}$$

But under open-circuit conditions, $Z_{22} = \infty$. Therefore,

4) $$\mathbf{E}_{2(\mathrm{oc})} = \frac{\pm jX_M \mathbf{E}_1}{Z_{11}} \qquad (10\text{-}19)$$

Equation (4) is based upon the assumed four-terminal network positive reference directions for the terminal currents and voltages shown in Fig. 10-8. These are the reference directions we have consistently been using with four-terminal networks in this and the preceding text. However, quite frequently the problem statement already includes current and/or voltage reference directions which are not in the standard form for a four-terminal network. In that case the sign of $\mathbf{E}_{2(\mathrm{oc})}$ in (4) is simply determined by noting the location of the dots and recognizing that the voltages at the dots follow in phase.

Now work through problem PS 10-7.

The Thévenin's equivalent impedance Z_{th} seen looking into the 2-2 terminals can be readily derived. However, it would seem intuitively apparent that, looking into the 2-2 terminals, we should see $z_{22} = R_2 + jX_{L2}$ in series with a reflected impedance term from the primary given by $Z_{\mathrm{ref}} = \dfrac{X_M{}^2}{Z_{11}}$. The procedure is identical with that used for expressing $Z_{\mathrm{in}(1\text{-}1)}$ except that the 2-2 terminals have now become the input terminals. Thus

5) $$Z_{\mathrm{th}} = z_{22} + \frac{X_M{}^2}{Z_{11}} = R_2 + jX_{L2} + \frac{X_M{}^2}{Z_{11}} \qquad (10\text{-}20)$$

and looking into the 2-2 terminals we have the Thévenin's equivalent circuit of Fig. 10-11.

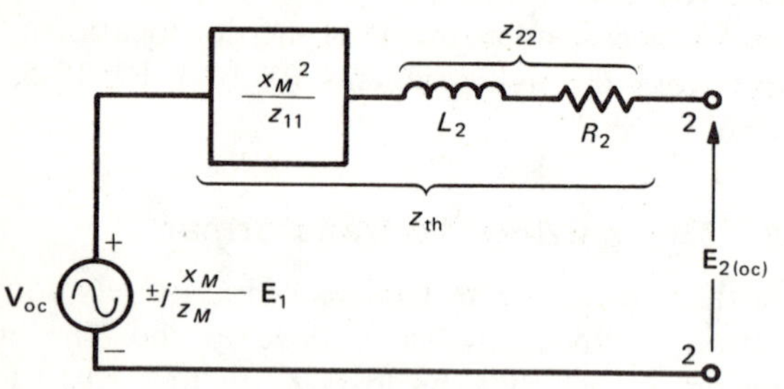

FIGURE 10-11

Now work through problem PS 10-8 and PS 10-9.

10-7 Inductive-conductive coupling

In circuits of the type shown in Fig. 10-12*a*, where the coupling is both conductive and inductive, the exact sign of M must be known if serious errors are to be avoided in computations. We shall show that the equivalent inductance of the two coupled coils in Fig. 10-12*a* is *not* $L_1 + L_2$, and that L_{eq} depends on the sign of M. This means, in problems of this nature, that the location of the dots must be known. The best procedure to use in handling circuits of this type is to draw in the equivalent self-induced and externally induced generators in place of each coil. Application of the dot rule is valid, but it can cause confusion in this type of circuit when there are multiple loops with many inductances between loops.

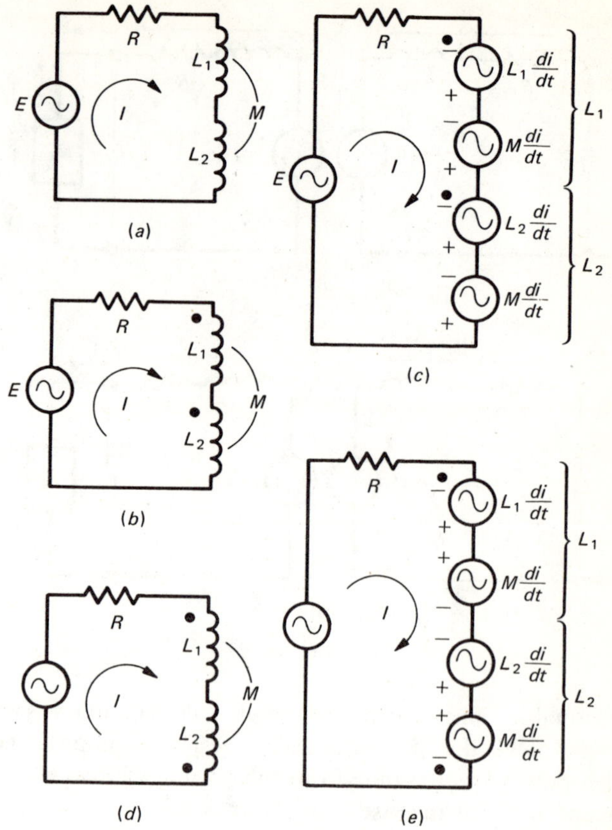

FIGURE 10-12

Suppose the dots were placed as in Fig. 10-12*b*. Following the instructions previously outlined, we have Fig. 10-12*c* as the equivalent of Fig. 10-12*b*. The voltage drop across the $M\,di/dt$ generator is of the same polarity as $L_1\,di/dt$. This is the same as if L_1 had been increased to $L_1 + M$. Similarly, L_2 appears as if it had been increased to $L_2 + M$. The equivalent circuit inductance is then

$$L_{eq} = L_1 + M + L_2 + M = L_1 + L_2 + 2M \qquad (10\text{-}21a)$$

If one of the dots is reversed, as in Fig. 10-12*d*, M will be negative. The equivalent circuit in this case is Fig. 10-12*e*. Notice that now the induced voltage in L_1, due to the same current in L_2, tends to buck the self-induced voltage in L_1. This is the same as if L_1 were decreased to $L_1 - M$. Similarly, L_2 appears as if it had been decreased to $L_2 - M$. The equivalent inductance in this circuit is therefore

$$L_{eq} = L_1 - M + L_2 - M = L_1 + L_2 - 2M \tag{10-21b}$$

If the coupling is tight, making k large, the difference in the equivalent inductance will depend markedly on the sign of M.

10-8 Measurement of mutual inductance

A method of determining the mutual inductance between two coupled coils is the following. Connect two terminals of the transformer of Fig. 10-13*a*, as in Fig. 10-13*b*.

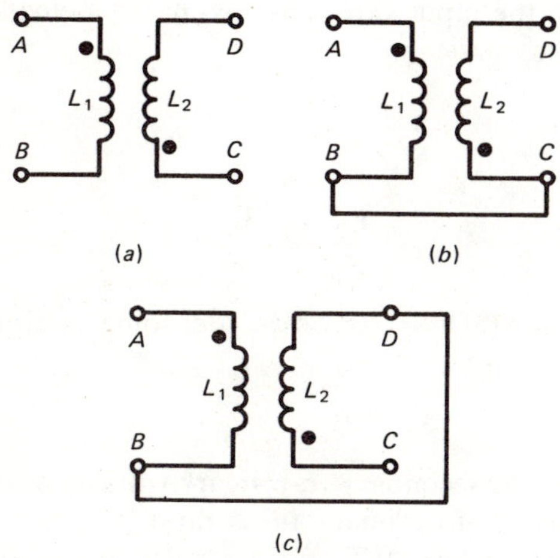

FIGURE 10-13

Measure the net inductance between terminals A and D with an impedance bridge. Call this equivalent inductance L'_{eq}. Now reverse one of the coils relative to the other, say, as in Fig. 10-13*c*, and measure the net inductance between terminals A and C. Call this equivalent inductance L''_{eq}. Now either L'_{eq} is greater than L''_{eq} or vice versa. Let us assume L'_{eq} is greater. As an exercise, the reader can then derive the relationships:

1) $$L'_{eq} = L_1 + L_2 + 2M$$

2) $$L''_{eq} = L_1 + L_2 - 2M$$

Solving for M in terms of L'_{eq} and L''_{eq}, we find, from Eq. (2),

$$L_1 + L_2 = L''_{eq} + 2M$$

Substituting $L''_{eq} + 2M$ for $L_1 + L_2$ in Eq. (1), we have

$$L'_{eq} = (L''_{eq} + 2M) + 2M \qquad \text{or} \qquad M = \frac{L'_{eq} - L''_{eq}}{4}$$

The self-inductance of each winding may be measured with the impedance bridge by leaving the other winding open. This assumes that the bridge frequency is low enough to make the effect of winding and interwinding capacitances negligible.

In circuits where both inductive and conductive coupling exist, the location of the dots must be known.

Now work through problem PS 10-10.

10-9 The ideal transformer

Up to this time, our discussion of the transformer has been perfectly general. We have made no restrictions concerning the tightness of coupling, efficiency of power transfer, or magnitudes of the source and load impedances. Neither have we overly concerned ourselves with the transformer turns ratio. In basic electronics the properties of a transformer are usually analyzed in terms of the turns ratio. For example, the secondary-to-primary voltage ratio is equal to the secondary-to-primary turns ratio. Too often the student loses sight of the fact that turns-ratio analysis is based upon an *ideal* transformer. As we shall soon see, any attempt to predict the voltage ratio in terms of the turns ratio could result in serious errors when the transformer is not ideal. However, many of the iron-core type of transformers used in power and audio work approximate ideal transformers and hence a discussion of this subject is in order.

First of all, what do we mean by an ideal transformer? An ideal transformer has 100 percent coupling. In other words, $k = 1$, and there is negligible leakage inductance and leakage flux. An ideal transformer has zero losses. This means negligible dc winding resistance. It also means negligible ac winding resistance and core losses at the frequency of interest. If the ideal transformer has no losses, it follows that the power P_1 put into the input side will all be dissipated in the load Z_L.

The ideal transformer also has infinite inductance windings, which implies that $X_{L1} \gg Z_S$, and $X_{L2} \gg Z_L$ in Fig. 10-14*a*. This might seem disconcerting, since it would seem that with an infinite primary winding reactance no current or power could be delivered to the transformer. However, this presumption ignores the fact that an infinitely small current and an infinitely large inductance may still yield a finite flux, so that transformer action can occur. Mathematically, we must exercise caution when dealing with situations where division or multiplication with zero or infinity is involved. Therefore, let us analytically investigate the input impedance (Z_{in}) seen by $\mathbf{E}_S$ and Z_S when looking into the 1-1 terminals in Fig. 10-14*a*. In general

1) $$Z_{\text{in}} = Z_{11} + \frac{X_M{}^2}{Z_{22}}$$
$$= R_1 + jX_{L1} + \frac{X_M{}^2}{Z_L + R_2 + jX_{L2}}$$

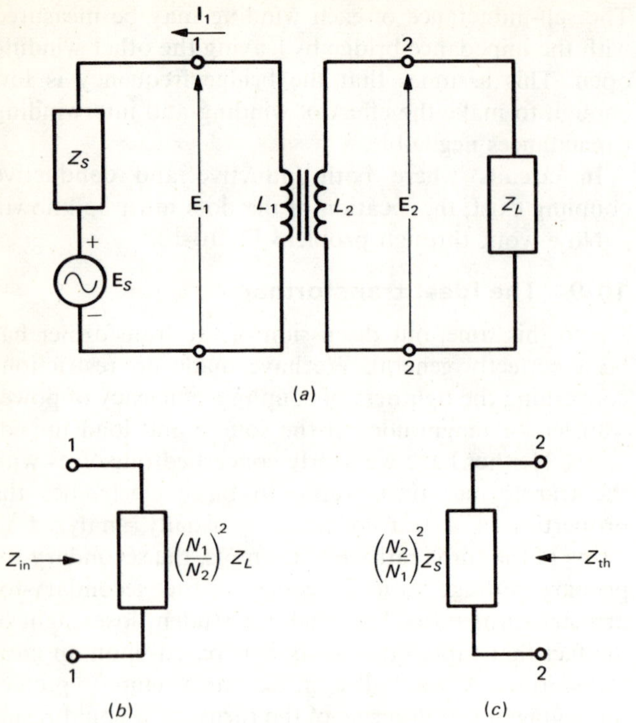

FIGURE 10-14

With negligible winding resistance, Eq. (1) reduces to

$$2) \qquad Z_{in} = jX_{L1} + \frac{X_M^2}{Z_L + jX_{L2}}$$

Rationalizing the second term in (2) yields

$$3) \qquad Z_{in} = jX_{L1} + \frac{X_M^2(Z_L - jX_{L2})}{Z_L^2 + X_{L2}^2}$$

Manipulating (3) so that we have a common denominator and simplifying yields

$$4) \quad Z_{in} = \frac{X_M^2 Z_L + j[X_{L1}(Z_L^2 + X_{L2}^2) - X_{L2}X_M^2]}{Z_L^2 + X_{L2}^2}$$

Now let us make the approximation $X_{L2} \gg Z_{L2}$ so that

$$5) \quad Z_{in} = \frac{X_M^2 Z_L + j(X_{L1}X_{L2}^2 - X_{L2}X_M^2)}{X_{L2}^2}$$

$$= \frac{X_M^2 Z_L + j(X_{L1}X_{L2} - X_M^2)}{X_{L2}^2}$$

Since

$$6) \qquad X_M = \omega k\sqrt{L_1 L_2}$$

and $k = 1$ in the ideal transformer, we have

$$7) \qquad X_M^2 = \omega^2 L_1 L_2 = \omega L_1 \omega L_2 = X_{L1}X_{L2}$$

Substituting (7) into (5) and recalling that inductance is proportional to the square of the number of turns, we obtain

$$8) \quad Z_{in} \cong \frac{X_{L1}}{X_{L2}} Z_L = \frac{L_1}{L_2} Z_L = \frac{N_1^2}{N_2^2} Z_L = \left(\frac{N_1}{N_2}\right)^2 Z_L \qquad (10\text{-}22)$$

where N_1 and N_2 is the number of primary and secondary winding turns respectively. Equation (8) indicates that, in the ideal transformer, the input impedance is just the load in the secondary reflected back as if it were multiplied by the square of the turns ratio N_1/N_2. In a similar manner it can be shown that looking back into the secondary we would see Z_S of Fig. 10-14*a* transformed to $\left(\frac{N_2}{N_1}\right)^2 Z_S$. These conclusions are summarized in Fig. 10-14*b* and *c*.

Since in an ideal (lossless) transformer, the power P_1 puts into the input side is all dissipated as load power P_L, we may write

$$9) \qquad P_1 = P_L$$

or

$$10) \qquad \frac{\mathbf{E}_1^2}{Z_{in}} = \frac{\mathbf{E}_2^2}{Z_L}$$

Substituting (8) into (10) yields, after some manipulation,

$$11) \qquad \frac{\mathbf{E}_2}{\mathbf{E}_1} = \frac{N_2}{N_1}$$

Note that the secondary-to-primary voltage ratio equals the secondary-to-primary turns ratio. If power in (10) were expressed as $\mathbf{I}^2 Z$ and a similar procedure were followed, there would result

$$12) \qquad \frac{\mathbf{I}_2}{\mathbf{I}_1} = \frac{N_1}{N_2}$$

Note that the turns ratio in this case is the reciprocal of that in (11).

In summary then, for the ideal transformer

$$P_1 = P_L \qquad (10\text{-}23a)$$

$$\frac{Z_L}{Z_{in}} = \left(\frac{N_2}{N_1}\right)^2 \qquad (10\text{-}23b)$$

$$\frac{\mathbf{E}_2}{\mathbf{E}_1} = \frac{N_2}{N_1} \qquad (10\text{-}23c)$$

$$\frac{\mathbf{I}_2}{\mathbf{I}_1} = \frac{N_1}{N_2} \qquad (10\text{-}23d)$$

10-10 Impedance transformation with the ideal transformer

Inspection of Figure 10-14*a*, *b*, and *c* reveals that the ideal transformer may be used as an impedance-changing (transforming) device. For example, suppose that in

Fig. 10-14a we want Z_S to look into a Z_{in} of a specific value. By choosing a transformer which approximates an ideal transformer and with the proper turns ratio, we can scale Z_L to the required Z_{in}. Conversely, if Z_S is fixed and we want the load to look into a specific Z_{th}, we again select the correct turns ratio according to Fig. 10-14c.

Frequently we encounter circuits in which an ideal transformer drives muliple loads, as in Fig. 10-15a. How do we analyze a circuit of this nature? Intuitively, we might sense that impedance reflected by Z_3 into the primary would appear in parallel with the reflected impedance of Z_2. The reason we sense this is that the more loads you hang on the transformer, the greater the primary current becomes, which implies that the total reflected impedance is decreasing. Intuition, however, can at times lead us astray. Therefore, let us develop a rigorous analysis of the circuit.

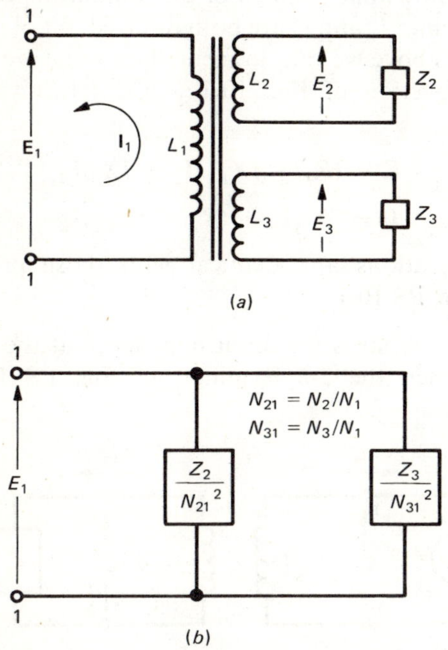

FIGURE 10-15

The power (volt-ampere product) put into the transformer primary is $E_{12}/Z_{1\text{-}1}$. Since the transformer is ideal, the primary power must equal the sum of the load powers (volt-ampere products). We can then write

$$\frac{E_1^2}{Z_{1\text{-}1}} = \frac{E_2^2}{Z_2} + \frac{E_3^2}{Z_3}$$

We can relate E_2 and E_3 to E_1 in terms of the turns ratios. Let $N_{21} = N_2/N_1$. and $N_{31} = N_3/N_1$; then

$$\frac{E_1^2}{Z_{1\text{-}1}} = \frac{(E_1 N_2/N_1)^2}{Z_2} + \frac{(E_1 N_3/N_1)^2}{Z_3}$$

$$\frac{E_1^2}{Z_{1\text{-}1}} = \frac{E_1^2 N_{21}^2}{Z_2} + \frac{E_1^2 N_{31}^2}{Z_3}$$

$$\frac{1}{Z_{1\text{-}1}} = \frac{1}{Z_2/N_{21}^2} + \frac{1}{Z_3/N_{31}^2} \tag{10-23}$$

Equation (10-23) is in the same form as that for two resistors in parallel:

$$\frac{1}{R_{\text{eq}}} = \frac{1}{R_1} + \frac{1}{R_2}$$

where R_1 is analogous to Z_2/N_{21}^2 and R_2 to Z_3/N_{31}^2. The equivalent circuit for Fig. 10-15a is, then, that of Fig. 10-15b.

Now work through problems PS 10-11 and PS 10-12.

PROBLEMS WITH SOLUTIONS

PS 10-1 Write loop equations for the circuit of Fig. PS 10-1a.

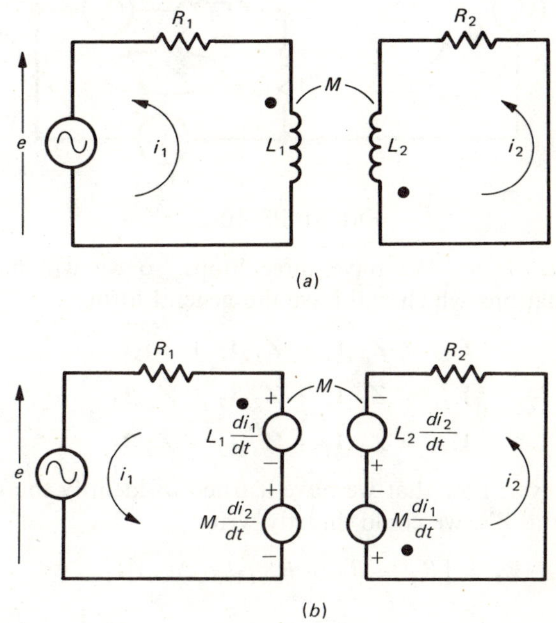

FIGURE PS 10-1

SOLUTION Each coil is replaced with a self-induced and an induced generator, as shown in Fig. PS 10-1b. Since the positive current directions are already specified in Fig. PS 10-1a, the polarity of the self-induced generators will be as shown in Fig. PS 10-1b. The self-induced generator of loop 1 drives the dot end positive as i_1 builds up. With the dot end of L_1 going positive because of i_1, the dot end of the $M\, di_1/dt$ induced generator in loop 2 is going positive. Similarly i_2 is developing a voltage across L_2, which drives the dot end positive in loop 2. The voltage induced in loop 1 by

i_2 is then driving the dot end of the $M\, di_2/dt$ generator in loop 1 positive. The loop equations are then

$$1) \qquad e = R_1 i_1 + L_1 \frac{di_1}{dt} + M \frac{di_2}{dt}$$

$$2) \qquad 0 = M \frac{di_1}{dt} + R_2 i_2 + L_2 \frac{di_2}{dt}$$

If the driving source is sinusoidal, we may in terms of rms values write

$$\mathbf{E} = (R_1 + jX_{L1})\mathbf{I}_1 + jX_M \mathbf{I}_2$$

$$0 = jX_M \mathbf{I}_1 + (R_2 + jX_{L1})\mathbf{I}_2$$

PS 10-2 Using the method of Eq. (10-6), write the loop equations for the circuit of Fig. PS 10-2.

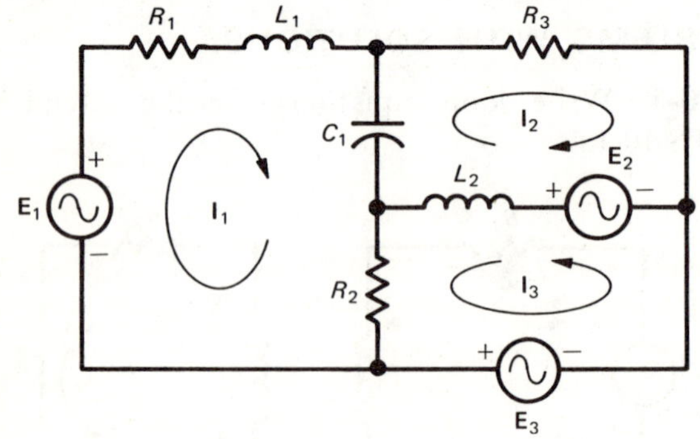

FIGURE PS 10-2

SOLUTION We have three loops, so we will have 3 equations which will have the general form

$$1a) \qquad \mathbf{E}_{11} = Z_{11}\mathbf{I}_1 + Z_{12}\mathbf{I}_2 + Z_{33}\mathbf{I}_3$$

$$2a) \qquad \mathbf{E}_{22} = Z_{21}\mathbf{I}_1 + Z_{22}\mathbf{I}_2 + Z_{23}\mathbf{I}_3$$

$$3a) \qquad \mathbf{E}_{33} = Z_{31}\mathbf{I}_1 + Z_{32}\mathbf{I}_2 + Z_{33}\mathbf{I}_3$$

However, now that we have learned to identify the **E**'s and the Z's, we could directly write

$$1b) \qquad -\mathbf{E}_1 = [R_1 + R_2 + j(X_{L1} - X_{C1})]\mathbf{I}_1 - jX_{C1}\mathbf{I}_2 + R_2\mathbf{I}_3$$

$$2b) \qquad -\mathbf{E}_2 = -jX_{C1}\mathbf{I}_1 + [R_3 + j(X_{L2} - X_{C1})]\mathbf{I}_2 + jX_{L2}\mathbf{I}_3$$

$$3b) \qquad \mathbf{E}_3 - \mathbf{E}_2 = R_2\mathbf{I}_1 + jX_{L2}\mathbf{I}_2 + (R_2 + jX_{L2})\mathbf{I}_3$$

PS 10-3 Rework problem PS 10-1 using the shortcuts we have developed for writing loop equations. Assume a sinusoidal voltage source.

SOLUTION By inspection of Fig. PS 10-1*a* we observe the following: There are two loops, so we will have two loop equations of the general form

$$\mathbf{E}_{11} = \mathbf{Z}_{11}\mathbf{I}_1 + \mathbf{Z}_{12}\mathbf{I}_2$$

$$\mathbf{E}_{22} = \mathbf{Z}_{21}\mathbf{I}_1 + \mathbf{Z}_{22}\mathbf{I}_2$$

The voltage $\mathbf{E}_{11} = +e$ since e looks like an aiding source to i_1. The voltage $\mathbf{E}_{22} = 0$ since there are no independent sources in loop 2. The self-impedance about loop 1 is $Z_{11} = R_1 + jX_{L1}$. The self-impedance about loop 2 is $Z_{22} = R_2 + jX_{L2}$. The loops are magnetically coupled due to a mutual inductance M. Hence, there will be a mutual impedance (reactance) between the loops given by $Z_{12} = Z_{21} = +jX_M$. We know that $Z_{12} = Z_{21}$ in passive networks with bilateral elements, and we know that the sign of the mutual impedance is positive, since i_1 and i_2 are both directed out of the dots.

All the above we can do mentally, so that we may by mere inspection of Fig. 10-1*a* write (effective values assumed)

$$\mathbf{E} = (R_1 + jX_{L1})\mathbf{I}_1 + jX_M\mathbf{I}_2$$

$$0 = jX_M\mathbf{I}_1 + (R_2 + jX_{L2})\mathbf{I}_2$$

These equations are identical with those previously derived in PS 10-1.

PS 10-4 What is the input impedance at the 1-1 terminals when the 2-2 terminals of Fig. PS 10-4 are shorted?

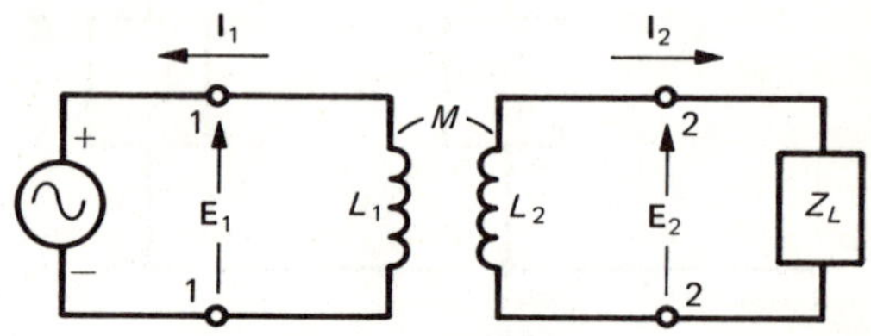

FIGURE PS 10-4

SOLUTION The general expression for Z_{in} as given by Eq. (10-17*b*) is

$$1) \qquad Z_{\text{in(1-1)}} = Z_{11} + \frac{X_M{}^2}{Z_{22}}$$

Equation (1) for Fig. PS 10-4 may be rewritten as

$$2) \qquad Z_{\text{in(1-1)}} = z_{11} + \frac{X_M{}^2}{z_{22} + Z_L}$$

With the 2-2 terminals shorted, $Z_L = 0$. Hence

$$3) \quad Z_{\text{in(1-1)}} = z_{11} + \frac{X_M{}^2}{z_{22}} = R_1 + jX_{L1} + \frac{X_M{}^2}{R_2 + jX_{L2}}$$

If $X_{L1} \gg R_1$ and $X_{L2} \gg R_2$, as is often the case,

$$4) \quad Z_{\text{in(1-1)}} = jX_{L1} + \frac{X_M{}^2}{jX_{L2}} = j\omega L_1 + \frac{\omega^2 M^2}{j\omega L_2}$$

$$= j\omega L_1 - j\omega \frac{M^2}{L_2} = j\omega L_1 - j\omega \frac{(k\sqrt{L_1 L_2})^2}{L_2}$$

$$= j\omega(L_1 - k^2 L_1)$$

or

5) $Z_{in(1\text{-}1)} = j\omega L_1(1 - k^2)$

Equation (5) indicates that shorting the secondary decreases the impedance seen looking into the primary. In the case of tightly coupled transformers such as those with ferromagnetic cores, $k \cong 1$ and hence the input impedance becomes very, very low. This is why shorting the secondary of a power transformer can blow fuses in the primary circuit.

PS 10-5 Assume Z_L in Fig. PS 10-4 is a pure capacitance C. What effect will this have upon the input impedance? What if Z_L were a pure inductance?

SOLUTION We can begin the analysis with (2) of the previous problem. Thus

1) $$Z_{in(1\text{-}1)} = R_1 + jX_{L1} + \frac{X_M{}^2}{R_2 + jX_{L2} - jX_C}$$

The effect of C is to introduce an X_C which tends to cancel X_{L2} in the denominator of the second (reflected impedance) term. Thus the reflected impedance increases until it reaches a maximum when X_C resonates with X_{L2}. If $X_C > X_{L2}$, Z_{22} is capacitive in nature so that the denominator phase angle is $-90° < \phi < 0°$. Since the negative denominator angle becomes positive in the numerator, the effect of a capacitive load is to try to make the reflected impedance look inductive. Conversely, an inductive load tends to make the reflected impedance look capacitive.

PS 10-6. A transformer has the following characteristics:

$L_1 = 3$ henries $\quad L_2 = 0.05$ henry $\quad k \approx 1$

$R_1 = 20$ ohms $\quad R_2 = 0.08$ ohm $\quad E = 115$ volts, 60 Hz

The generator source impedance is negligible, and the secondary load resistance (R_L) is 42 ohms. What is the primary current in Fig. PS 10-6*a*.

SOLUTION $\omega = 2\pi f = 377$ rad/sec

$$M = k\sqrt{L_1 L_2} \cong \sqrt{0.15} = 0.387 \text{ H}$$

$$Z_{12} = j\omega M = j146\ \Omega$$

$$Z_{11} = R_1 + j\omega L_1 = 20 + (j377\ \Omega)(3\text{ H})$$
$$= 20 + j1{,}131\ \Omega$$

$$Z_{22} = 42.08 + (j377\ \Omega)(0.05\text{ H})$$
$$= 42.08 + j18.85 = 46.1\angle 24.1°\ \Omega$$

$$Z_{ref} = \frac{\omega^2 M^2}{Z_{22}} = \frac{(3.77)^2 \times 10^4 \times (0.387)^2}{46.1\angle 24.1°}$$
$$= 462\angle -24.1° = 422 - j189\ \Omega$$

Note that the inductive impedance of the secondary loop reflects into the primary as a capacitive impedance having a reactive component of $-j189$ ohms. The reflected impedance appears in series with the primary-loop self-impedance. The equivalent circuit that the generator sees is shown in Fig. PS 10-6*b*. The input impedance is

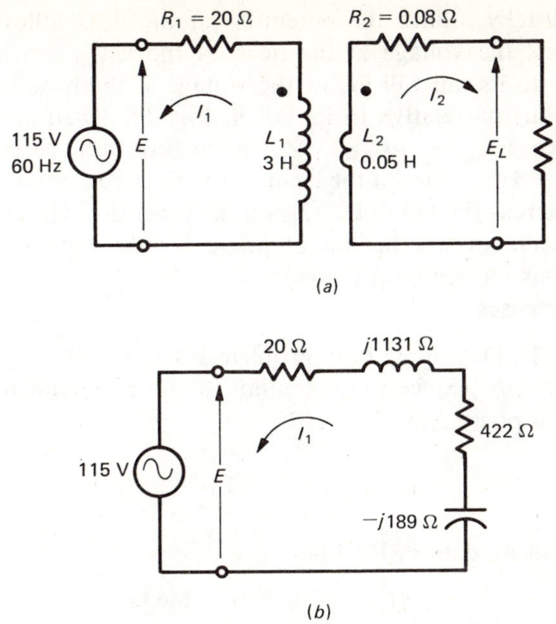

FIGURE PS 10-6

$$Z_{in} = Z_{11} + Z_{ref} = 20 + j1{,}131 + 422 - j192$$
$$= 441 + j942 = 1{,}040\angle 64.9°\ \Omega$$

and the rms input current is

$$\mathbf{I}_1 = \frac{115\angle 0°\text{ V}}{1.04\angle 64.9°\text{ k}\Omega} = 111\text{ mA}\angle -64.9°$$

PS 10-7 The open-circuit secondary voltage in a transformer as given by Eq. (10-19) is $\mathbf{E}_{2(oc)} = \pm jX_M\mathbf{E}_1/Z_{11}$. Determine if the plus or minus sign is to be used in each of the four cases in Fig. PS 10-7.

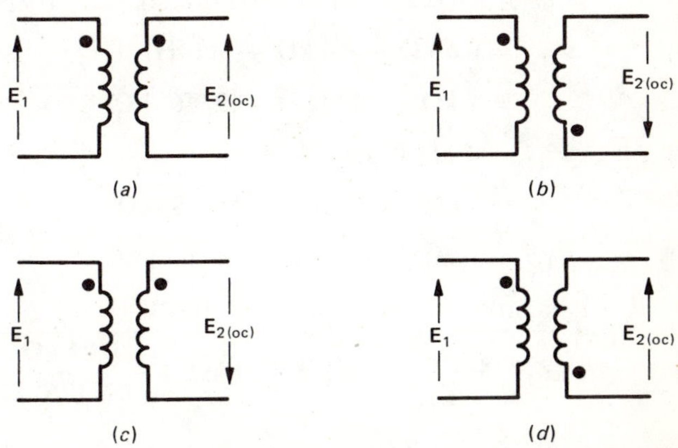

FIGURE PS 10-7

SOLUTION Since the potentials at the dots follow in phase, the voltage at the head of the $\mathbf{E}_{2(oc)}$ arrow relative to its tail will follow the voltage at the head of the $\mathbf{E}_1$ arrow relative to its tail in Fig. PS 10-7*a* and *b*. Hence, $\mathbf{E}_{2(oc)} = +jX_M\mathbf{E}_1/Z_{11}$ in each of these cases. In Fig. PS 10-7*c* and *d* the head of the $\mathbf{E}_1$ arrow is on a dot, whereas the tail of $\mathbf{E}_{2(oc)}$ is on the other dot. Hence, these voltages will be out of phase for the reference directions chosen, and therefore $\mathbf{E}_{2(oc)} = -jX_M\mathbf{E}_1/Z_{11}$ in these cases.

PS 10-8 Determine $\mathbf{I}_2$ in problem PS 10-6.
SOLUTION Since the potentials at the dots tend to follow in phase,

1) $$\mathbf{E}_{L(oc)} = +\frac{jX_M}{Z_{11}}\mathbf{E}_1$$

where (using data in PS 10-6)

$$X_M = \omega M = 377(0.387) = 146\ \Omega$$

$$Z_{11} = 20 + j1131\ \Omega = 1131\underline{/89^\circ}\ \Omega$$

$$\mathbf{E} = \mathbf{E}_1 = 115\underline{/0^\circ}\text{ V at 60 Hz}$$

Therefore plugging these values into (1) yields

2) $$\mathbf{E}_{L(oc)} = 14.8\underline{/1^\circ}\text{ V}$$

From Fig. 10-11 we see that

3) $$Z_{th} = R_2 + jX_{L2} + \frac{X_M{}^2}{Z_{11}} = 0.08 + j18.85 + \frac{(146)^2}{1131\underline{/89^\circ}}$$

$$= 0.409 + j0.01\ \Omega = 0.409\underline{/1.4^\circ}\ \Omega$$

Therefore $\mathbf{I}_2$ is given by

4) $$\mathbf{I}_2 = \frac{\mathbf{E}_{2(oc)}}{Z_{th} + R_L} = \frac{14.8\underline{/1^\circ}\text{ V}}{42.4 + j0.01} \cong 0.349\underline{/1^\circ}\text{ A}$$

PS 10-9 Determine I_1 and I_2 in Fig. PS 10-9*a*.
SOLUTION

1) $$M = 0.1\sqrt{(1\text{H})(4\text{H})} = 0.2\text{H}$$

2) $$\omega = 2\pi(159.2\text{ Hz}) = 1000\text{ rad/sec}$$

3) $$Z_{22} = 0.6\text{ k}\Omega + 0.4\text{ k}\Omega + j\omega(4\text{H})$$

$$= 1\text{ k}\Omega + j4\text{ k}\Omega = 4.12\underline{/76^\circ}\text{ k}\Omega$$

4) $$Z_{11} = 1\text{ k}\Omega + j\omega(1\text{H})$$

$$= 1\text{ k}\Omega + j1\text{ k}\Omega = 1.41\underline{/45^\circ}\text{ k}\Omega$$

5) $$X_M{}^2 = (\omega M)^2 = (0.2\text{ k}\Omega)^2 = .04\text{ k}\Omega^2$$

6) $$\mathbf{I}_1 = \frac{100\text{ V}}{Z_{11} + \dfrac{X_M{}^2}{Z_{22}}} = \frac{100\text{ V}}{1\text{ k}\Omega + j1\text{ k}\Omega + \dfrac{0.04\text{ k}\Omega^2}{4.12\underline{/76^\circ}\text{ k}\Omega}}$$

$$= 71\underline{/-44.7^\circ}\text{ mA}$$

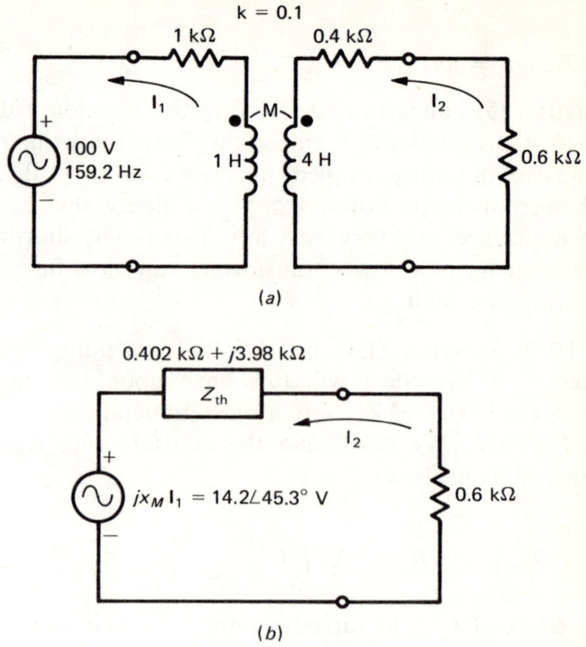

FIGURE PS 10-9

Thevenizing looking into the secondary winding we see the equivalent circuit of Fig. PS 10-9*b* where

7) $$\mathbf{E}_{2(oc)} = +jX_M\mathbf{I}_1$$

$$= j(0.2\text{ k}\Omega)(71\underline{/-44.7^\circ}\text{ mA}) = 14.2\underline{/45.3^\circ}\text{ V}$$

8) $$Z_{th} = z_{22} + \frac{X_M{}^2}{Z_{11}}$$

$$= 0.4\text{ k}\Omega + j\omega(4\text{H}) + \frac{0.04\text{ k}\Omega^2}{1.41\underline{/45^\circ}\text{ k}\Omega}$$

$$= 0.420\text{ k}\Omega + j3.98\text{ k}\Omega$$

9) $$\mathbf{I}_2 = \frac{\mathbf{E}_{2(oc)}}{Z_{th} + 0.6\text{ k}\Omega} = \frac{14.2\underline{/45.3^\circ}\text{ V}}{4.11\underline{/75.6^\circ}\text{ k}\Omega}$$

$$= 3.45\underline{/-30.3^\circ}\text{ mA}$$

Thus the current in the secondary winding lags the applied (input) voltage by 30.3°.

PS 10-10 Find the circuit current in Fig. PS 10-10*a*, if $R_1 = 1$ kilohm, $R_2 = 1$ kilohm, $L_1 = 1$ henry, $L_2 = 2$ henries, $M = 0.5$ henry, $E = 100$ volts/100 Hz. What is the coefficient of coupling between the two coils?
SOLUTION Replace L_1 with two generators—one for self-induced voltage due to i_1 in L_1 and another for the voltage induced in L_1 by the same current i_1 in coil L_2. Similarly, L_2 is replaced with two generators. By observing the manner in which the self-inductances are polarized, we can determine the polarity of the induced generators via the dots. This leads to Fig. PS 10-10*b*.

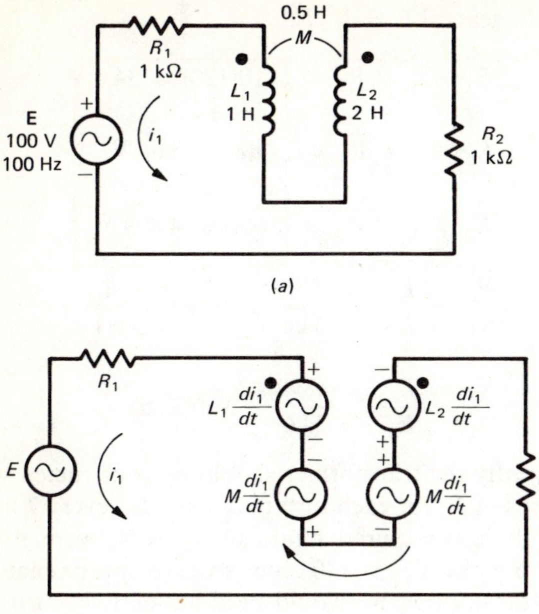

FIGURE PS 10-10

The loop equation is then

$$E = i_1 R_2 + L_2 \frac{di_1}{dt} - M \frac{di_1}{dt} - M \frac{di_1}{dt}$$

$$+ L_1 \frac{di_1}{dt} + i_1 R_1$$

$$= i_1(R_1 + R_2) + (L_1 + L_2 - 2M) \frac{di_1}{dt}$$

Applying the dot rule, we see that the current goes into one dot and out the other. Therefore, the self and mutual terms will have opposite signs, and M will be negative.

Since **E** is a sinusoidal generator, we can write

$$\mathbf{E} = (R_1 + R_2)\mathbf{I}_1 + j\omega(L_1 + L_2 - 2M)\mathbf{I}_1$$

$$= [R_1 + R_2 + j\omega(L_1 + L_2 - 2M)]\mathbf{I}_1 = Z\mathbf{I}_1$$

where E and I_1 are effective values.

$$Z = 2\ \Omega + j2\pi 100(2\text{ H})$$

$$= 2.36\underline{/32.1^\circ}\text{ k}\Omega$$

$$\mathbf{I} = \frac{\mathbf{E}\underline{/0^\circ}}{Z\underline{/\phi^\circ}} = \frac{100\text{ V}}{2.36\underline{/32.1^\circ}\text{ k}\Omega} = 42.3\text{ mA}\underline{/-32.1^\circ}$$

$$k = \frac{M}{\sqrt{L_1 L_2}} = \frac{0.5}{\sqrt{1 \times 2}} = \frac{0.5}{1.41} = 0.354\text{, or } 35.4\%$$

PS 10-11 The turns ratio of the transformer in Fig. PS 10-11 has been chosen for maximum power transfer. Find the turns ratio N, Z_{in}, I_1, E_1, I_2, E_2, and the power developed in the loudspeaker (P_L).

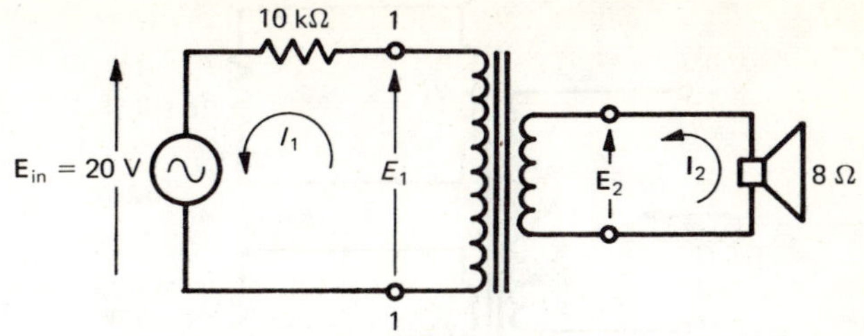

FIGURE PS 10-11

SOLUTION To find N, we will make use of the fact that the turns ratio has been chosen for maximum power transfer. This means that the reflected impedance of the load as it appears between the 1-1 terminals must match (equal) the 10-kilohm impedance. Therefore,

$$\frac{Z_L}{N^2} = 10\text{ k}\Omega$$

or

$$N^2 = \frac{8}{10{,}000}$$

Therefore

$$N = \frac{N_2}{N_1} = \frac{\sqrt{8}}{\sqrt{10{,}000}} = 2.83 \times 10^{-2}$$

or

$$N_1 : N_2 = 35:1$$

$$Z_{in} = 10\text{ k} + \frac{Z_L}{N^2} = 10\text{ k} + 10\text{ k} = 20\text{ k}\Omega$$

$$I_1 = \frac{20\text{ V}}{20\text{ k}\Omega} = 1\text{ mA}$$

$$E_1 = I_1 \frac{Z_L}{N^2} = 1\text{ mA} \times 10\text{ k}\Omega = 10\text{ V}$$

$$I_2 = \frac{I_1}{N} = \frac{1\text{ mA}}{2.83 \times 10^{-2}} = 35.4\text{ mA}$$

$$E_2 = E_1 N = 10\text{ V} \times 2.83 \times 10^{-2} = 0.283\text{ V}$$

$$P_L = E_2 I_2 = 0.283\text{ V} \times 35.4\text{ mA} = 10\text{ mW} = E_1 I_1{}^*$$

$$= 10\text{ V} \times 1\text{ mA} = 10\text{ mW}$$

PS 10-12 An audio frequency ouput transformer must deliver 4 watts to an 8-ohm load and 6 watts to a 4-ohm load as shown in Fig. PS 10-12. The impedance seen looking into the primary is to be 120 ohms. Determine

*Actually $P_L = E_2 I_2 \cos \phi_2$, but since the load is assumed purely resistive in the frequency range of interest the power factor is 0° and hence $\cos \phi_2 = 1$.

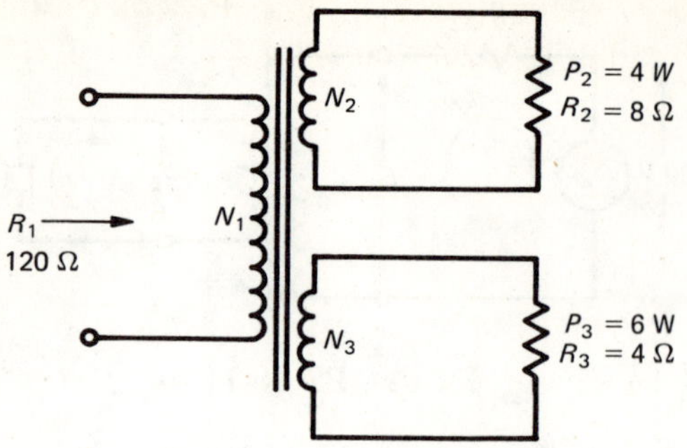

FIGURE PS 10-12

the required turns ratios and voltage that must be developed across the primary (N_1) winding. Assume the transformer is ideal. No dots are shown as we have no interest in phase relationships.

SOLUTION

$$P_1 = P_2 + P_3$$

$$P_1 = 4\text{ W} + 6\text{ W} = 10\text{ W}$$

Since in general $P = E^2/R$,

$$E_1 = \sqrt{P_1 R_1} = \sqrt{10(120)} = 34.6\text{ V}$$

$$E_2 = \sqrt{P_2 R_2} = \sqrt{4(8)} = 5.66\text{ V}$$

$$E_3 = \sqrt{P_3 R_3} = \sqrt{6(4)} = 4.899\text{ V}$$

$$\frac{N_2}{N_1} = \frac{E_2}{E_1} = \frac{5.66}{34.6} = 0.164 \cong \frac{1}{6.1}$$

$$\frac{N_3}{N_1} = \frac{E_3}{E_1} = \frac{4.899}{34.6} = 0.142 \cong \frac{1}{7}$$

Apparently the transformer should be constructed with 6.1 turns of N_1 for each turn of N_2, and for every 7 turns of N_1 there is required 1 turn of N_3. If N_1 were 10,000 turns to make $X_{L1} \gg 120$ ohms so as to approximate an ideal transformer, we would need about 1640 turns on N_2 and 1420 turns on N_3.

PROBLEMS WITH ANSWERS

PA 10-1 An air-core transformer has a primary winding inductance L_1 of 12 mH, secondary winding inductance L_2 of 60 mH, primary winding $Q_1 = 48$ from 500 kHz to 800 kHz, secondary winding $Q_2 = 32$ from 500 kHz to 800 kHz, and a coefficient of coupling $k = 0.47$. What is the mutual inductance between the windings in this part of the frequency spectrum?

ANSWER $M = 12.6$ mH

PA 10-2 Write the loop equations for the circuit of Fig. PA 10-2.

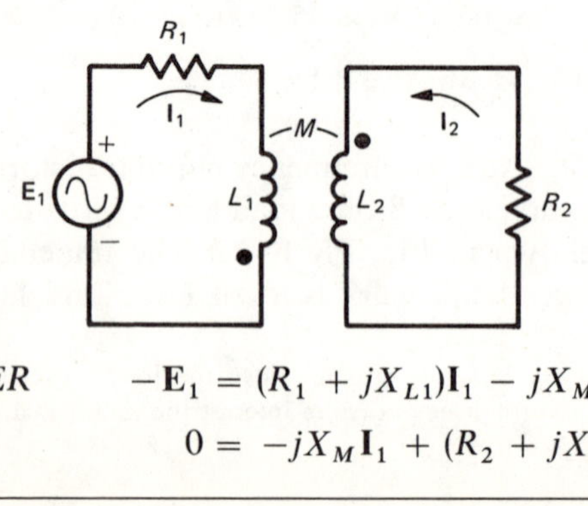

ANSWER

$$-\mathbf{E}_1 = (R_1 + jX_{L1})\mathbf{I}_1 - jX_M\mathbf{I}_2$$

$$0 = -jX_M\mathbf{I}_1 + (R_2 + jX_{L2})\mathbf{I}_2$$

PA 10-3 Determine $Z_{11}, Z_{12}, \ldots, Z_{21}, Z_{22}, \ldots,$ in the circuit of Fig. PA 10-3 and write the loop equations.

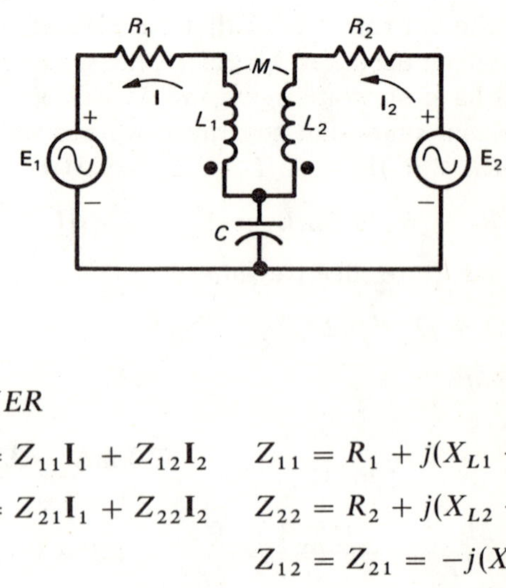

ANSWER

$$\mathbf{E}_1 = Z_{11}\mathbf{I}_1 + Z_{12}\mathbf{I}_2 \qquad Z_{11} = R_1 + j(X_{L1} - X_C)$$

$$-\mathbf{E}_2 = Z_{21}\mathbf{I}_1 + Z_{22}\mathbf{I}_2 \qquad Z_{22} = R_2 + j(X_{L2} - X_C)$$

$$Z_{12} = Z_{21} = -j(X_M + X_C)$$

PA 10-4 What are the z parameters of the tapped transformer (autotransformer) shown in Fig. PA 10-4?

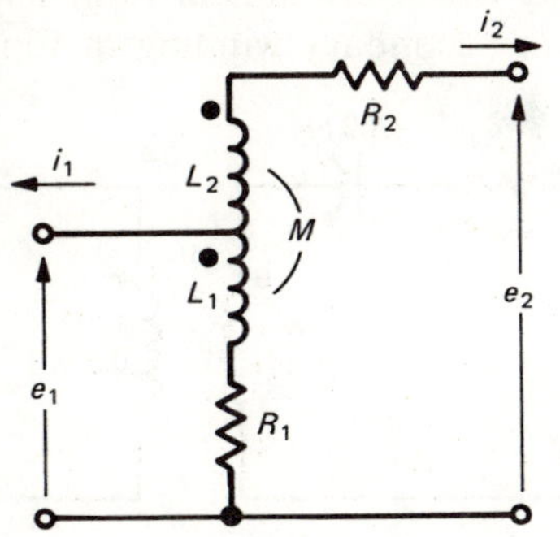

ANSWER

$$z_{11} = R_1 + jX_{L1},$$

$$z_{12} = z_{21} = R_1 + jX_M,$$

$$z_{22} = R_1 + R_2 + j(X_{L1} + X_{L2} + 2X_M)$$

PA 10-5 Determine $\mathbf{I}_1$ in Fig. PA 10-5 by applying Thévenin's theorem.

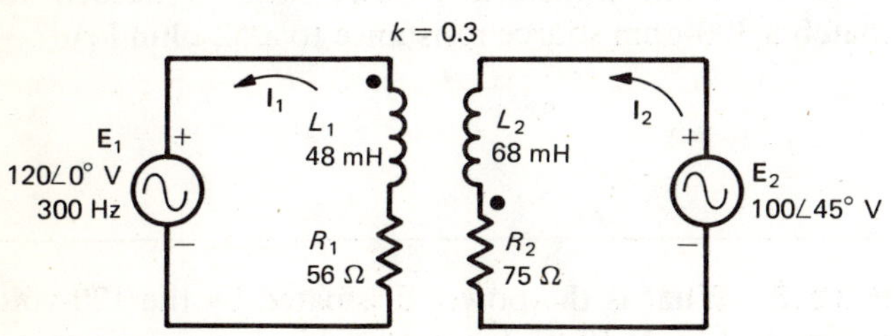

ANSWER $\mathbf{I}_1 = 1.37\underline{/-57.04°}$ A

PA 10-6 Determine $\mathbf{I}_2$ in Fig. PA 10-5 by applying Thévenin's theorem,

ANSWER $\mathbf{I}_2 = 0.941\underline{/-14.8°}$ A

PA 10-7 Determine **I** in Fig. PA 10-7. Hint: Use branch rather than loop currents.

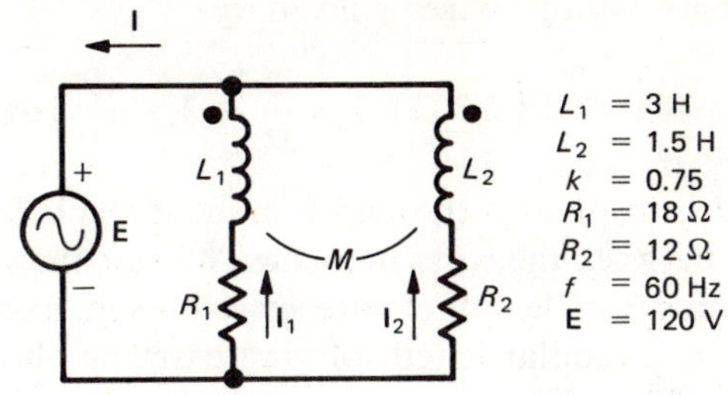

ANSWER $\mathbf{I} = 0.213\underline{/-88.6°}$ A

PA 10-8 Determine the required turns ratio in Fig. PA 10-8 so that the impedance looking into N_1 matches source E_S of internal resistance R_S. The power in R_2 and R_3 is to be 0.5 and 0.1 watts. Assume an ideal transformer.

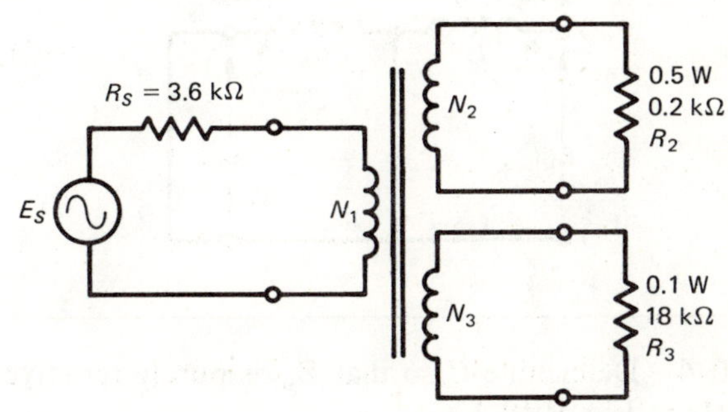

ANSWER $N_1/N_3 = 1.095$, $N_1/N_2 = 4.65$

PROBLEMS WITHOUT ANSWERS

P 10-1 The mutual inductance of two straight and equal length parallel wires is given by

$$M = 0.005081\left(2.303 \log_{10} \frac{2l}{D} - 1 + \frac{D}{l}\right) \mu\text{H}$$

where M is in μH, l is the length in inches, and D is the distance between the wires in inches. What is the voltage induced in a 3-in. length of wire spaced 0.8 in. from, and parallel to, a similar length of wire carrying 50 mA of current at 456 mHz.

P 10-2 What is the resonant frequency and Q of the circuit in Fig. P 10-2?

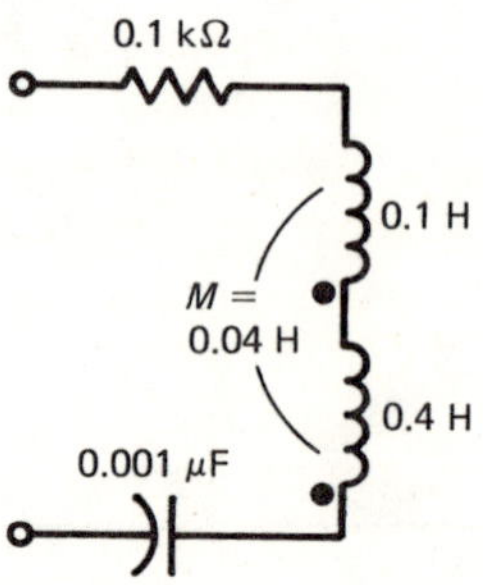

P 10-3 Derive an expression for the impedance seen looking into the 1-1 terminals. Will any $\mathbf{I}_2$ flow if an $\mathbf{E}_1$ is applied?

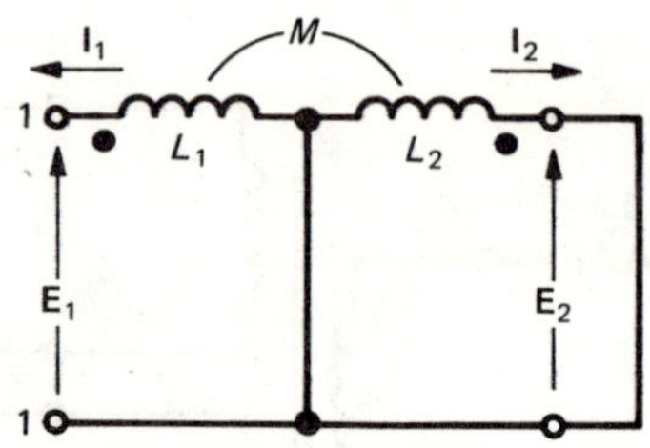

P 10-4 Determine C so that Z_{in} is purely resistive at 1 MHz in Fig. P 10-4.

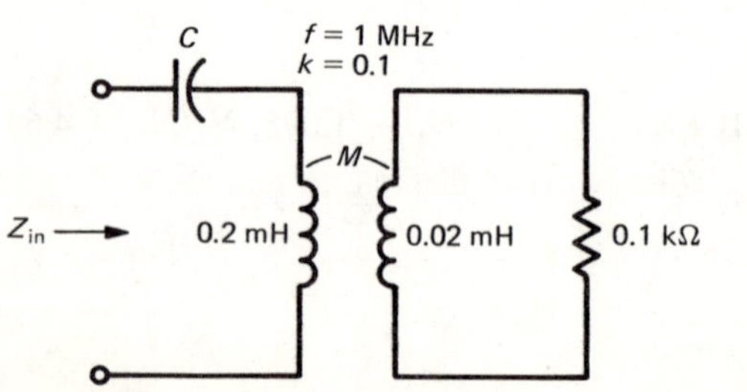

P 10-5 Plot the voltage gain $\mathbf{E}_S/\mathbf{E}_g$ versus frequency for a few points which are indicative of selectivity seen looking into the secondary winding of Fig. P 10-5.

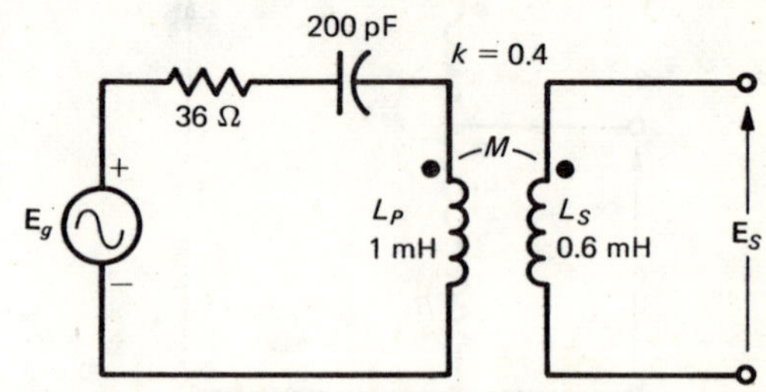

P 10-6 Determine $\mathbf{I}_2$ in Fig. P 10-6.

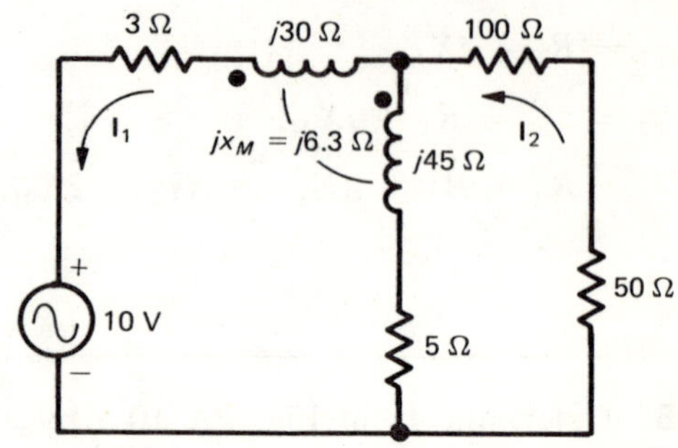

P 10-7 What transformer-turns ratio is needed to match a 300-ohm source resistance to a 52-ohm load?

P 10-8 What is the power dissipated by the 120-volt source in Fig. P 10-8? Assume an ideal transformer.

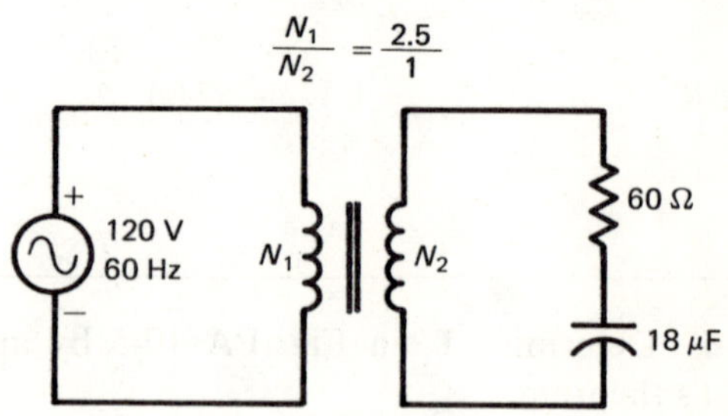

P 10-9 Estimate a minimum value of L_1 to approximate an ideal transformer in the previous problem.

11 the Bode plot

A need frequently arises to examine the frequency response (both amplitude and phase) of a circuit or amplifier. This is particularly true in more advanced courses where attention is focused on the behavior of feedback amplifiers or control systems. Hence it is advisable to introduce the subject at an elementary level so that we will be on firmer ground in the future.

11-1 The transfer function

Most any circuit or system that we might consider has some kind of input and output. Generally we are interested in output voltage versus input voltage or voltage gain as a function of frequency, but this need not be the case. We could just as well be interested in current gain, or output voltage versus input current, and so on. When dealing with control systems we might be interested in motor speed versus armature voltage, or height of liquid in a tank versus a potentiometer setting, etc. In any event, the output versus input relationship as a function of frequency is, for our purposes, called the transfer function. The transfer functions of most concern to us are exclusively electrical in nature and generally a voltage gain.

There are a few different methods of plotting transfer functions versus frequency. One of the simplest and most useful is called the Bode plot, after a famous scientist who did some outstanding research in feedback theory. It turns out that to conveniently construct a Bode plot, the transfer function is best expressed in certain algebraic forms which are readily conveyed by example.

11-2 The low-pass filter

The first circuit we will develop the transfer function for is the low-pass filter of Fig. 11-1. We seek the voltage

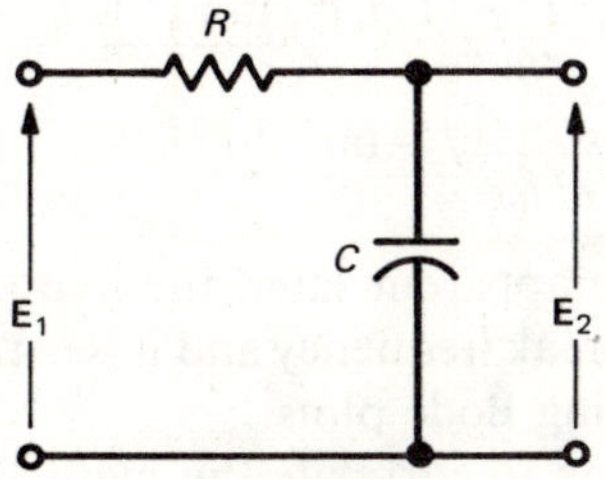

$$\frac{E_2}{E_1} = \frac{1}{1 + j\dfrac{f}{f_h}}$$

$$f_h = \frac{1}{RC}$$

FIGURE 11-1

gain $A_v = \mathbf{E}_2/\mathbf{E}_1$ as a function of frequency. Clearly we have a voltage divider here and hence

1) $$\mathbf{A}_v = \frac{\mathbf{E}_2}{\mathbf{E}_1} = \frac{-jX_C}{R - jX_C} = \frac{1/j\omega C}{R + \dfrac{1}{j\omega C}}$$

Remember, we can transfer j from numerator to denominator and vice versa simply by changing its sign, since this corresponds to the similar transfer of an angle in polar form.

Finding a common denominator and simplifying (1), we obtain

2) $$\mathbf{A}_v = \frac{1}{1 + j\omega RC} = \frac{1}{1 + j\omega\tau_h}$$

where

3) $$\tau_h = RC$$

Now it turns out that the product RC is called a time constant and it has the dimensions of seconds, since[1]

4) $$RC = \frac{\text{V}}{\text{A}} \times \frac{\text{charge}}{\text{V}} = \frac{\not{\text{V}}}{\not{\text{A}}} \times \frac{\not{\text{A}} - \text{sec}}{\not{\text{V}}} = \text{sec}$$

Now it is convenient to reason that the time constant τ_h is the period corresponding to a fictitious radian frequency ω_h so that

5) $$\omega_h = \frac{1}{\tau_h}$$

or

6) $$\tau_h = \frac{1}{\omega_h}$$

The use of the h subscript will be clarified later. Substituting (6) into (2) yields, for the low-pass filter,

7) $$\mathbf{A}_V = \frac{\mathbf{E}_2}{\mathbf{E}_1} = \frac{1}{1 + j\dfrac{\omega}{\omega_h}} = \frac{1}{1 + j\dfrac{f}{f_h}} \qquad (11\text{-}1a)$$

Notice how simple this expression is and, in particular, note its form. In polar form we should express (7) as

8) $$\mathbf{A}_V = \frac{\mathbf{E}_2}{\mathbf{E}_1} = \frac{1}{\sqrt{1 + (f/f_h)^2}\,\underline{/\tan^{-1} f/f_h}}$$

$$= \frac{1}{\sqrt{1 + (f/f_h)^2}}\,\underline{/-\tan^{-1} f/f_h} \qquad (11\text{-}1b)$$

For reasons to become apparent later, the frequency f_h is called the corner or break frequency and it is extremely important in constructing Bode plots.

It might be interesting to correlate the mathematical form of 11-1a (or 11-1b) with a qualitative analysis of the circuit. Surely at low frequencies X_C is so much larger than R that the magnitude of $E_2 \cong E_1$. Correspondingly in (11-1a) for f approaching zero, or more significantly for $f \ll f_h$, the j term approaches zero and $A_v \cong 1\underline{/0^\circ}$.

[1] In a similar manner, the time constant L/R in an inductive circuit has the dimensions of seconds, since

$$\frac{L}{R} = \frac{\text{H}}{\Omega} = \frac{\dfrac{\text{V}}{\text{A/sec}}}{\text{V/A}} = \text{sec}$$

As f increases, we would expect A_V to decrease and exhibit *some* phase lag since $\mathbf{E}_2$ is taken off the capacitor and capacitor voltage lags capacitor current by 90°. Inspection of Eq. (11-1b) [or Eq. (11-1a)] indicates that when f, the frequency of the input signal E_1, is the same as f_h, the quantity $f/f_h = 1$ and hence the magnitude of $A_v = 1/\sqrt{2} = 0.707$ and the phase angle is $\theta = -\tan^{-1} 1 = -45°$.

As f becomes very large, and in particular for $f \gg f_h$, the j term in Eq. (11-1a) approaches infinity, so that $A_v \cong 1/j\infty = -j0$, which means the gain magnitude approaches zero and the gain phase is $-90°$. This is to be expected, since at high frequencies $X_C \ll R$, so that the circuit impedance to $\mathbf{E}_1$ is essentially resistive and hence the current tends to be in phase with $\mathbf{E}_1$. However, this approximately zero-degree phase angle of the current is added to the 90° phase lag ($-90°$) of the capacitor to yield the phase of $\mathbf{E}_2$. Hence $\mathbf{E}_2$ lags $\mathbf{E}_1$ by approximately 90° at $f \gg f_h$. Of course, with X_C approaching zero in the high-frequency spectrum ($f \gg f_h$), $X_C \ll R$ and hence the magnitude of E_2 approaches zero. Thus the gain magnitude approaches zero.

Now work through problems PS 11-1, PS 11-2, and PS 11-3.

11-3 The high-pass filter

Another circuit commonly encountered is the high-pass filter of Fig. 11-2. This circuit is commonly used as a coupling network in many amplifier circuits where $\mathbf{E}_1$

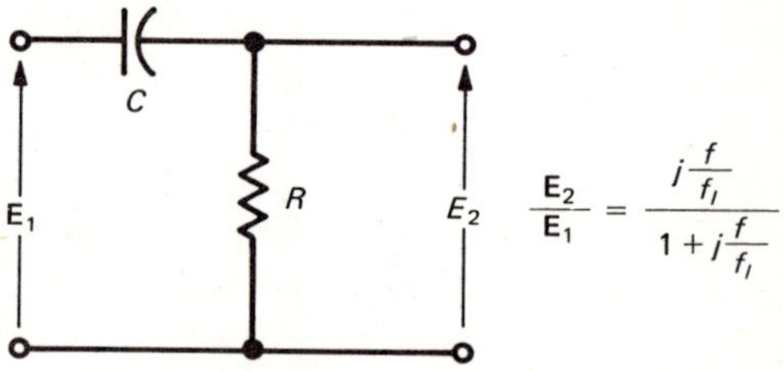

High-pass filter

FIGURE 11-2

contains a dc as well as an ac component. Usually we want only the ac component to appear at the output. The capacitor C, also called a blocking capacitor, assures this, since only a time-varying voltage can cause capacitor current to flow.[2]

We can develop the voltage transfer function by the voltage divider relationship and writing

1) $$\mathbf{A}_v = \frac{\mathbf{E}_2}{\mathbf{E}_1} = \frac{R}{R - jX_C} = \frac{1}{1 - j\dfrac{X_C}{R}} = \frac{1}{1 - j\dfrac{1}{\omega CR}}$$

[2] Recall that $i_c = C\dfrac{de_c}{dt}$.

If we let

2) $$\omega_l = \frac{1}{CR}$$

we can rewrite (1) as

3) $$\mathbf{A}_v = \frac{1}{1 - j\dfrac{\omega_l}{\omega}} = \frac{1}{1 - j\dfrac{f_l}{f}} \qquad (11\text{-}2a)$$

Now Eq. (3) is perfectly valid, but it is more desirable (also for reasons to become apparent later) to get rid of the negative denominator term. This may be accomplished by multiplying the numerator and denominator of (3) by jf/f_l. If you do this and rearrange terms, there results

4) $$\mathbf{A}_v = \frac{j\dfrac{f}{f_l}}{1 + j\dfrac{f}{f_l}} \qquad (11\text{-}2b)$$

which may also be written as

5) $$\mathbf{A}_v = \frac{f/f_l}{\sqrt{1 + (f/f_l)^2}} \angle 90° - \tan^{-1} f/f_l \qquad (11\text{-}2c)$$

Although (4) appears more complicated than (3), it is somewhat easier to Bode plot, which after all is our goal.

To correlate (5) with a qualitative analysis of Fig. 11-2, we may reason as follows. In the low-frequency spectrum, say for $f \ll f_l$, the numerator magnitude $\cong 0$ and the denominator magnitude $\cong 1$. Hence for $f \ll f_l$, $|A_v| \cong 0$. Physically we should expect this, since $X_C \gg R$ at low frequencies. The question is what constitutes low frequencies? The answer is that f_l in the high-pass or f_h in the low-pass filter forms a convenient spot frequency to separate the low and high frequency spectrum.

Since at low frequencies $X_C \gg R$, the current leads $\mathbf{E}_1$ by almost 90°; hence the output voltage which is taken off the resistor should be in phase with the current. Inspection of the phase portion of (5) indicates that for $f \ll f_l$ the tangent is $\cong 0°$ and hence the net angle is $90° - 0°$ or $\cong 90°$.

At $f = f_l$ where f_l is the corner or break frequency, $|A_v| = 1/\sqrt{2}$ and the phase is $90° - 45° = 45°$.

In the high-frequency spectrum ($f \gg f_l$) we see from Eq. (4) that the reactive term in the denominator is predominant, and hence $\mathbf{A}_v$ approaches $1\angle 0°$. This is in agreement with the physical facts, for $X_C \ll R$, $|E_2| \cong |E_1|$. Furthermore, the current in this case is essentially determined by R and hence is almost in phase with $\mathbf{E}_1$. Since $\mathbf{E}_2$ is taken off the resistor which has no phase angle associated with it, it follows that $\mathbf{E}_2$ is essentially in phase with $\mathbf{E}_1$.

Now work through problems PS 11-4, PS 11-5, and PS 11-6.

Note that, in all the problems, we have tried to express the transfer function so that it contains terms of the form jx and/or $1 + jx$. This will be possible whenever we deal with circuits containing a single-energy storage element such as just one capacitor or one inductor.

11-4 The decibel

The transfer functions encountered in practice will in general contain factors in the numerator and denominator. Therefore evaluation of the transfer function at a particular frequency requires the process of multiplication and/or division, neither of which is conveniently performed by graphic techniques. Since a graphic display of the transfer function is our goal, we seek an alternative approach.

You will recall that with the aid of logarithms, the operations of multiplication and division are converted to addition and subtraction, since

1) $$\log ab = \log a + \log b \qquad (11\text{-}3a)$$

2) $$\log \frac{a}{b} = \log a - \log b \qquad (11\text{-}3b)$$

and raising a number to a power converts to multiplication

3) $$\log a^x = x \log a \qquad (11\text{-}3c)$$

Now graphic addition and subtraction is no problem since we need only count divisions or lay off distances with a pair of dividers. For this reason, among others, the vertical scale in our graphic display will be expressed in a logarithmic unit called the decibel, abbreviated dB. The decibel is formally defined as a unit of power gain given by

4) $$A_p = 10 \log \frac{P_2}{P_1} \text{ dB} \qquad (11\text{-}4a)$$

Equation (11-4*a*) is read as, "the power gain A_p, expressed in dB, is equal to 10 times the log of the power ratio P_2/P_1." A more common form of expressing Eq. (11-4*a*) is shown in Eq. (11-4*b*):

5) $$A_{p(\text{dB})} = 10 \log \frac{P_2}{P_1} \qquad (11\text{-}4b)$$

Although Eq. (11-4*a*) is in better mathematical form, Eq. (11-4*b*), which is read and interpreted in the same way, is more prevalent and hence both forms will be used here.

Symbols other than A are commonly used to designate the gain or ratio between two quantities. For example, the symbol K or G is often used, and as a further refinement the symbol G is used if $P_2 > P_1$ or α if $P_2 < P_1$. We will, however, stick to the symbol A irrespective of whether P_2 or P_1 is the greater number.

Since electrical power is given by $P = E^2/R$ or I^2R, we may express Eq. (4) as

$$6)\quad A_p = 10 \log \frac{P_2}{P_1}\text{ dB} = 10 \log \frac{E_2{}^2/R_2}{E_1{}^2/R_1}\text{ dB}$$

$$= 10 \log \frac{E_2{}^2R_1}{E_1{}^2R_2}\text{ dB}$$

$$= 10 \log \left(\frac{E_2}{E_1}\right)^2\text{ dB} + 10 \log \frac{R_1}{R_2}\text{ dB}$$

$$= 20 \log \frac{E_2}{E_1}\text{ dB} + 10 \log \frac{R_1}{R_2}\text{ dB}$$

Similarly

$$7)\quad A_p = 10 \log \frac{P_2}{P_1}\text{ dB} = 10 \log \frac{I_2{}^2R_2}{I_1{}^2R_1}\text{ dB}$$

$$= 10 \log \left(\frac{I_2}{I_1}\right)^2 \frac{R_2}{R_1}\text{ dB}$$

$$= 10 \log \left(\frac{I_2}{I_1}\right)^2\text{ dB} + 10 \log \frac{R_2}{R_1}\text{ dB}$$

$$= 20 \log \frac{I_2}{I_1}\text{ dB} + 10 \log \frac{R_2}{R_1}\text{ dB}$$

Equations (6) and (7) are particularly useful if one wishes to determine power gain (or loss) by measuring E and R, or I and R, which is usually easier to do than a direct measurement of power. However, many electronic circuits are primarily designed from a standpoint of voltage or current gain, and hence we drop the resistive terms in (5) and (6) so that the definition of voltage and current gain in decibels are respectively,

$$8)\qquad A_v = 20 \log \left|\frac{E_2}{E_1}\right|\text{ dB} = 20 \log |A_v|\text{ dB} \quad (11\text{-}5a)$$

$$9)\qquad A_i = 20 \log \left|\frac{I_2}{I_1}\right|\text{ dB} = 20 \log |A_1|\text{ dB} \quad (11\text{-}5b)$$

The ratio of the output over the input is correctly called gain when two quantities are of similar units. However, it is common practice to refer to the output-input ratio as gain or system gain even when they are different dimensionally. For example, we might consider the gain of an optical system as being the output light intensity of a lamp versus applied voltage. It should also be clear that gain may be a number more or less than unity. If the gain is less than unity, it represents attenuation or loss within the network or system.

Note that if the gain is less than unity, the gain expressed in dB will be negative, since the logarithm of a fraction is negative. For example,

$$A = 20 \log \tfrac{1}{2}\text{ dB} = 20\,(\log 1 - \log 2)\text{ dB}$$

Since the logarithm of 1 to any base is zero, the above reduces to

$$A = 20\,(0 - \log 2)\text{ dB} = -20 \log 2\text{ dB}$$

$$= -20\,(0.301)\text{ dB} = -6.02\text{ dB}$$

Now work through problems PS 11-7, PS 11-8, and PS 11-9.

11-5 The Bode plot

A most useful means of displaying the amplitude and phase response of a circuit or system is to plot the magnitude (amplitude response) in decibels versus frequency as one curve and the phase versus frequency as a separate curve. The same frequency axis is used for both curves, and semilog graph paper is used so that a wide frequency range can be displayed. Curves with this type of display are known as Bode plots, and they are most useful in electronic circuit and control system analysis.

To familiarize ourselves with Bode plots, we will first construct the Bode plots of some basic transfer function factors. For the moment we are not concerned with how these functions are physically realized or even if they can be. We just want to learn the mechanics of constructing them. Once these basic forms are mastered, we can move on to more complicated forms which are essentially combinations of the basic forms.

Bode also showed that the phase and amplitude response for most transfer functions are uniquely related and that the slopes of the amplitude response are indicative of the phase response. Since it is usually easier to measure amplitude rather than phase, it behooves us to see how we can predict the phase shift by observing the amplitude response.

11-6 Basic forms

The first transfer function form (factor) we will consider is that of $A = k$ where A is the general symbol we are using to denote a transfer function and k is some arbitrary constant, a number which is independent of frequency. Thus if $A = 100$, we have A in dB given by $A = 20 \log 100\text{ dB} = 40\text{ dB}$. Similarly if $A = 0.2$, we have $A = 20 \log 0.2\text{ dB} \cong -14\text{ dB}$. These functions Bode plot as straight lines, as shown in Fig. 11-3. Note that there is no phase shift when the amplitude response is flat (A = a constant).

Next assume a transfer function of the form

$$1)\qquad \mathbf{A} = \left(j\frac{\omega}{\omega_c}\right)^n \qquad (11\text{-}6a)$$

where ω_c is some arbitrary constant which we will refer

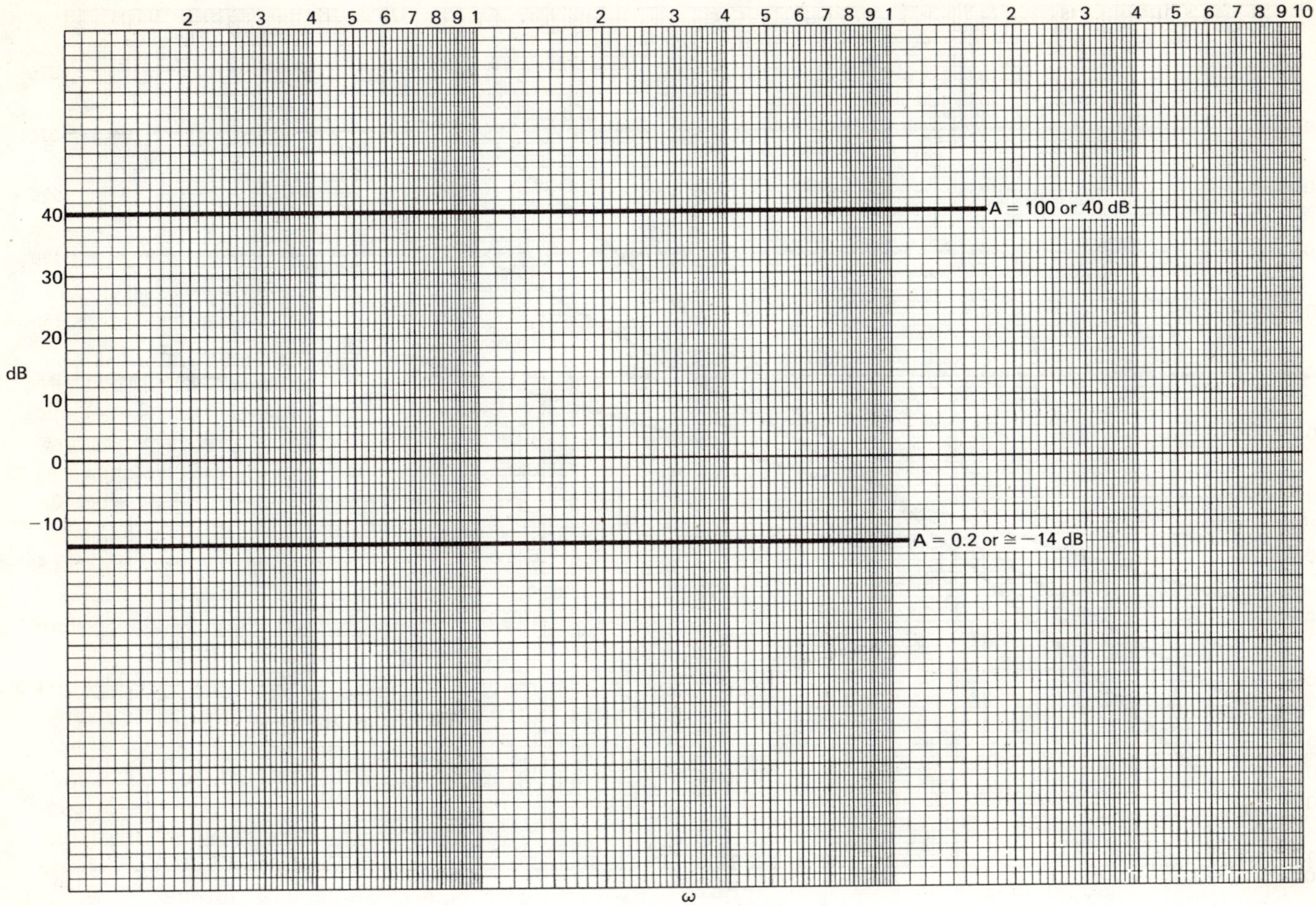

Bode plot of a constant
A = k

FIGURE 11-3

to as the corner frequency, and n is an integral exponent 1, 2, 3, 4 , . . . , n.[3] To Bode plot this function, let us first consider $n = 1$. In this case the function is simply

$$\text{2)} \qquad \mathbf{A} = j\frac{\omega}{\omega_c}$$

Clearly the j indicates a constant phase shift of 90° irrespective of the ratio ω/ω_c. Hence the phase response plots as a straight line (dashed) at 90 ° as shown in Fig. 11-4 for $n = 1$.

The only way we can plot the magnitude (in dB) of A is to assume various values of ω relative to our arbitrary ω_c. Constructing a table of values we have

ω/ω_c	0.1	0.5	1	2	5	10
dB = 20 log (ω/ω_c)	−20	−6.02	0	6.02	13.98	20
θ	←——— 90° ———→					

[3]A function of this form is sometimes called a derivative function.

These values are plotted for the $n = 1$ curve in Fig. 11-4, with the dashed line for the phase and solid line for the amplitude.

Note and memorize the following, as evidenced in Fig. 11-4. For a transfer function of the form in Eq. (1) with $n = 1$, the slope of the Bode plot is positive and a constant equal to 20 dB/decade ≅ 6 dB/octave where a decade is a frequency ratio of 10, and an octave is a frequency span of 2. The plot crosses 0 dB at $\omega = \omega_c$. The phase angle associated with this slope is 90°.

Now let $n = 2$ in Eq. (1) so that the specific function we are dealing with is

$$\text{3)} \qquad \mathbf{A} = \left(j\frac{\omega}{\omega_c}\right)^2 = j^2\left(\frac{\omega}{\omega_c}\right)^2 = -1\left(\frac{\omega}{\omega_c}\right)^2 = \left(\frac{\omega}{\omega_c}\right)^2 \underline{/180^\circ}$$

Clearly the phase angle in this case is a constant equal to 180°. The magnitude and phase of (3) may also be tabulated in order to sketch the Bode plot. Thus we obtain

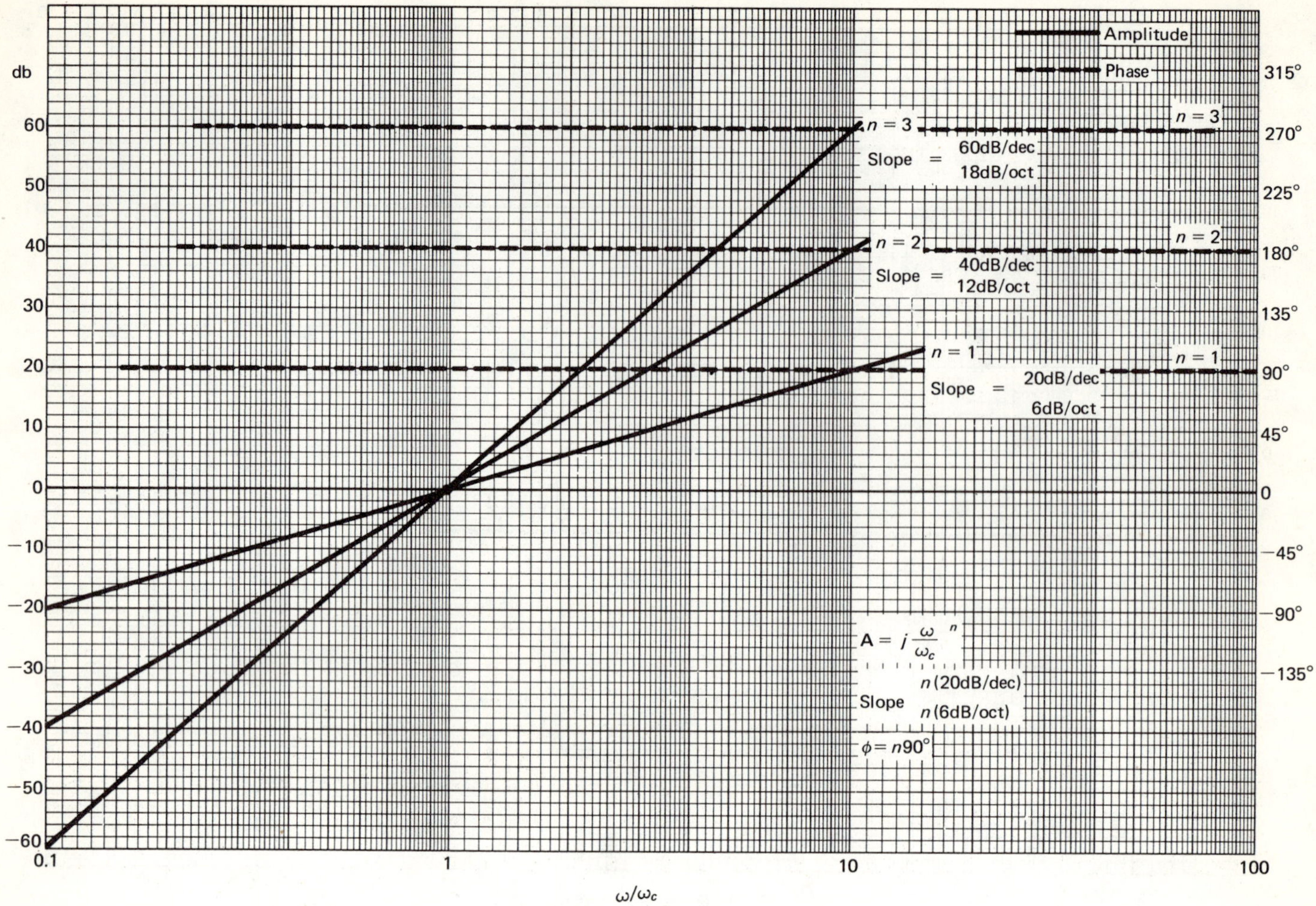

FIGURE 11-4

ω/ω_c	0.1	0.5	1	2	5	10
dB $= 40 \log (\omega/\omega_c)$	−40	−12.04	1	12.04	27.96	40
θ	←———		180°		———→	

Note that because this time the transfer function is the square of the previous case ($n = 1$), the slope is exactly twice as much and the phase angle is doubled. Therefore for $n = 2$ and a transfer function of the form given in Eq. (1), you should think in terms of a positive slope of 40 dB/dec or about 12 dB/oct and associate a 180° phase shift with this slope. These results are also shown in Fig. 11-4.

A Bode plot for $n = 3$ is also shown in Fig. 11-4. As might be expected, this curve also goes through 0 dB at $\omega = \omega_c$ with a positive slope equal to 60 dB/dec or approximately 18 dB/oct and the associated phase angle is 270°.

As you can see from Fig. 11-4, the slope in general is given by

$$\text{slope} = n\,(20 \text{ dB/dec}) \cong n\,(6 \text{ dB/oct}) \qquad (11\text{-}6b)$$

and the phase is given by

$$\phi = n90° \qquad (11\text{-}6c)$$

Another commonly encountered form of transfer function term is[4]

1) $$\mathbf{A} = \frac{1}{\left(j\dfrac{\omega}{\omega_c}\right)^n} \qquad (11\text{-}7a)$$

Inspection of (1) reveals that an increase of ω causes the denominator to increase, since ω_c is some fixed value, and hence it causes A to decrease. Therefore, the amplitude response should have a negative slope. We also see that for any value of n, the amplitude must be 0 dB at $\omega = \omega_c$. Since the j represents 90° of phase in the denominator of A, the net phase angle in the denominator must be n 90°. Consequently, the net phase of A must be $-n$ 90°.

Although we will not take the time to derive the data

[4] This transfer function is sometimes referred to as the integral function.

for plotting A, the results are given in Fig. 11-5. The magnitude response from (1) is as follows:

2) $$\mathbf{A} = \frac{1}{(\omega/\omega_c)^n}$$

3) $$A_{dB} = 20\left[\log 1 - n \log \frac{\omega}{\omega_c}\right]$$

Since the log 1 to any base is zero, we have

4) $$A_{dB} = -n\, 20 \log \frac{\omega}{\omega_c}$$

Inspection of Fig. 11-5 indicates that except for being negative, the slopes and phase for various values of n are identical with Fig. 11-4. Therefore, for a function of the form in (1),

$$\text{slope} = -n\,(20\text{ dB/dec}) \text{ or } -n\,(6\text{ dB/oct}) \quad (11\text{-}7b)$$

$$\phi = -n\,90° \quad (11\text{-}7c)$$

Now work through problems PS 11-10 and PS 11-11.

The next form of a transfer function factor we might familiarize ourselves with is

1) $$\mathbf{A} = \left(1 + j\frac{\omega}{\omega_c}\right)^n$$

Let us first consider the case where $n = 1$ and calculate the magnitude and phase. This yields the data in the following table:

ω/ω_c	0.1	0.5	1	2	5	10
ϕ	5.71°	26.6°	45°	63.4°	78.7°	84.3°
A	1.00	1.12	1.414	2.24	5.1	10.0
A_{dB}	0.043	0.969	3.01	6.99	14.1	20.0

A Bode plot of the above data is in the solid line curves of Fig. 11-6. Figure 11-6 also shows some dashed lines which represent straight-line approximations to the actual curves. The beauty of the Bode plot (among other things) lies in the ease with which the straight-line approximations to the actual characteristics may be constructed. For example, consider the upper magnitude response curve. Note that it may be approximated by the two dashed straight lines which intersect at ω_c. We

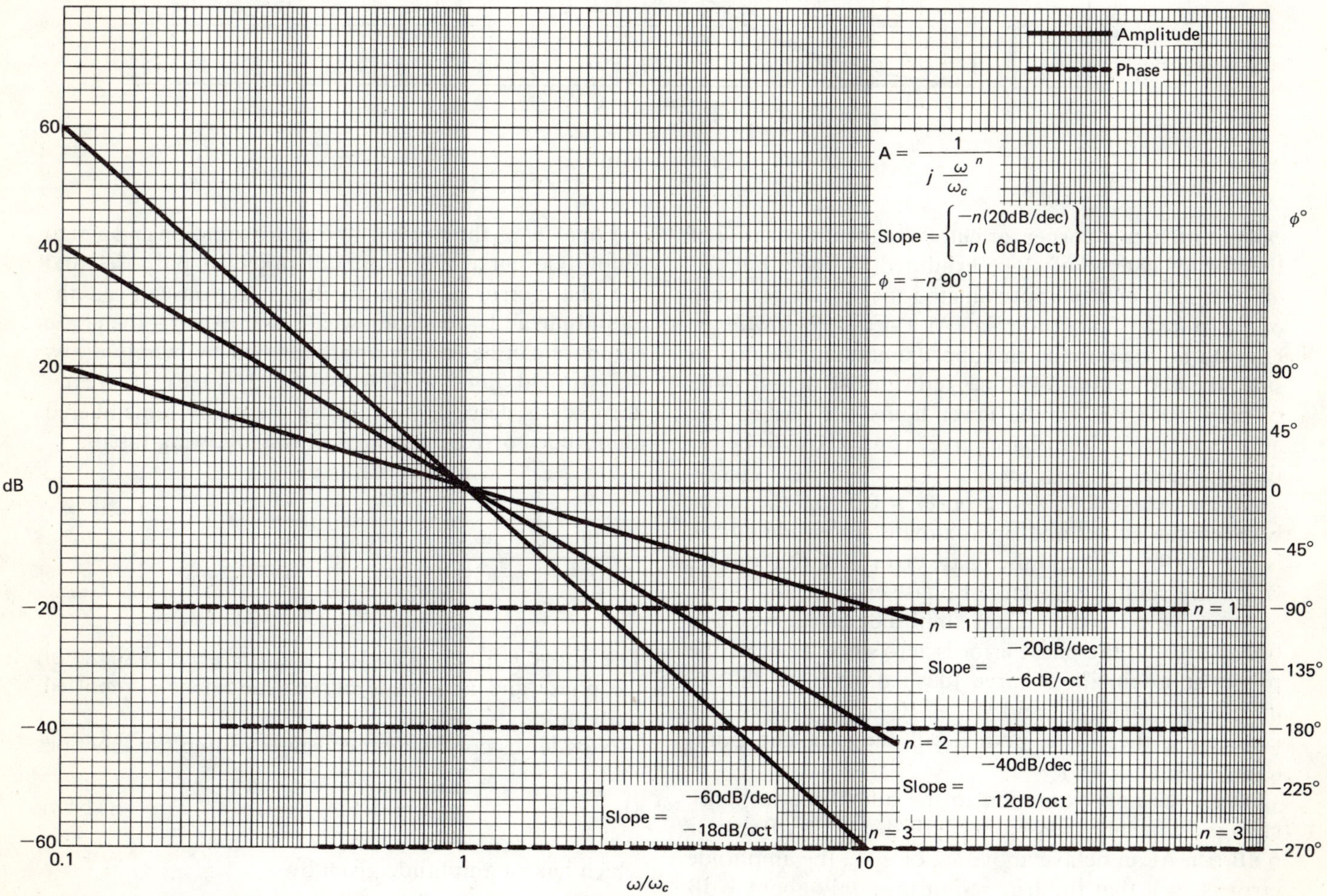

FIGURE 11-5

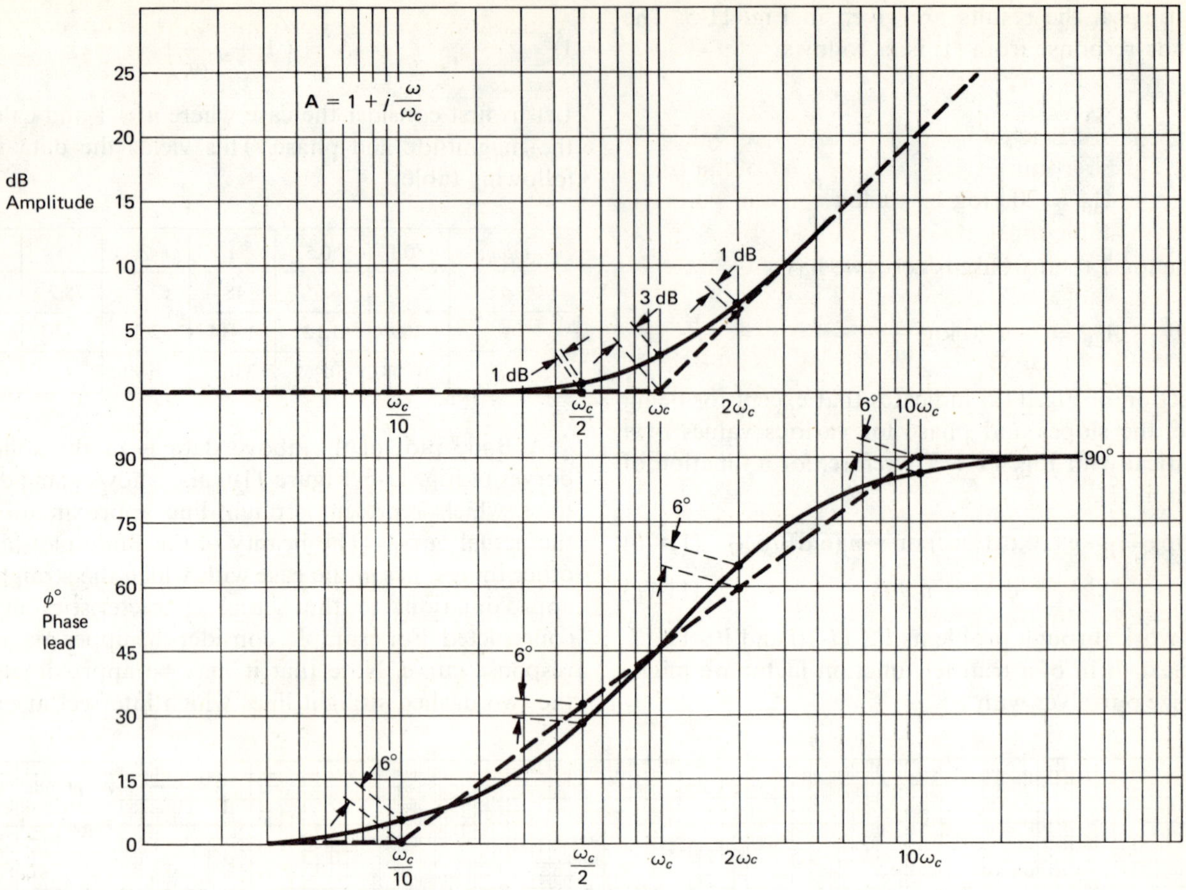

FIGURE 11-6

call ω_c the break, corner, or cutoff frequency because of the abrupt change of slope exhibited by these lines at ω_c. For $\omega < \omega_c$ the approximated magnitude response of this transfer function is flat. At a decade below ω_c or $\omega = 0.1\ \omega_c$ the error in this approximation is negligible since the above table of values indicates an actual value of only 0.043 dB. In the lower figure the dashed-line approximation to the phase curve is only about 6° below the true value.

At an octave below ω_c, or $\omega = 0.5\ \omega_c$, the error between the actual and approximate amplitude curves is only about 1 dB and the phase error is only about 6°, which is essentially the largest phase error at any point.

At ω_c we have the largest amplitude error, which is only 3 dB, and the phase error is zero since we draw the phase approximation curve (dashed line) so that it intersects 0° at $\omega_c/10$ and 90° at 10 ω_c. A straight line connecting these two points will intersect the point ($\omega/\omega_c = 1$, $\phi = 45°$).

At ω_c the approximation to the amplitude response changes slope abruptly from 0° dB/dec to 20 dB/dec or 6 dB/oct. At an octave above ω_c, or 2 ω_c, the amplitude error is such that the true response is only about 1 dB high and the phase error is about 6° low.

For $\omega \gg \omega_c$ the straight-line approximation and actual amplitude curve merge for all practical purposes and the phase error is again about 6°. As you can see, the amplitude and—to a somewhat lesser extent—the phase response can be readily approximated by merely drawing straight lines with well-defined slopes. If greater accuracy is desired, we can use the easily remembered amplitude correction factors of the following table:

ω/ω_c	0.1	0.5	1	2	10
dB	0	1	3	1	0
$\phi°$	6	6	0	6	6

The above correction factors and the general shape of the phase and amplitude curves should be committed to memory.

Now let us consider other values of n in the function

1) $$\mathbf{A} = \left(1 + j\frac{\omega}{\omega_c}\right)^n \tag{11-8a}$$

which has an amplitude given by

2) $$A_{\text{dB}} = n\ 20 \log \sqrt{1 + (\omega/\omega_c)^2} \tag{11-8b}$$

and a phase angle

3) $$\phi = n \tan^{-1} \frac{\omega}{\omega_c} \qquad (11\text{-}8c)$$

From (2) we see that the amplitude in dB and phase in degrees will be n times the values we just obtained for plotting Fig. 11-6. Therefore, the correction factors in the previous table must also be multiplied by n.

A Bode plot of straight-line approximations to (1) for different values of n is shown in Fig. 11-7. Since the amplitude slope for $\omega < \omega_c$ was zero, it will still be zero for any value of n. However, for $\omega > \omega_c$ the slope must be multiplied by the appropriate n. Thus, for $n = 3$, the slope for $\omega > \omega_c$ is 3 (20 dB/dec) = 60 dB/dec or 18 dB/oct. The corresponding phase asymptotically approaches 3 (90°) = 270°. For $n = 3$ at $\omega = \omega_c$, the phase is n (45°) = 135° and the amplitude error is n (3 dB) = 9 dB.

Now work problem PS 11-12.

Another commonly encountered form of transfer function is

1) $$\mathbf{A} = \frac{1}{\left(1 + j\dfrac{\omega}{\omega_c}\right)^n} \qquad (11\text{-}9a)$$

You will notice that this expression is the reciprocal of Eq. (11-8a), which we just studied. The amplitude response is given by

2) $$A_{\text{dB}} = -n\, 20 \log \sqrt{1 + (\omega/\omega_c)^2} \qquad (11\text{-}9b)$$

The phase response is given by

$$\phi = -n \tan^{-1} \frac{\omega}{\omega_c} \qquad (11\text{-}9c)$$

A Bode plot for $n = 1$ is shown in Fig. 11-8 and, as you can see, the only difference in the amplitude response in this case and that of Fig. 11-6 is that for $\omega > \omega_c$ the gain decreases. The slopes for $\omega > \omega_c$ in Eq. (11-9a) for any value of n are simply the negative of the slopes for Eq. (11-8a). The correction factors are also identical in magnitude except that now they lie below the straight-line approximations. The phase characteristic and its correction factors for any value of n is also identical with the single exception that we now have phase lag rather than lead. Thus for $\omega \gg \omega_c$ and $n = 1$, the phase angle approaches $-90°$.

Now work through problem PS 11-13.

Transfer functions occasionally contain factors of the form

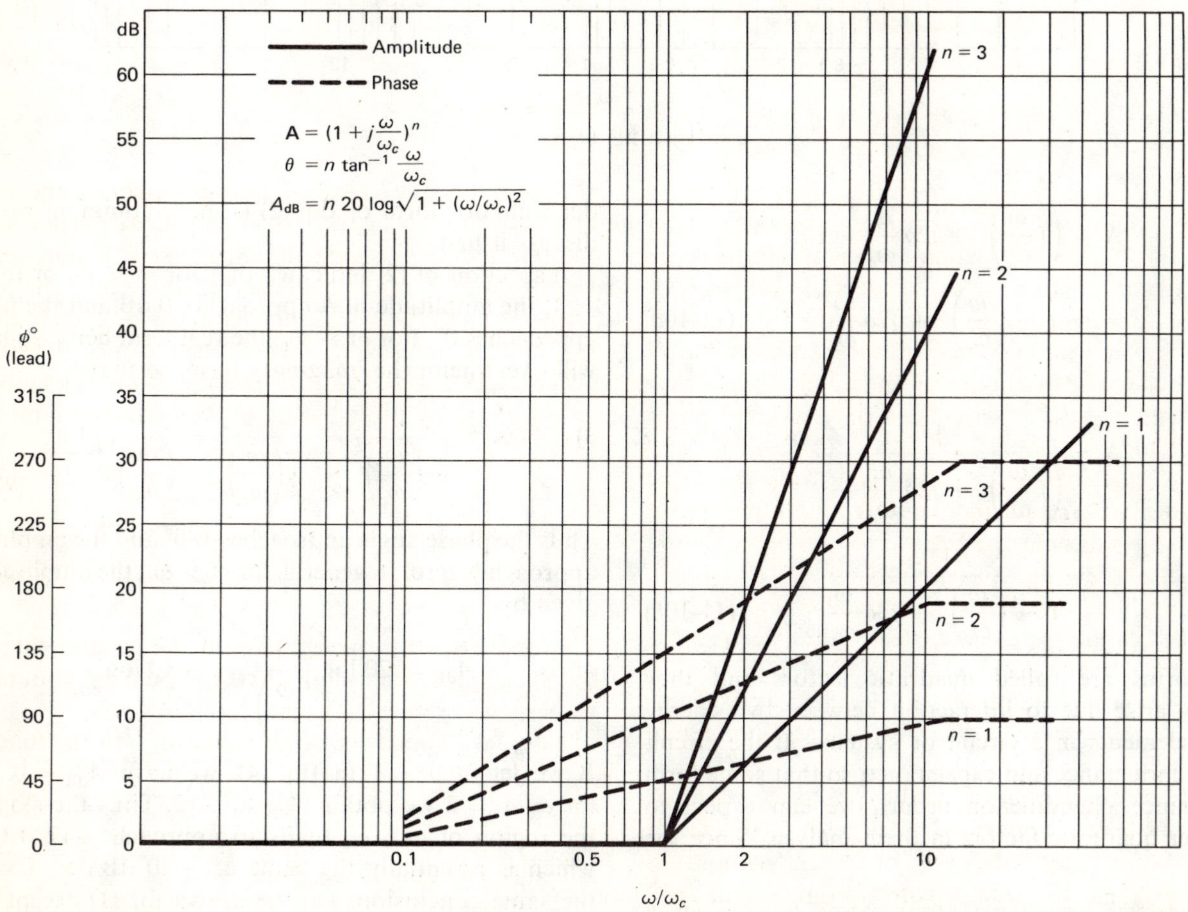

FIGURE 11-7

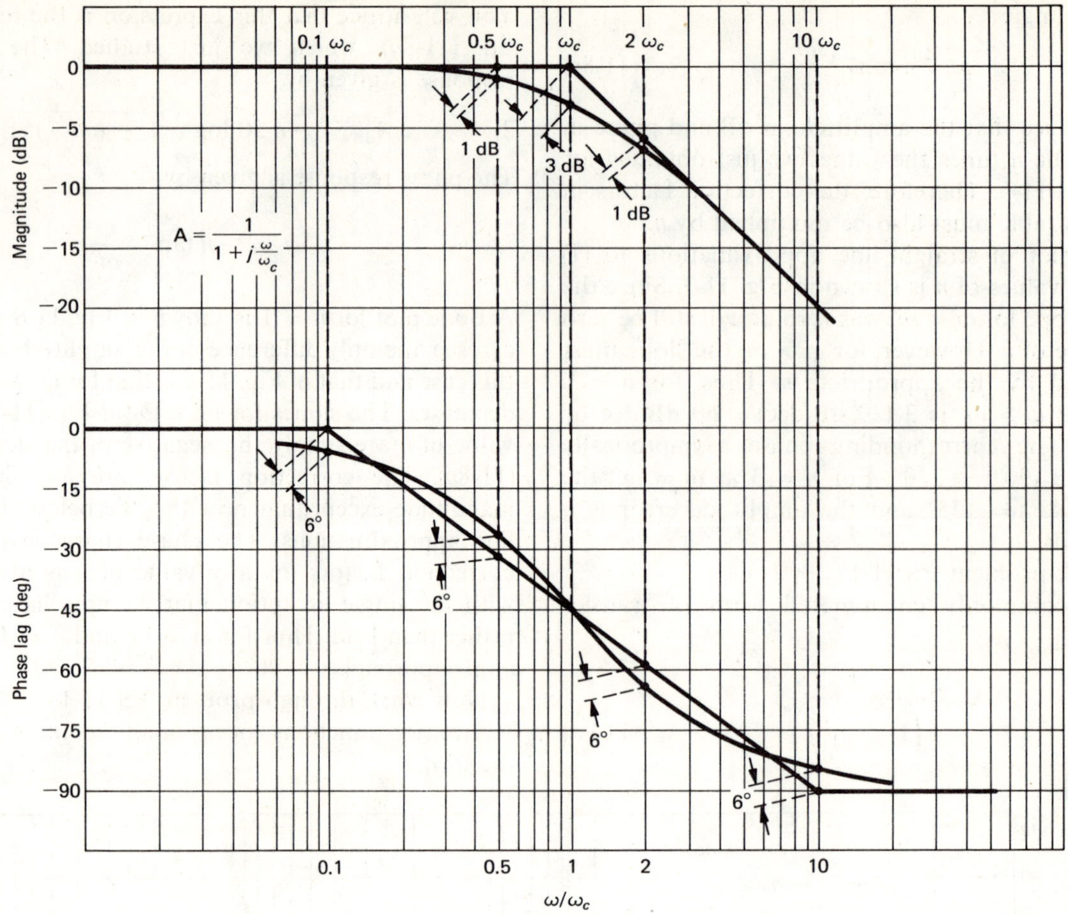

FIGURE 11-8

$$1)\qquad \mathbf{A} = \left(j\frac{\omega}{\omega_c}\right)^2 + j2\delta\frac{\omega}{\omega_c} + 1$$

$$= 1 - \left(\frac{\omega}{\omega_c}\right)^2 + j2\delta\frac{\omega}{\omega_c} \qquad (11\text{-}10a)$$

or

$$2)\qquad \mathbf{A} = \frac{1}{\left(j\frac{\omega}{\omega_c}\right)^2 + j2\delta\frac{\omega}{\omega_c} + 1}$$

$$= \frac{1}{1 - \left(\frac{\omega}{\omega_c}\right)^2 + j2\delta\frac{\omega}{\omega_c}} \qquad (11\text{-}10b)$$

These forms are called quadratic factors and they generally arise due to interaction between two energy storage elements in a circuit or system.[5] If the circuit contains inductance and capacitance, so that some form of resonance phenomenon occurs, we can expect to encounter quadratic factors in their analysis. Since the denominator form of Eq. (2) is more common, we will discuss it first.

Inspection of (2) indicates that for $\omega \ll \omega_c$ or $(\omega/\omega_c) \ll 1$, the amplitude of **A** approaches 0 dB and the phase approaches 0°. For $\omega \gg \omega_c$, the real term being squared will overwhelm the imaginary term, so that

$$3)\qquad \mathbf{A} \cong \frac{1}{-\left(\frac{\omega}{\omega_c}\right)^2} = -\frac{1}{\left(\frac{\omega}{\omega_c}\right)^2} = \frac{1}{\left(\frac{\omega}{\omega_c}\right)^2}\underline{/180^\circ}$$

Thus the phase angle approaches 180° and the amplitude approaches zero. In general, for $\omega \gg \omega_c$, the amplitude is given by

$$4)\qquad A_{dB} = 20\log\frac{1}{\left(\frac{\omega}{\omega_c}\right)^2} = -40\log\frac{\omega}{\omega_c}$$

If we let $\omega/\omega_c = 1$ in Eq. (4), we have $A_{dB} = 0$, and for $\omega/\omega_c = 2$ we obtain $A_{dB} \cong -12$. Thus the slope in the region of $\omega \gg \omega_c$ tends to approach -12 dB/oct, which is essentially the same as -40 dB/dec. Exactly the same conclusions can be drawn for (1) except that there the amplitude response rises and the phase leads

[5] The energy storage elements can be two capacitors, two inductors, or an inductor and capacitor.

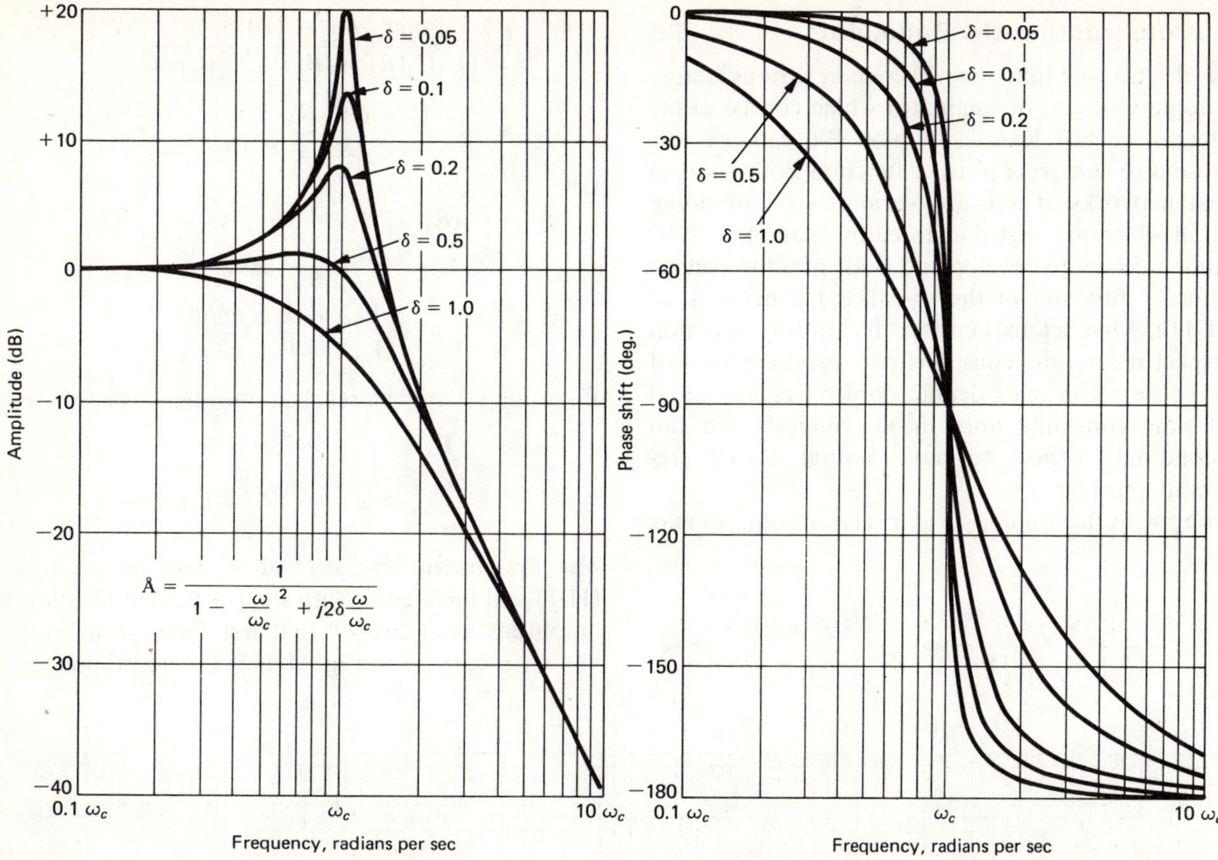

FIGURE 11-9

as ω increases, whereas in (2) the amplitude decreases and the phase lags. In both cases the phase asymptotically approaches 180°, though from different directions.

Now what happens at $\omega = \omega_c$ in (2)? Clearly the function reduces to

5) $$\mathbf{A} = \frac{1}{j2\delta} = \frac{1}{2\delta}\angle -90^\circ$$

Thus the phase at $\omega = \omega_c$ is -90° in (2) or $+90^\circ$ in (1). The corresponding amplitude in (2) is

6) $$A_{\text{dB}} = -20 \log 2\delta \qquad \text{(11-10c)}$$

Inspection of (6) indicates that if $\delta = \frac{1}{2}$, $\log 2\delta = \log 1 = 0$, so that $A_{\text{dB}} = 0$. If $\delta < \frac{1}{2}$, say $\delta = \frac{1}{5}$ for example, $\log 2\delta$ will be the log of a number less than 1 which is a negative quantity, and in this case

$$A_{\text{dB}} = -20 \log 2(\tfrac{1}{5}) = -20(-0.3979) \cong +7.96 \text{ dB}$$

Thus we see that the amplitude response actually rises above the 0 dB level for $\delta < \frac{1}{2}$ in (2). For $\delta > \frac{1}{2}$, say $\delta = 1$, the log is positive and hence the number of dB is now negative. Both the phase and magnitude of (1) and (2) strongly depend upon the value of δ in the vicinity of $\omega = \omega_c$, and it is probably best to plot a few points for an accurate representation. However, as an aid to sketching denominator quadratic factors, we have Fig. 11-9. Note that as δ (sometimes called damping factor) increases, the peak or resonant point shifts below ω_c.

The same curves may be used for numerator quadratic factors if you remember that for $\omega \gg \omega_c$ the amplitude curve slopes upwards at 40 dB/dec; and at ω_c, instead of a resonant rise, we may see a resonant dip as shown in Fig. 11-10*a*. The phase response is also similar except that it approaches 180° of phase lead for $\omega \gg \omega_c$. Fig. 11-10*b* illustrates the phase response in this case.

Now work through problem PS 11-14.

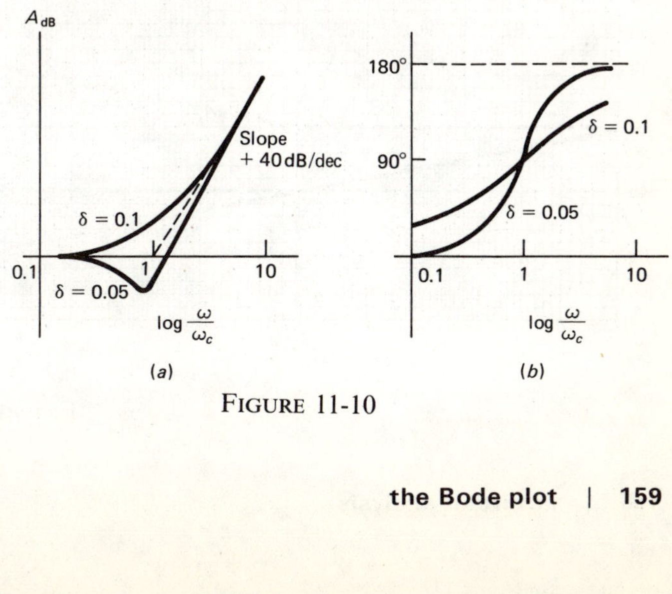

FIGURE 11-10

11-7 Constructing the Bode plot

In general, a transfer function will contain various factors in the numerator and denominator which consist of the basic forms we just learned to plot. We can use this background to construct transfer function Bode plots of practical networks. There are various ways of doing this, all of which are best illustrated by example.

Assume we wish to develop and Bode plot the voltage gain transfer function of the so-called lag network of Fig. 11-11*a*. First let us develop the transfer function and inspect it for time constants (*RC* products or *L*/*R* ratios) whose reciprocals denote corner frequencies. If several time constants arise in the analysis, we can correspondingly expect to have several corner frequencies to consider.

Starting with the voltage divider relationship, we may derive

$$\frac{\mathbf{E}_o}{\mathbf{E}_{in}} = \frac{R_2 - jX_{C_2}}{R_1 + R_2 - jX_{C_2}} = \frac{R_2 + 1/j\omega C_2}{R_1 + R_2 + 1/j\omega C_2}$$

$$= \frac{(j\omega C_2 R_2 + 1)(1/j\omega C_2)}{[j\omega C_2(R_1 + R_2) + 1](1/j\omega C_2)}$$

$$= \frac{1 + j\omega C_2 R_2}{1 + j\omega C_2(R_1 + R_2)}$$

Let

$$\tau_1 = C_2(R_1 + R_2) = 1/\omega_1, \tau_2 = C_2R_2 = 1/\omega_2,$$

$$\omega_2 = 10\omega_1$$

1) $$\frac{\mathbf{E}_o}{\mathbf{E}_{in}} = \frac{1 + j\omega/\omega_2}{1 + j\omega/\omega_1}$$

Equation (1) may be resolved into two factors

2) $$\frac{\mathbf{E}_o}{\mathbf{E}_{in}} = \left(1 + j\frac{\omega}{\omega_2}\right)\frac{1}{1 + j\dfrac{\omega}{\omega_1}}$$

The first factor was studied in conjunction with Eq. (11-8) and the second with Eq. (11-9). A Bode plot of the amplitude response of the first factor (numerator) is shown as curve *N* in Fig. 11-11*b*. The amplitude response

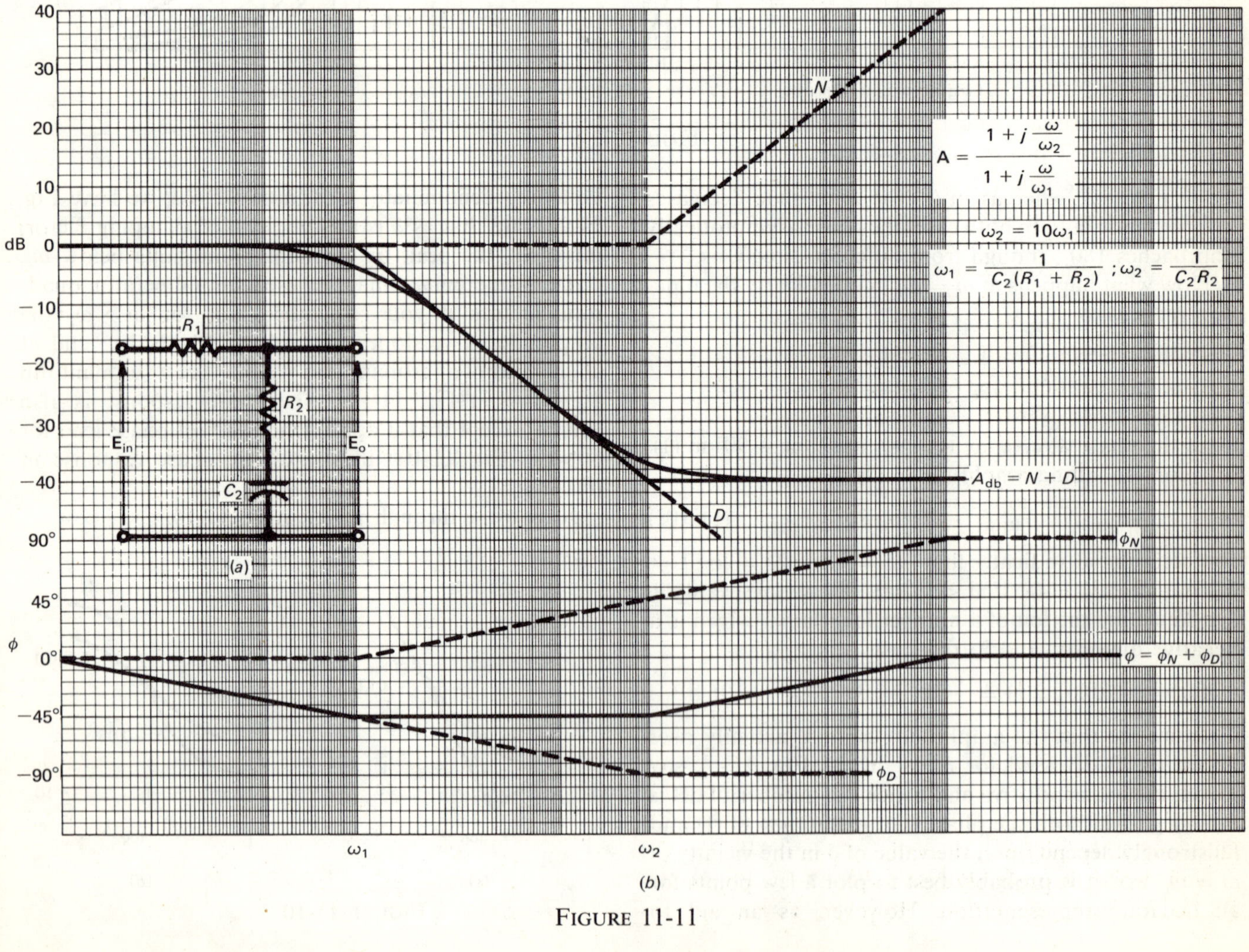

FIGURE 11-11

of the second factor is shown as curve D. Since the amplitude scale is logarithmic (dB), multiplication of the amplitude of these factors is equivalent to addition. Hence, the composite amplitude response is simply the point-by-point addition of curves N and D as shown in curve $A_{dB} = N + D$. A pair of dividers is most useful in this regard. Correction factors can be inserted so that a smooth curve can be drawn, but generally the straight-line approximations are adequate.

The phase of the first and second factors has also been approximated with straight-line segments as ϕ_N and ϕ_D respectively. The phase angles may also be algebraically added with a pair of dividers or by counting squares in a manner similar to the amplitude response.

With a little practice you will in most cases be able to construct the amplitude response of **A** without having to construct the individual factors. For example, with $\omega_2 > \omega_1$ in (1), we know the first break (corner) occurs at ω_1 as we go from the low to the high-frequency range. Since ω_1 is in the denominator, A_{dB} will break downward at ω_1 with a slope of 20 dB/dec and continue doing so until ω_2, whereupon the numerator starts climbing at 20 dB/dec. Thus the positive numerator slope cancels the negative denominator slope so that the slope of A_{dB} levels off at 0 dB/dec.

Although it is somewhat trickier, the phase response may be similarly reasoned. To a first approximation the denominator phase breaks at $0.1\omega_1$ and decreases at about 45°/dec. Thus at ω_1 the denominator angle is 45°, but note the numerator phase is coming into play, since $\omega_1 = 0.1\omega_2$ and the numerator phase is increasing at 45°/dec. Thus at ω_1 the numerator and denominator phase *slopes* cancel, so that the phase remains constant at the $-45°$ level it acquired at ω_1. This holds true until ω_2, at which point the denominator phase has gone through its maximum possible shift of $-90°$, so that the *slope* of the denominator phase angle is zero. The numerator phase slope, however, continues to climb at 45°/dec for a decade above ω_2. This pulls the net phase slope (ϕ) up at 45°/dec above ω_2 until $10\omega_2$, at which point the numerator has gone through its maximum possible phase shift of $\pm 90°$. Thus for $\omega > \omega_2$ both numerator and denominator phase slopes are zero and the net phase slope levels off at 0° due to the $-90°$ in the denominator canceling the $+90°$ due to the numerator.

From a physical viewpoint we can reason in Fig. 11-11*a* that at low frequencies C_2 looks like an open circuit and hence $\mathbf{E}_o = \mathbf{E}_{in}$ or $A_{dB} = 0$. As ω increases, C_2 comes into play so that it starts to load R_1, which decreases $\mathbf{E}_o$. This accounts for the break at ω_1. As ω increases further, C_2 eventually looks like a short and $\mathbf{E}_o = \mathbf{E}_{in} R_2/(R_1 + R_2)$. This accounts for ω_2, where C_2 fades out of the picture.

Now work through problems PS 11-15, PS 11-16, PS 11-17.

PROBLEMS WITH SOLUTIONS

PS 11-1 What is the corner frequency of the low-pass filter in Fig. PS 11-1? What is the voltage gain transfer function at the corner frequency and at 1/10, 1/2, twice, and ten times the corner frequency?

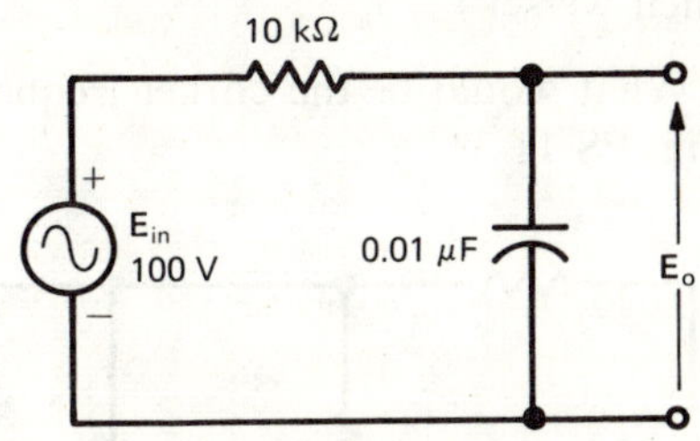

FIGURE PS 11-1

SOLUTION The time constant is

1) $\tau_h = RC = 10\ \text{k}\Omega(0.01\ \mu\text{F}) = 1 \times 10^{-4}$ sec

The corresponding *radian* corner frequency is

2) $$\omega_h = \frac{1}{\tau_h} = 10^4 \text{ rad/sec}$$

The corresponding corner frequency is

3) $$f_h = \frac{\omega_h}{2\pi} = 1592 \text{ Hz}$$

The general expression for the transfer function by Eq. (11-1) is

4) $$A_V = \frac{\mathbf{E}_o}{\mathbf{E}_{in}} = \frac{1}{1 + j\dfrac{f}{f_h}}$$

Plugging in the given values of $f/f_h = 0.1, 0.5, 1, 2$, and 10 into (4), we have

f/f_h	0.1	0.5	1	2	10
$\lvert A_V \rvert$	0.995	0.894	0.707	0.447	0.0995
$\angle\theta$	$-5.71°$	$-26.6°$	$-45°$	$-63.4°$	$-84.3°$

PS 11-2 Prove that the corner frequency of a low-pass filter such as shown in Fig. 11-1 is that frequency at which $X_C = R$.

SOLUTION Applying the voltage divider relationship to Fig. 11-1, we have

1) $$\mathbf{A}_v = \frac{\mathbf{E}_2}{\mathbf{E}_1} = \frac{-jX_C}{R - jX_C}$$

Dividing numerator and denominator by $-jX_C$ and rearranging terms, we obtain

2) $$\mathbf{A}_v = \frac{1}{1 - \dfrac{R}{-jX_C}} = \frac{1}{1 + j\dfrac{R}{X_C}}$$

If we let $R = X_C$ in Eq. (2),

3) $$\mathbf{A}_v = \frac{1}{\sqrt{2}\,\underline{/45^\circ}} = 0.707\underline{/-45^\circ}$$

which is precisely what we showed in the previous problem when we set $f = f_h$.

PS 11-3 What would be the corner frequency of the circuit in Fig. PS 11-3?

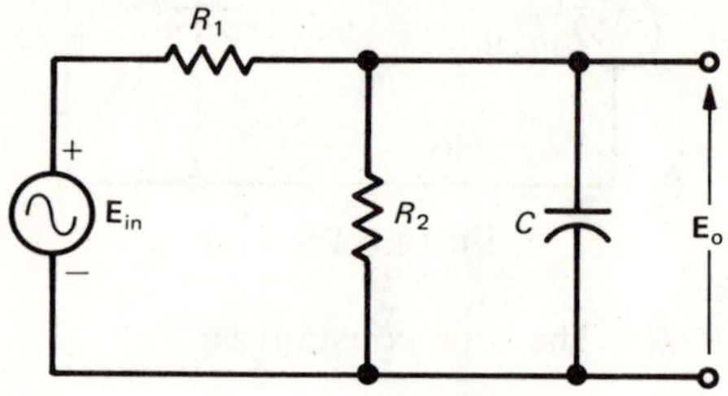

FIGURE PS 11-3

SOLUTION By inspection we see that R_1 and R_2 may be thevenized with the voltage source to simplify the equivalent circuit that the capacitor looks back into. Thus, by analogy with Fig. 11-1, intuition, or both, $\tau_h = R_{th}C = (R_1 \| R_2)C$. The radian corner frequency is then $\omega_h = 1/\tau_h$.

PS 11-4 What is the corner frequency of the high-pass filter in Fig. PS 11-4? What is the voltage gain transfer function at the corner frequency and at 1/10, 1/2, twice, and ten times the corner frequency?

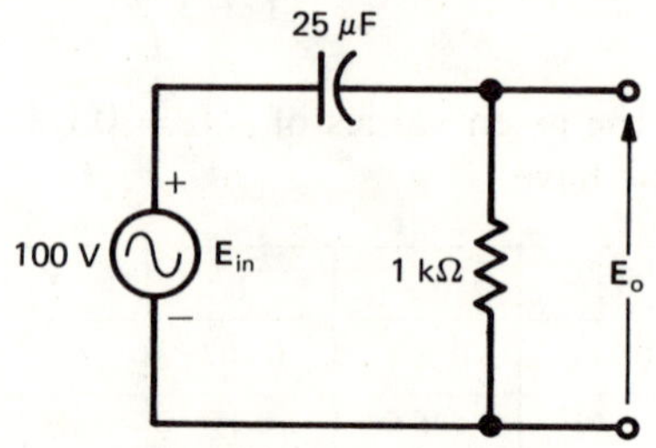

FIGURE PS 11-4

SOLUTION The time constant from Eq. (11-2) is

1) $\tau_l = RC = 1\ \text{k}\Omega(25\ \mu\text{F}) = 25 \times 10^{-3}$ sec

The corresponding *radian* corner frequency is

2) $$\omega_l = \frac{1}{\tau_l} = 40\ \text{rad/sec}$$

And the corresponding corner frequency is

3) $$f_l = \frac{\omega_l}{2\pi} = 6.37\ \text{Hz}$$

The general expression for the voltage transfer function by Eq. (11-2*b*) is

4) $$\mathbf{A}_v = \frac{j\dfrac{f}{f_l}}{1 + \dfrac{f}{f_l}}$$

Setting f/f_l equal to 0.1, 0.5, 1, 2, and 10 yields the following values

f/f_l	0.1	0.5	1	2	10
$\lvert A_V \rvert$	0.9950	0.447	0.707	0.894	0.995
$\underline{/\theta}$	84.3°	63.4°	45°	26.6°	5.71°

PS 11-5 Determine the voltage transfer function of the coupling network shown in Fig. PS 11-5. Resistor R_1 is analogous to the output resistance of a voltage source driving a high-pass filter.

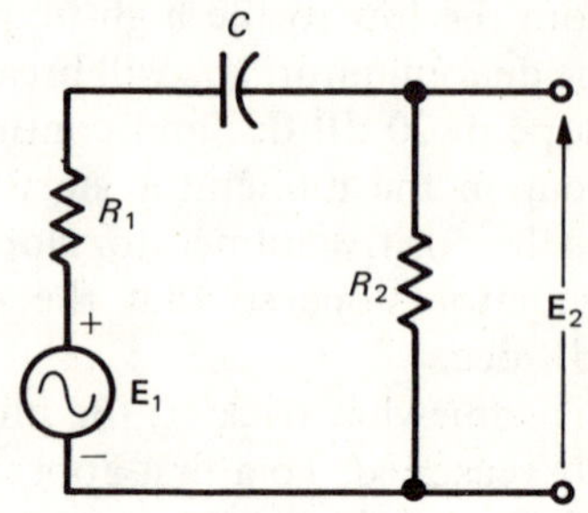

FIGURE PS 11-5

SOLUTION Applying the voltage divider relationship we have

1) $$\mathbf{A}_v = \frac{\mathbf{E}_2}{\mathbf{E}_1} = \frac{R_2}{R_1 + R_2 - jX_C}$$

Dividing numerator and denominator by $R_1 + R_2$ yields

2) $$\mathbf{A}_v = \frac{\dfrac{R_2}{R_1 + R_2}}{\dfrac{R_1 + R_2}{R_1 + R_2} - j\dfrac{X_C}{R_1 + R_2}}$$

$$= \frac{R_2}{R_1 + R_2} \times \frac{1}{1 - j\dfrac{1}{\omega C(R_1 + R_2)}}$$

If we let the corner frequency be

3) $$\omega_l = \frac{1}{C(R_1 + R_2)}$$

we can rewrite (2) as

4) $$\mathbf{A}_v = \frac{R_1}{R_1 + R_2} \times \frac{1}{1 - j\dfrac{\omega_l}{\omega}}$$

Multiplying numerator and denominator of (4) by $j\omega/\omega_l$ and introducing f yields

5) $$\mathbf{A}_v = \frac{R_1}{R_1 + R_2} \times \frac{j\dfrac{f}{f_l}}{1 + j\dfrac{f}{f_l}}$$

Note that for $f \gg f_l$ we just see the R_1, R_2 voltage divider, since $X_C \cong 0$.

PS 11-6 Prove that the corner frequency is the frequency at which $X_C = R_1 + R_2$ in Fig. PS 11-5.

SOLUTION If we divide numerator and denominator of Eq. (1) by $R_1 + R_2$ in the solution of the previous problem, we obtain

$$\mathbf{A}_v = \frac{R_2}{R_1 + R_2} \times \frac{1}{1 - j\dfrac{X_C}{R_1 + R_2}}$$

When $X_C = R_1 + R_2$ the above equation reduces to

$$\mathbf{A}_v = \frac{R_2}{R_1 + R_2} \times \frac{1}{1 - j1} = 0.707\left(\frac{R_2}{R_1 + R_2}\right)\underline{/45^\circ}$$

This is exactly the same result that would be obtained by setting $f = f_l$ in (5) of the previous problem. Thus when $f = f_l$, $X_C = R_1 + R_2$.

PS 11-7 Express the voltage gain of the circuit in Fig. PS 11-7 in dB at $f = f_r$, the resonant frequency.

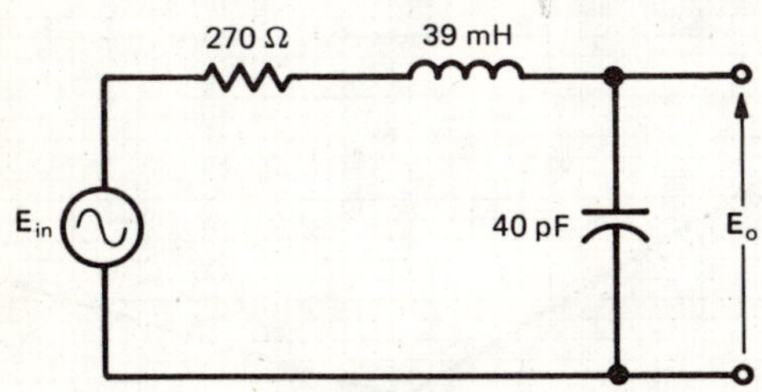

FIGURE PS 11-7

SOLUTION

1) $$f_r = \frac{1}{2\pi\sqrt{LC}} = 127.4\ \text{kHz}$$

2) $$\omega_r = 800.6 \times 10^3\ \text{rad/sec}$$

3) $$X_L = \omega_r L = 31.22\ \text{k}\Omega$$

4) $$Q = \frac{X_L}{R} = \frac{31.22\ \text{k}\Omega}{0.27\ \text{k}\Omega} = 115.6$$

5) $$\mathbf{E}_o = Q\mathbf{E}_{in}$$

6) $$\mathbf{A}_v = \frac{\mathbf{E}_o}{\mathbf{E}_{in}} = Q$$

Therefore

7) $$A_v = 20 \log \frac{\mathbf{E}_o}{\mathbf{E}_{in}}\ \text{dB} = 20 \log 115.6\ \text{dB}$$

$$= 20(2.0631)\ \text{dB} = 41.26\ \text{dB}$$

PS 11-8 Express the voltage gain of the network in Fig. PS 11-8 in dB.

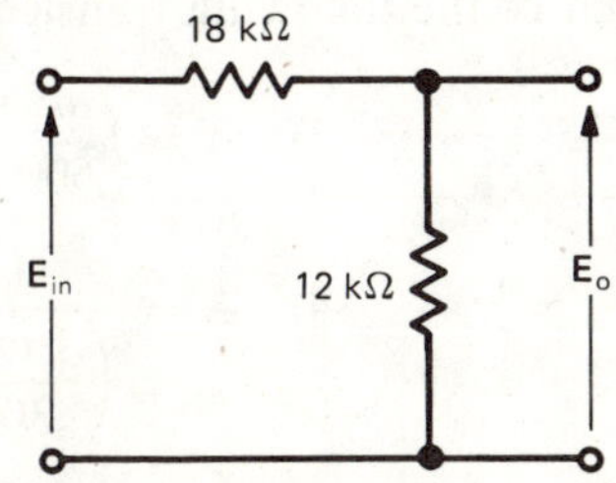

FIGURE PS 11-8

SOLUTION

$$A_v = 20 \log \frac{\mathbf{E}_o}{\mathbf{E}_{in}}\ \text{dB} = 20 \log \frac{12\ \text{k}\Omega}{12\ \text{k}\Omega + 18\ \text{k}\Omega}\ \text{dB}$$

$$= 20 \log (0.4)\ \text{dB} = 20(-0.3979)\ \text{dB}$$

$$= -7.9588\ \text{dB}$$

PS 11-9 The voltage gain in dB of the four-terminal network in Fig. PS 11-9 is $A_v = -5.67$ dB. What is the voltage gain ratio E_2/E_1?

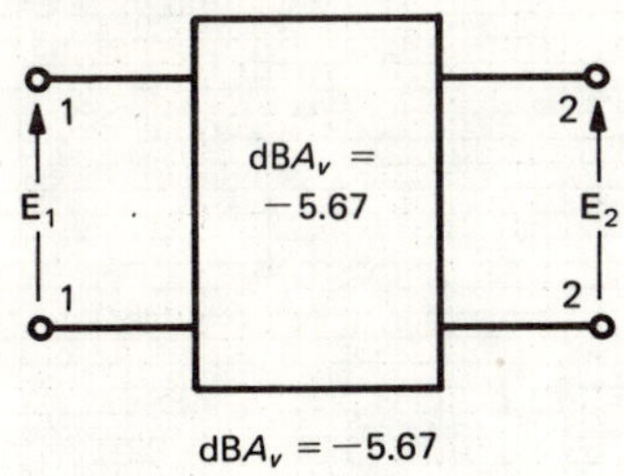

FIGURE PS 11-9

SOLUTION

1) $$A_v = 20 \log \frac{E_2}{E_1}\ \text{dB}$$

2) $$-5.67\ \text{dB} = 20 \log \frac{E_2}{E_1}\ \text{dB}$$

3) $$\frac{E_2}{E_1} = \log^{-1}\left(\frac{-5.67}{20}\right) = \log^{-1}(-0.2835)$$

Therefore we seek the number which represents the ratio E_2/E_1 and whose logarithm is -0.2835 (the notation $\log^{-1}$ implies the antilog). You will recall that in log tables positive mantissa values only are usually tabulated, and hence to obtain a positive mantissa we add and subtract 1, so that

4) $$\frac{E_2}{E_1} = \log^{-1}(1 - 0.2835 - 1) = \log^{-1}(0.7165 - 1)$$

Looking in a table of logarithms (or on a slide rule) for the number corresponding to 0.7165, we have 5.206 and with a characteristic of -1.*

5) $$\frac{E_2}{E_1} = 0.5206$$

PS 11-10 Construct a Bode plot for the amplitude response for each of the following transfer functions:

1) $\mathbf{A} = 3200$ 3) $\mathbf{A} = j\frac{\omega}{50}$

2) $\mathbf{A} = 0.004$ 4) $\mathbf{A} = \frac{1}{\left(j\frac{\omega}{300}\right)^2}$

Determine the phase response in each case by inspection of the above transfer functions.

SOLUTION

1) $$A = 20 \log 3200 = 70.1 \text{ dB}$$

*Negative mantissas for negative logarithms may be conveniently evaluated on a slide rule with the L and CI scales.

The phase angle is always zero, since **A** equals a constant. Remember that a constant amplitude response implies zero phase shift.

2) $$A = 20 \log 0.004 \cong -48 \text{ dB}$$

The phase angle is again zero at all ω.

3) The corner frequency is $\omega_c = 50$ rad/sec. Since $n = 1$, this Bode plot will have a positive slope of 20 dB/dec (or 6 dB/oct) and pass through 0 dB at $\omega = \omega_c$. The phase shift is a constant 90°.

4) The corner frequency is $\omega_c = 300$ rad/sec. Hence the Bode plot will pass through 0 dB at $\omega_c = 300$. Since the ω term is in the denominator and $n = 2$, the slope will be negative and equal to -40 dB/dec (or -12 dB/oct). For $n = 2$, $\phi = -n\,90° = -2(90°) = -180°$.

The resultant curves are plotted in Fig. PS 11-10.

PS 11-11 What is the transfer function corresponding to the amplitude response curves labeled *a*, *b*, *c*, *d* in Fig. PS 11-11?

SOLUTION Curve *a* crosses 0 dB at a corner frequency $f_c = 4000$ Hz and a positive slope $\cong$ 6 dB/oct.

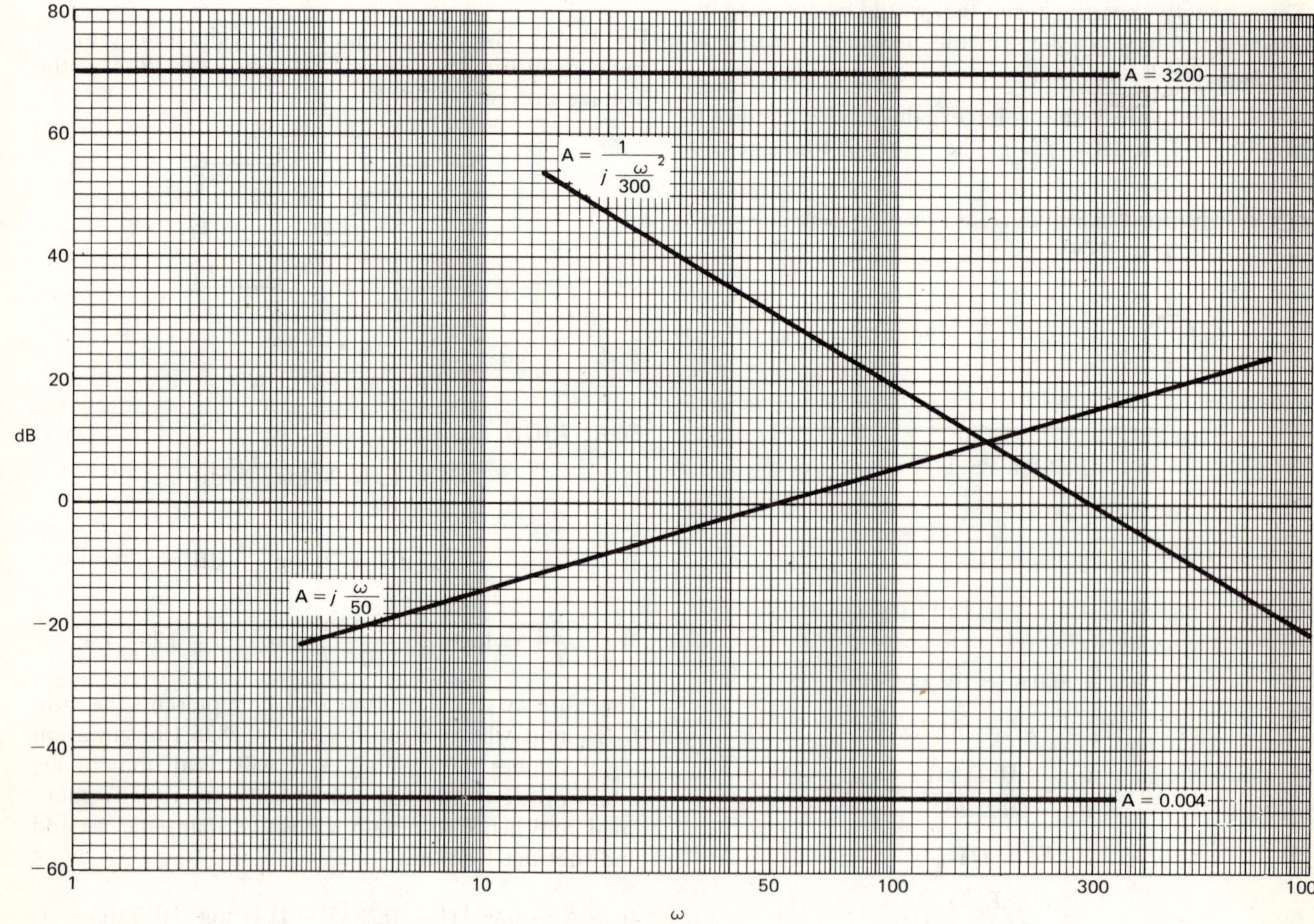

FIGURE PS 11-10

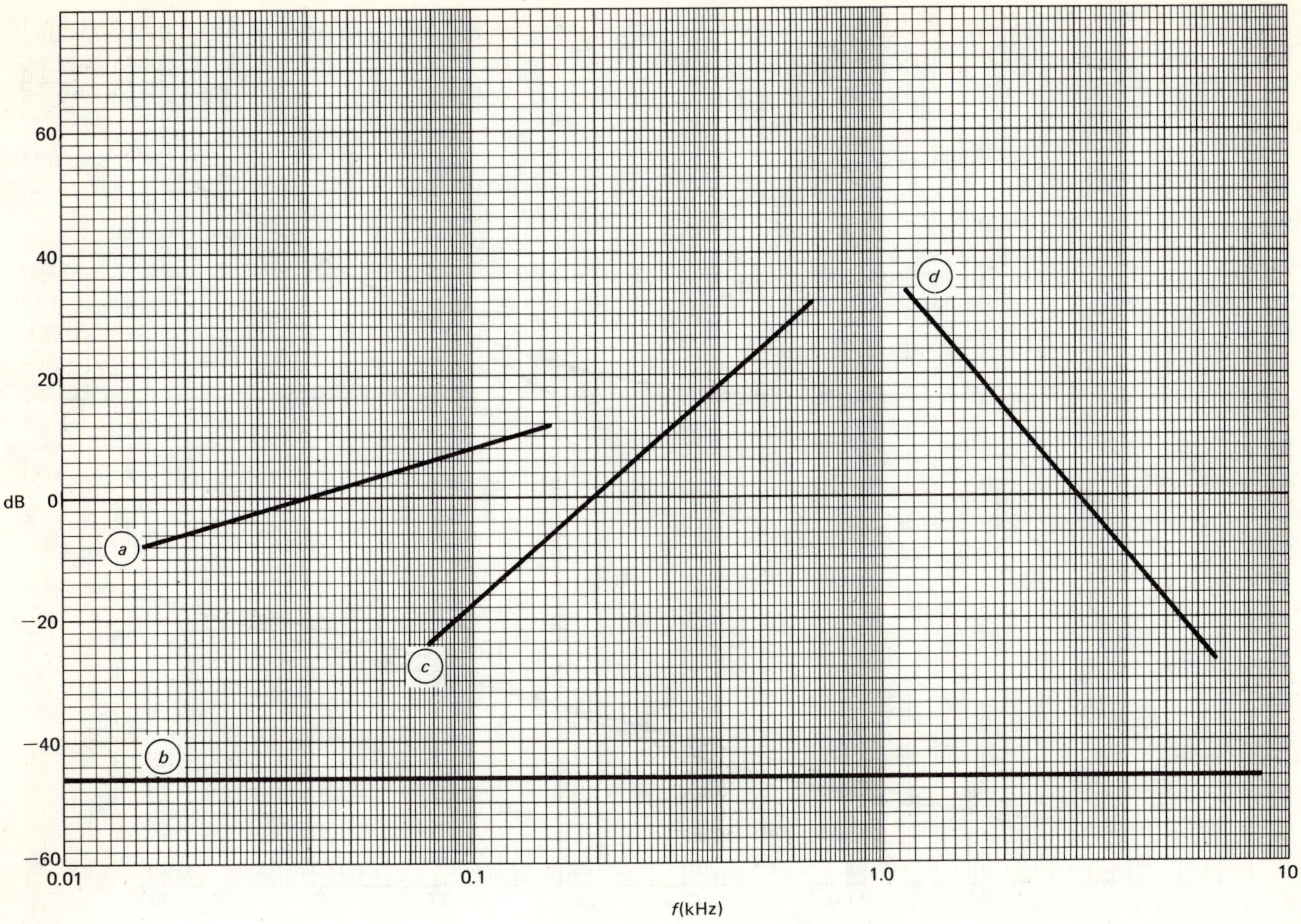

FIGURE PS 11-11

Since the 6 dB/oct slope implies 90° of phase shift,

$$\mathbf{A}_a = j\frac{f}{40}$$

Curve b is constant at -43 dB. Therefore

$$-43 \text{ dB} = 20 \log \mathbf{A}_b$$

$$\mathbf{A}_b = \log^{-1}\left(-\frac{43 \text{ dB}}{20}\right) = 0.00708$$

Curve c has $f_c = 200$ Hz and a positive slope of 18 dB/oct. Hence the phase shift is 3(90°) as $n = 3$. Therefore

$$\mathbf{A}_c = \left(j\frac{f}{200}\right)^3$$

Curve d has $f_c = 3000$ Hz and a negative slope of 24 dB/oct. Therefore

$$\mathbf{A}_d = \frac{1}{\left(j\dfrac{f}{3000}\right)^4}$$

PS 11-12 Construct a Bode plot for the function $\mathbf{A} = \left(1 + j\dfrac{f}{500 \text{ Hz}}\right)^2$.

First show the straight-line approximations for the amplitude and phase response and then insert the correction factors so that a smooth curve may be constructed.

SOLUTION The amplitude response will be flat from $f = 0$ Hz until $f = f_c = 500$ Hz. For $f > f_c$, the slope rises at 40 dB/dec or 12 dB/oct. Since $n = 2$, the amplitude correction factors are 2, 6, and 2 dB at $0.5 f_c$, f_c, and $2 f_c$ respectively. The phase corrections are approximately $+12°$ and $-12°$ at $0.1 f_c$ and $10 f_c$ respectively and $-12°$ and $+12°$ at $0.5 f_c$ and $2 f_c$ respectively. The phase curve will go through 90° at f_c and approach 0° and 180° asymptotically for $f \ll f_c$ and $f \gg f_c$ respectively. The resultant plot is shown in Fig. PS 11-12.

PS 11-13 Construct a Bode plot of the transfer function $\mathbf{A} = \dfrac{1}{\left(1 + j\dfrac{f}{500 \text{ Hz}}\right)^3}$.

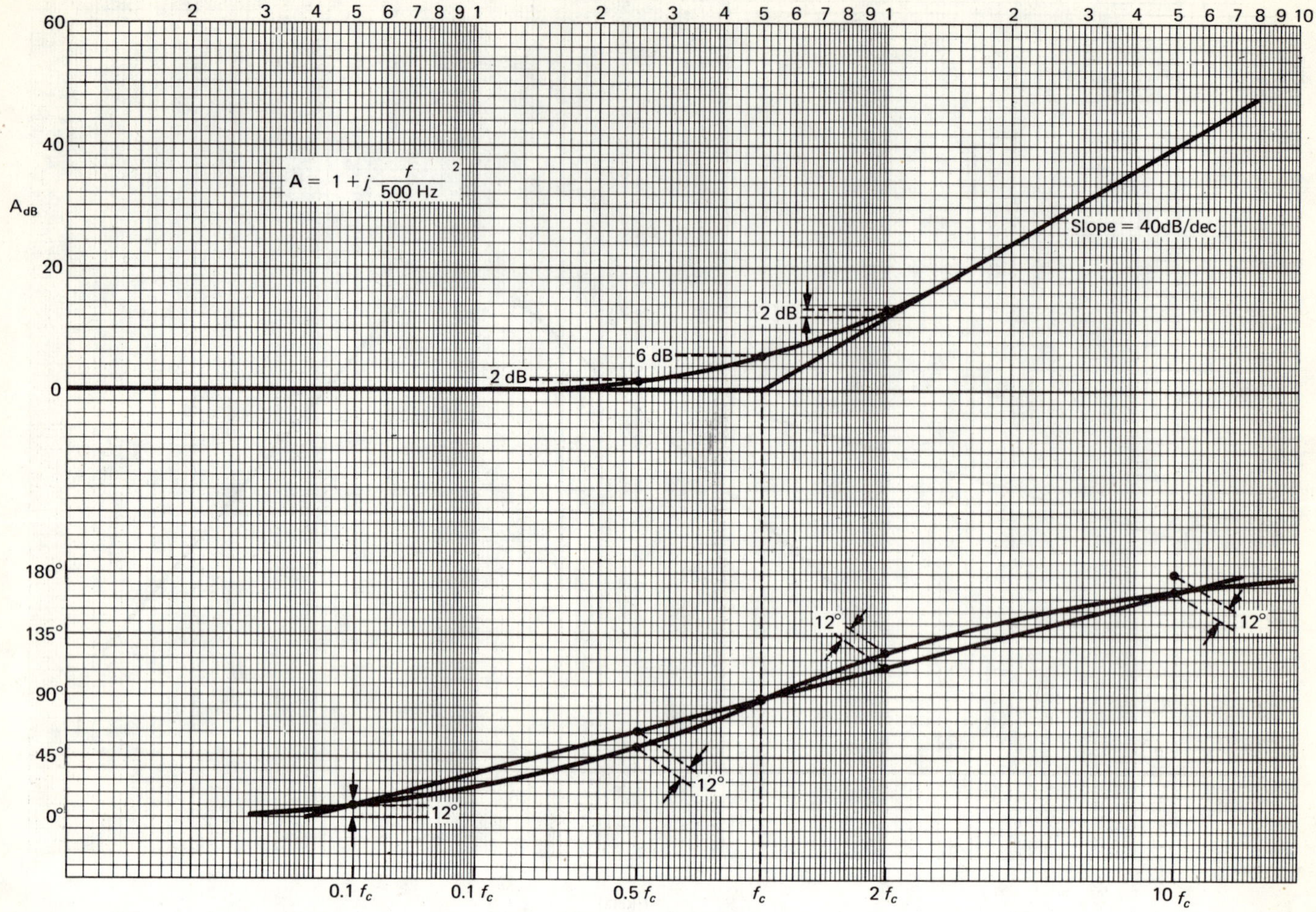

FIGURE PS 11-12

SOLUTION The approximated amplitude response will be flat for $f > f_c$ until f_c. At f_c the slope changes from zero to -60 dB/dec. Correction factors for sketching the actual response are 3, 9, and 3 dB at 0.5 f_c, f_c, and 2 f_c respectively and $+18°$ and $-18°$ at 0.5 f_c and 2 f_c respectively. The phase curve will go through $-135°$ at f_c and approach 0° and $-270°$ asymptotically for $f \ll f_c$ and $f \gg f_c$ respectively. The resultant plot is shown in Fig. PS 11-13.

PS 11-14 Determine the voltage gain transfer function and construct a Bode plot for the circuit of Fig. PS 11-14*a*.

SOLUTION First we must develop the transfer function. Thus

1) $$\mathbf{A}_v = \frac{\mathbf{E}_o}{\mathbf{E}_{in}} = \frac{-jX_C}{R + j(X_L - X_C)}$$

$$= \frac{1/j\omega C}{R + j\left(\omega L - \dfrac{1}{\omega C}\right)}$$

$$= \frac{1}{j\omega CR + j^2(\omega^2 LC - 1)}$$

Now the trick is to algebraically wrestle the final form of (1) into the quadratic form of Eq. (11-10*b*), repeated here as (2).

2) $$\mathbf{A} = \frac{1}{1 - \left(\dfrac{\omega}{\omega_c}\right)^2 + j2\delta\dfrac{\omega}{\omega_c}}$$

If we let

3) $\omega_c^2 = \dfrac{1}{LC} = 83.333 \times 10^6$ (rad/sec)^2 or $\omega = 9129$ rad/sec and therefore obtain

4) $$C = \frac{1}{\omega_c^2 L}$$

and then substitute (3) and (4) into (1), we have

5) $$\mathbf{A}_v = \frac{1}{j\dfrac{\omega R}{\omega_c^2 L} - \dfrac{\omega^2}{\omega_c^2} - 1}$$

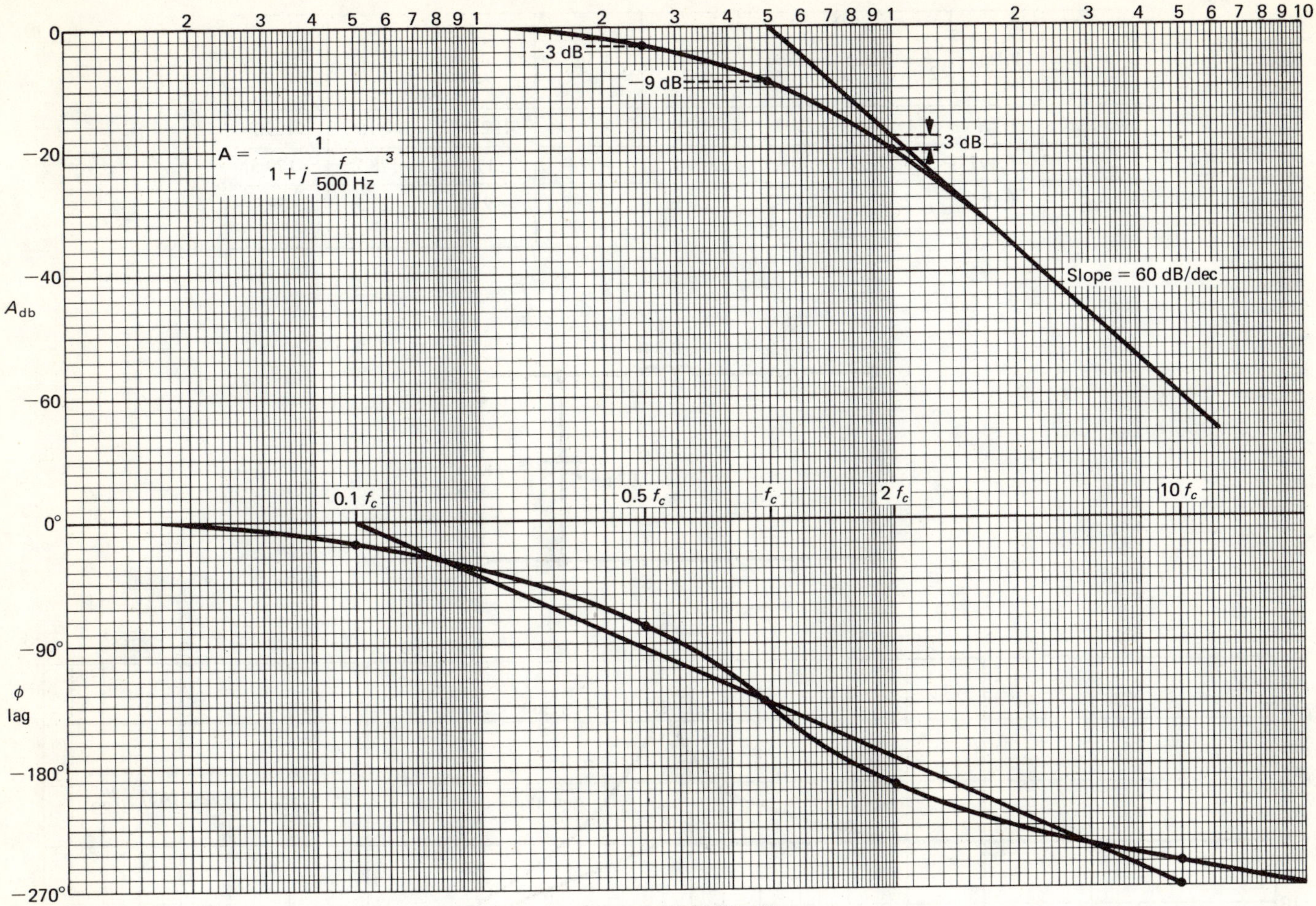

FIGURE PS 11-13

$$= \frac{1}{1 - \left(\frac{\omega}{\omega_c}\right)^2 + j\frac{R}{\omega_c L} \times \frac{\omega}{\omega_c}}$$

Since

6) $$Q = \frac{\omega_c L}{R} = 11.65$$

we may substitute (6) into (5) to obtain

7) $$\mathbf{A}_v = \frac{1}{1 - \left(\frac{\omega}{\omega_c}\right)^2 + j\frac{1}{Q}\frac{\omega}{\omega_c}}$$

Equation (7) is in the same form as (2), with

8) $$2\delta = \frac{1}{Q}$$

or

9) $$\delta = \frac{1}{2Q} = 0.0429 \cong 0.043$$

In constructing the Bode plot, we know we will have for $\omega \ll \omega_c$ a straight line along 0 dB until ω_c, at which point it breaks downward with a slope of −40 dB/dec for $\omega \gg \omega_c$. The response in the vicinity of ω_c is determined by the damping factor δ. Turning to Fig. 11-9 we see that for $\delta = 0.043$ we are off the plot. Hence we turn to Eq. (11-10*c*).

10) $$\text{dB peak} = -20 \log 2\delta = 21.3 \text{ dB}$$

which yields the height of the resonant peak at ω_c. For a more accurate plot, we might as well evaluate the transfer function. Thus we have

11) $$\mathbf{A} = \frac{1}{1 - \left(\frac{\omega}{9129}\right)^2 + j2(0.0429)\frac{\omega}{9129}}$$

$$= \frac{1}{1 - \frac{\omega^2}{83.33 \times 10^6} + j(9.399 \times 10^{-6})\omega}$$

and plugging in various values of ω yields the following table of data, which is plotted in Fig. PS 11-14*b*:

	$0.1\omega_c$	$\frac{\omega_c}{2} = 0.5\omega_c$	$0.8\omega_c$	ω_c	$1.5\omega_c$	$2\omega_c$	$4\omega_c$	$10\omega_c$
ω rad/sec	912.9	4564	7303	9129		18258	36516	91290
ϕ	$-0.497°$	$-3.27°$	$-10.8°$	$-90°$	$-174°$	$-177°$	$-178.7°$	$-179.5°$
A_{dB}	0.087	2.48	8.72	21.3	-1.98	-9.56	-23.5	-39.9

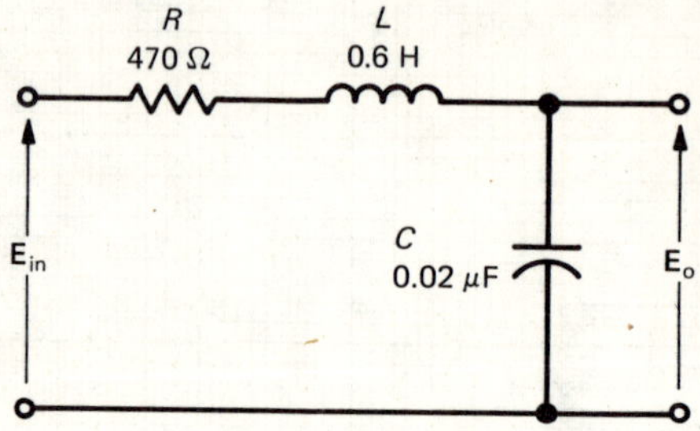

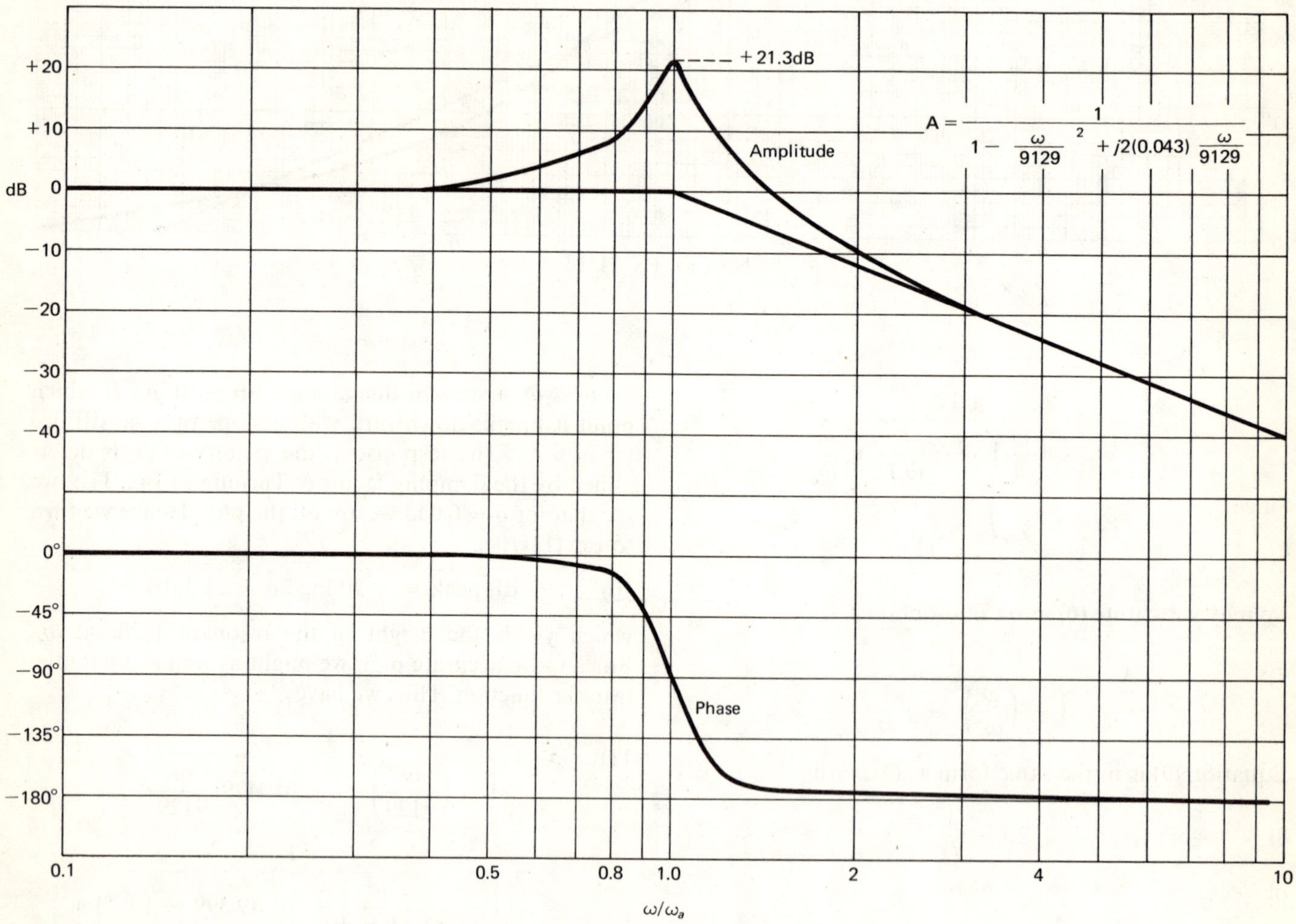

(b)

FIGURE PS 11-14

PS 11-15 Develop the transfer function and construct a Bode plot of the amplitude response of the network shown in Fig. PS 11-15*a*. Assume $R_1 = R_2$.

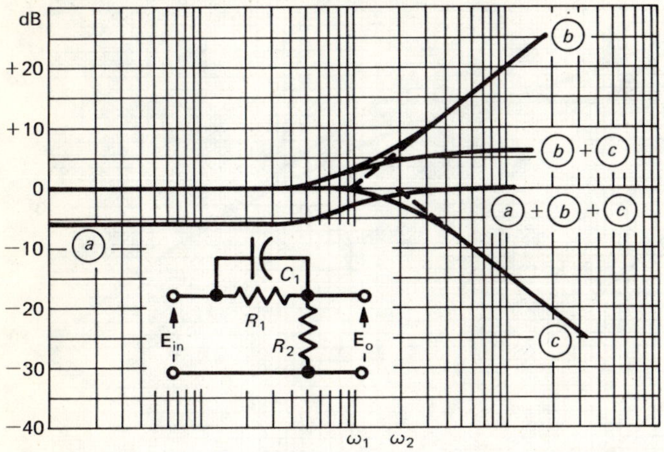

FIGURE PS 11-15

SOLUTION

1. First develop the transfer function.

1) $$\frac{\mathbf{E}_o}{\mathbf{E}_{in}} = \frac{R_2}{R_2 + \dfrac{R_1(-jX_{C_1})}{R_1 - jX_{C_1}}}$$

$$= \frac{R_2}{R_2 + \dfrac{R_1}{j\omega C_1}\dfrac{1}{R_1 + 1/j\omega C_1}}$$

$$= \frac{R_2}{R_2 + \dfrac{R_1}{j\omega C_1}\dfrac{1}{(j\omega R_1 C_1 + 1)/j\omega C_1}}$$

$$= \frac{R_2}{R_2 + \dfrac{R_1}{1 + j\omega R_1 C_1}}$$

$$= \frac{R_2}{\dfrac{R_2 + j\omega R_1 R_2 C_1 + R_1}{1 + j\omega R_1 C_1}}$$

$$= \frac{R_2(1 + j\omega R_1 C_1)}{R_1 + R_2 + j\omega R_1 R_2 C_1}$$

Dividing numerator and denominator by $R_1 + R_2$ yields

$$= \frac{R_2}{R_1 + R_2}\,\frac{1 + j\omega R_1 C_1}{1 + j\omega R_{eq} C_1}$$

where

$$R_{eq} = \frac{R_1 R_2}{R_1 + R_2}$$

Now let $R_1 C_1 = \tau_1 = 1/\omega_1$ and $R_{eq} C_1 = \tau_2 = 1/\omega_2$:

2) $$\frac{\mathbf{E}_o}{\mathbf{E}_{in}} = \frac{R_2}{R_1 + R_2}\,\frac{1 + j\omega/\omega_1}{1 + j\omega/\omega_2}$$

2. Analyze the transfer function to ascertain what contributions the various factors will make in the Bode plot. Equation (2), the transfer function, is of the form

$$\frac{\mathbf{E}_o}{\mathbf{E}_{in}} = a\frac{b}{c}$$

where $a = R_2/(R_1 + R_2)$, $b = 1 + j\omega/\omega_1$, and $c = 1 + j\omega/\omega_2$. Therefore,

$$\text{dB} = 20\,(\log a + \log b - \log c)$$

The log a term will contribute a component which is independent of frequency, while the b and c terms are frequency-dependent. We have shown in previous examples that the b term will break at ω_1 and slope upward at 20 dB per decade, while the c term breaks at ω_2 and slopes downward at 20 dB per decade. Inspection of the circuit indicates that at low frequencies, where X_{C_1} is very large compared with R_1, the output will equal $[R_2/(R_1 + R_2)]E_{in}$. At frequencies where X_{C_1} is small compared with R_2, the output will approach the input.

3. Construct the Bode plot. For the purpose of sketching, let us assume that $R_2 = R_1$. Then $R_2/(R_1 + R_2) = \frac{1}{2}$, or -6 dB. This fixes the ratio of τ_1 to τ_2:

$$\tau_1 = R_1 C_1$$

$$\tau_2 = \frac{R_1 R_2}{R_1 + R_2} C_1 \quad \text{or} \quad R_1(\tfrac{1}{2})C_1 = 0.5 R_1 C_1$$

$$\frac{\tau_1}{\tau_2} = \frac{R_1 C_1}{0.5 R_1 C_1} = 2 = \frac{\omega_2}{\omega_1}$$

The Bode plot is sketched in Fig. PS 11-15*b*.

PS 11-16 Bode plot the transfer function

$$\mathbf{A} = \frac{10\left(1 + j\dfrac{\omega}{20}\right)}{\left(1 + j\dfrac{\omega}{5}\right)\left(1 + j\dfrac{\omega}{100}\right)}$$

SOLUTION In the very low frequency spectrum or for ω approaching zero, we see that $\mathbf{A} = 10$, or 20 dB. This means we have a network capable of providing an amplification of ten times down to dc or zero frequency, as shown in Fig. PS 11-16. Electronic amplifiers can do this. As ω increases, the first break occurs at $\omega = 5$ rad/sec, and since this is in the denominator, the gain (amplitude) rolls off at -20 dB/dec from the 20-dB level. The next break at $\omega = 20$ rad/sec is in the numerator, which tries to pull $\mathbf{A}$ up at $+20$ dB/dec. However, this is canceled by the previous -20 dB/dec slope, so that the net slope is now zero and it stays that way until

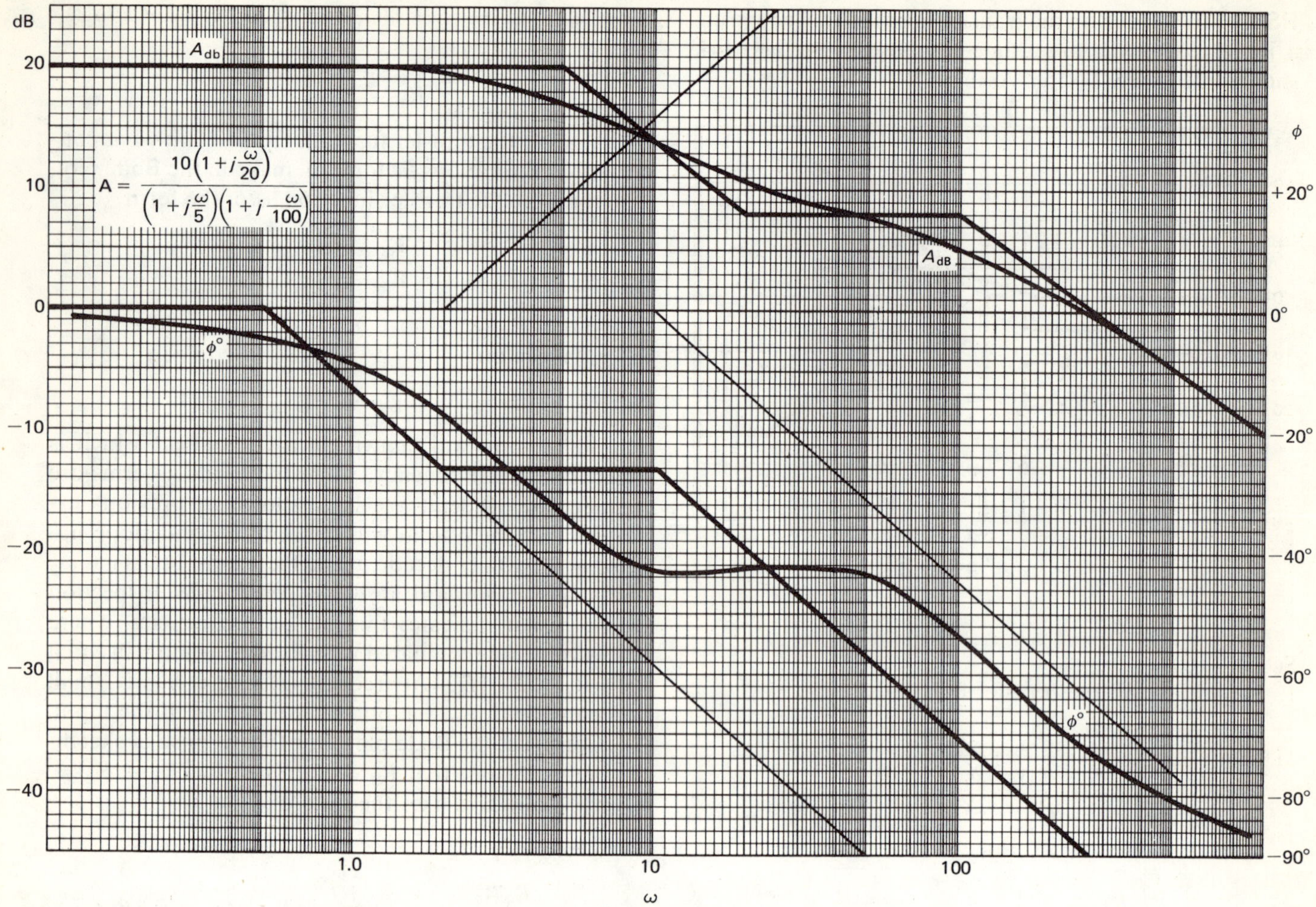

FIGURE PS 11-16

$\omega = 100$, at which it breaks down again at -20 dB/dec and remains so. If correction factors are inserted or A_{dB} is actually computed, the smoothed amplitude response curve results.

The phase response is approximated by lines sloping plus or minus 45°/dec, depending upon whether or not the term is in the numerator or denominator. The phase approximations start at 0° at 0.1 ω_c, go through 45° at ω_c, and terminate at $\pm 90°$ at 10 ω_c. The resultant phase approximation is illustrated by the stepped phase curve. For more accuracy, the various correction factors can be inserted or the actual phase angle of **A** may be computed to yield the smoothed phase curve of Fig. PS 11-16.

PS 11-17 Bode plot the transfer function

$$1) \qquad \mathbf{A} = \frac{j\dfrac{\omega}{\omega_1}}{\left(1 + j\dfrac{\omega}{\omega_1}\right)^2}$$

SOLUTION The above transfer function may be rewritten as

$$2) \qquad \mathbf{A} = \frac{j\dfrac{\omega}{\omega_1}}{1 + j\dfrac{\omega}{\omega_1}} \; \frac{1}{1 + j\dfrac{\omega}{\omega_1}}$$

Inspection of (2) indicates that the first factor is that of a high-pass *RC* type filter and the second factor is that of a low-pass filter with identical corner frequencies. Therefore, in the frequency range $\omega < \omega_1$, we have the low-pass response climbing at 20 dB/dec while the high-pass response is flat at 0 dB. At $\omega = \omega_1$, the high-pass response breaks to zero slope while the low-pass response breaks downward at -20 dB/dec. This yields the approximated amplitude response of Fig. PS 11-17.

The approximate phase response may be sketched by first drawing the high-pass filter phase response as a line starting at $\omega/\omega_1 = 0.1$, sloping down 45°/dec from $+90°$

to 0° at $\omega/\omega_1 = 10$, and then sketching the low-pass filter phase with the same slope and ω values but starting at 0° and terminating at $-90°$. The approximate phase may then be sketched by algebraically adding both the high- and low-pass phase curves. Should greater accuracy be required, correction factors may be inserted or actual values computed in (2). The actual phase and magnitude are also shown in Fig. PS 11-17.

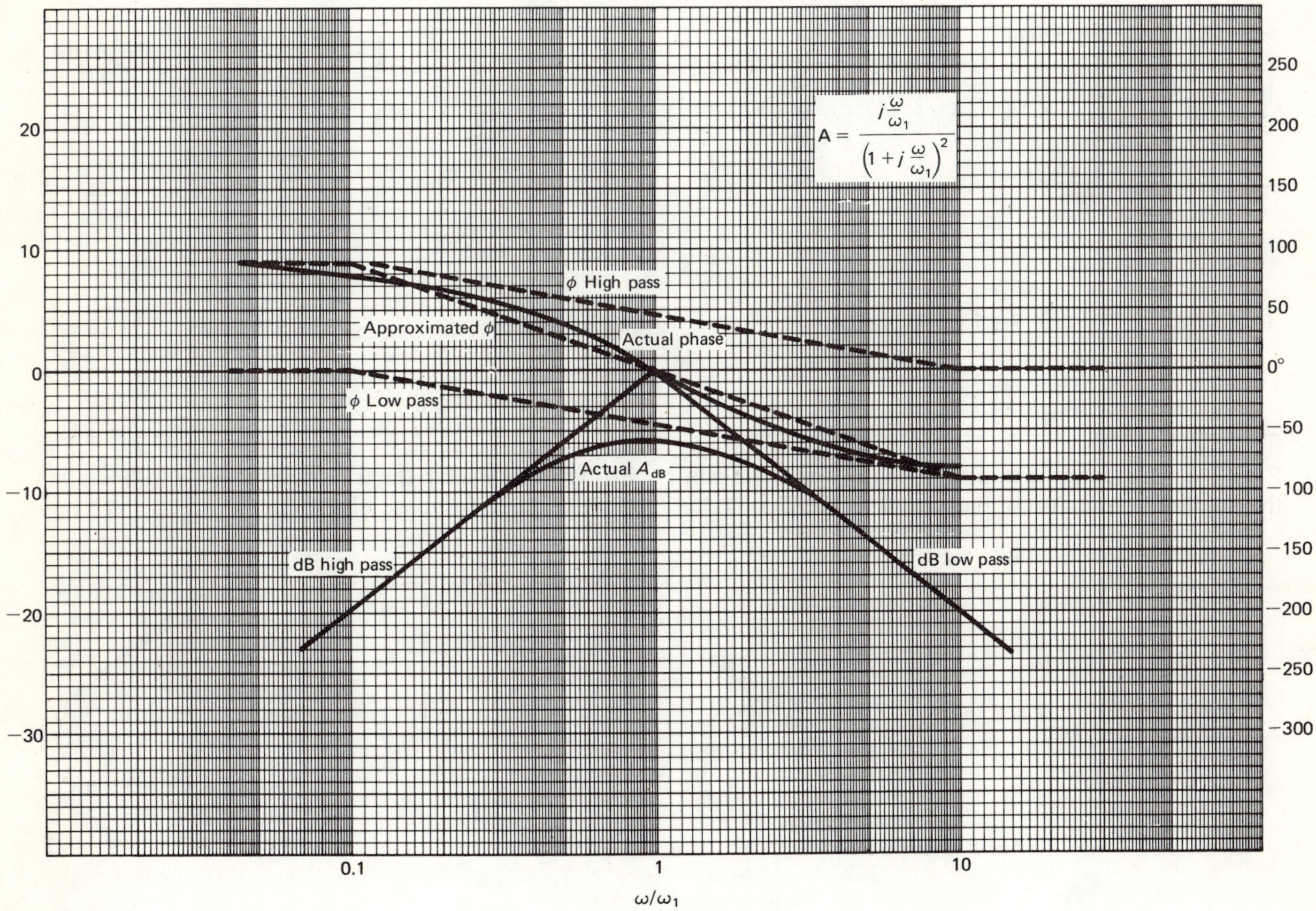

FIGURE PS 11-17

PROBLEMS WITH ANSWERS

PA 11-1 Develop the transfer function and Bode plot the amplitude response of the circuit in Fig. PA 11-1*a*.

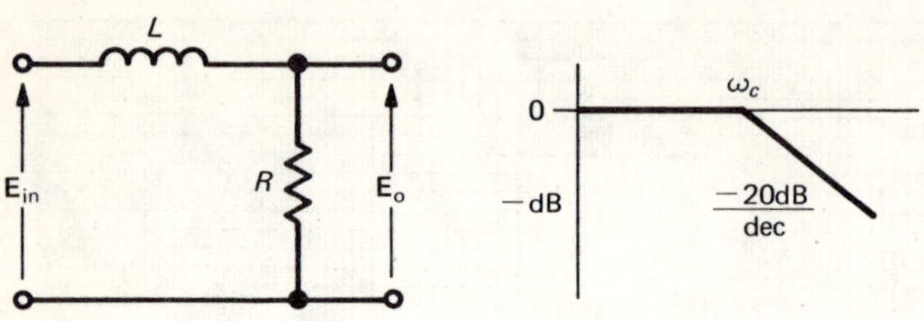

(*a*) (*b*)

ANSWER See Fig. PA 11-1*b*.

PA 11-2 Develop the transfer function and Bode plot the amplitude response of the circuit in Fig. PA 11-2*a*.

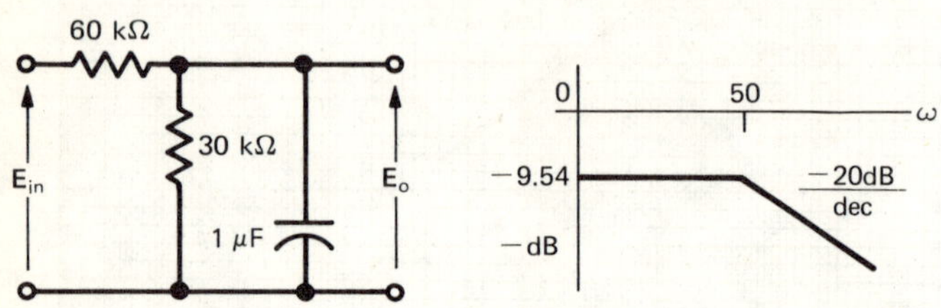

(*a*) (*b*)

ANSWER See Fig. PA 11-2*b*.

PA 11-3 Develop the transfer function and Bode plot the amplitude response of the circuit in Fig. PA 11-3*a*.

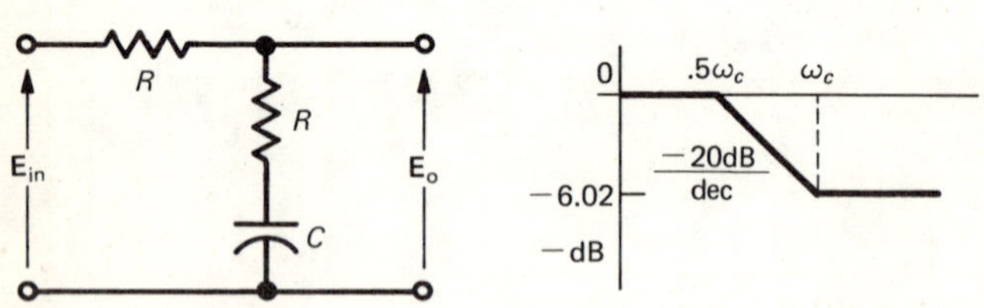

(*a*) (*b*)

ANSWER See Fig. PA 11-3*b*.

PA 11-4 Develop the transfer function and Bode plot the amplitude response of the circuit in Fig. PA 11-4*a*.

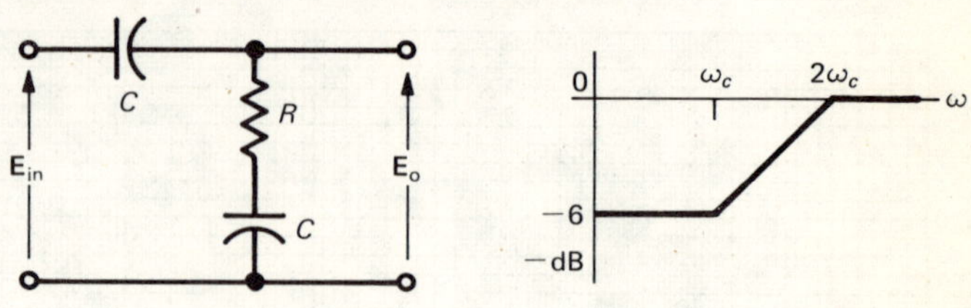

(*a*) (*b*)

ANSWER See Fig. PA 11-4*b*.

PA 11-5 Develop the transfer function and Bode plot the amplitude response of the circuit in Fig. PA 11-5*a*.

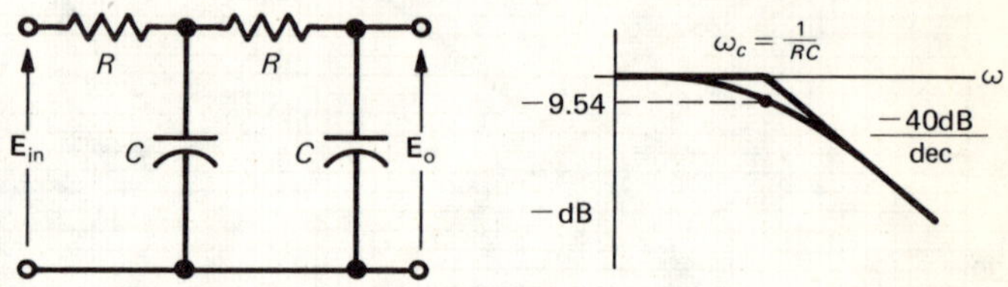

(*a*) (*b*)

ANSWER See Fig. PA 11-5*b*.

PA 11-6 Develop the transfer function and Bode plot the amplitude response of the circuit in Fig. PA 11-6*a*.

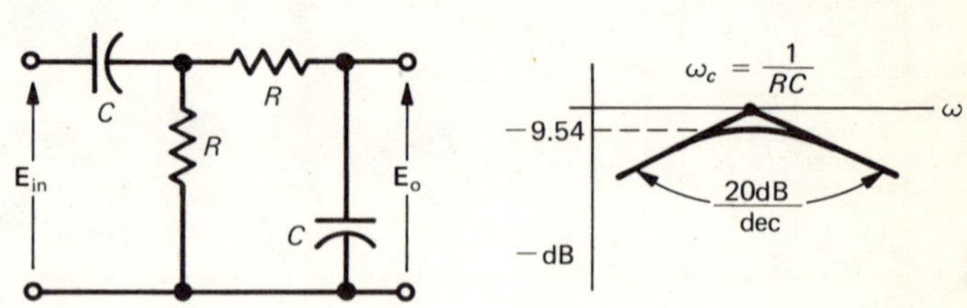

(*a*) (*b*)

ANSWER See Fig. PA 11-6*b*.

PA 11-7 Evaluate **A** at $\omega = \omega_c$ if $\delta = \frac{1}{2}$ in the following transfer function:

$$\mathbf{A} = \frac{\left(1 + j\dfrac{\omega}{\omega_c}\right)^2 100}{\left(1 + j\dfrac{\omega}{\omega_c}\right)\left[1 - \left(\dfrac{\omega}{\omega_c}\right)^2 + j2\delta\dfrac{\omega}{\omega_c}\right]}$$

ANSWER 43 dB $\underline{/-45^\circ}$

PA 11-8 Bode plot $\mathbf{A} = \left[\dfrac{1}{1 + j\omega(0.5 \times 10^{-6})}\right]^3$. Show both phase and amplitude.

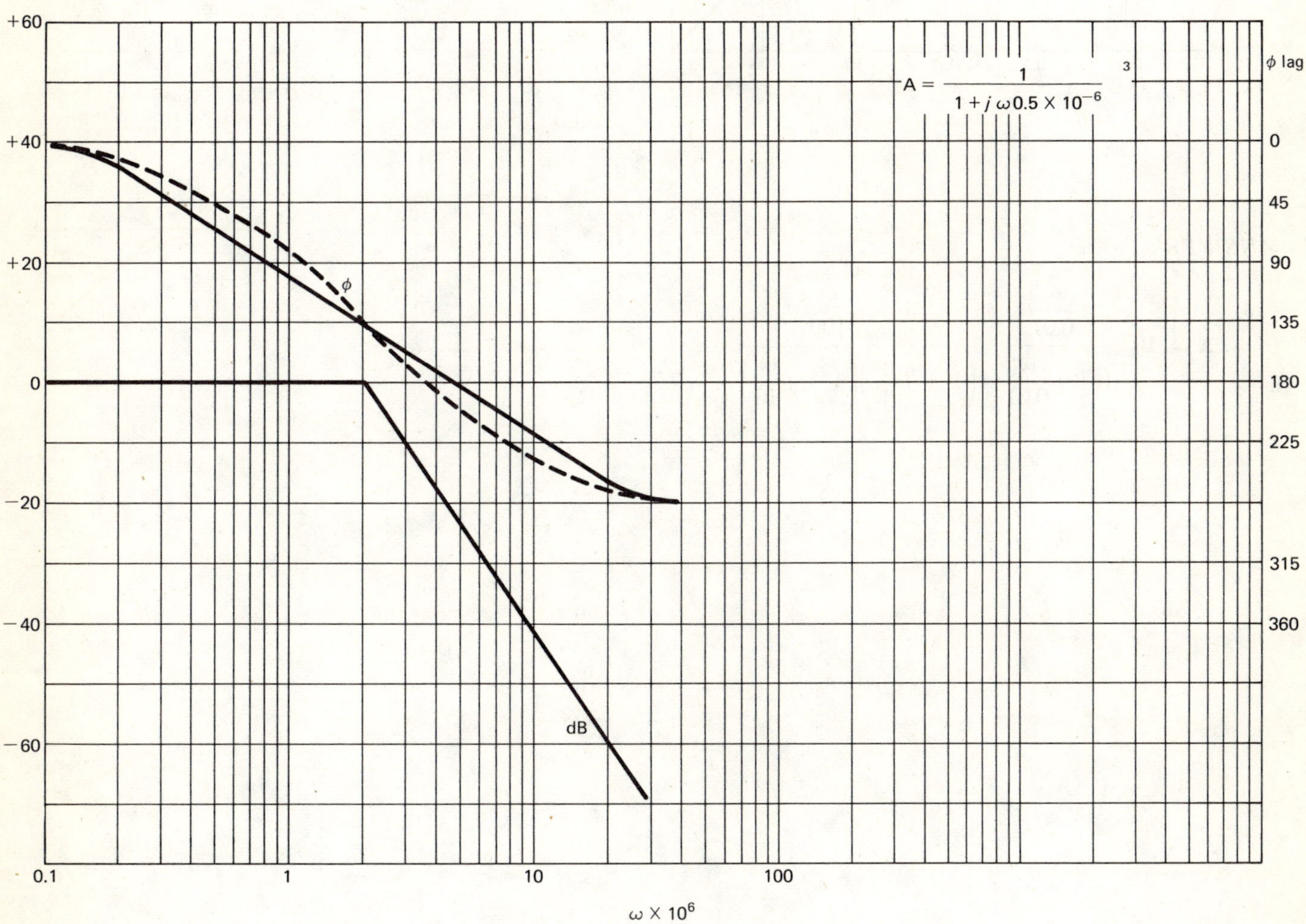

ANSWER See Fig. PA 11-8.

PA 11-9 Estimate the transfer function applicable to Fig. PA 11-9.

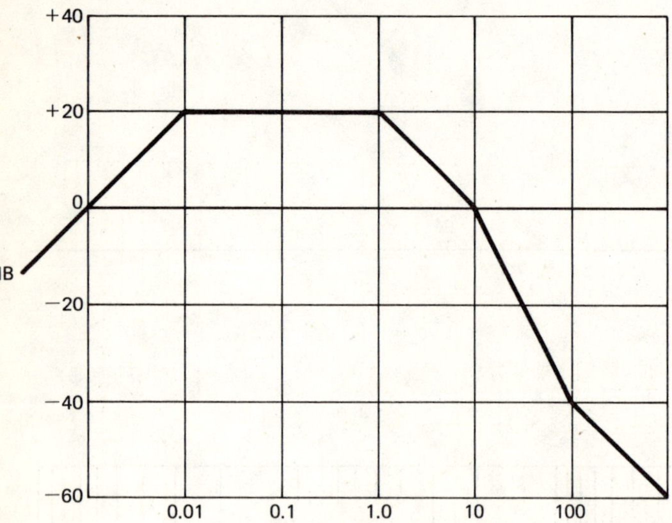

ANSWER

$$\mathbf{A} = 10\,\frac{j\dfrac{\omega}{0.01}}{1 + j\dfrac{\omega}{0.01}}\;\frac{1 + j\dfrac{\omega}{100}}{\left(1 + j\dfrac{\omega}{1}\right)\left(1 + j\dfrac{\omega}{10}\right)}$$

PA 11-10 Bode plot

$$\mathbf{A} = \frac{1\left(1 + j\frac{\omega}{1}\right)\left(1 + j\frac{\omega}{0.1}\right)}{\left(j\frac{\omega}{1}\right)^3\left(1 + j\frac{\omega}{100}\right)^2}$$

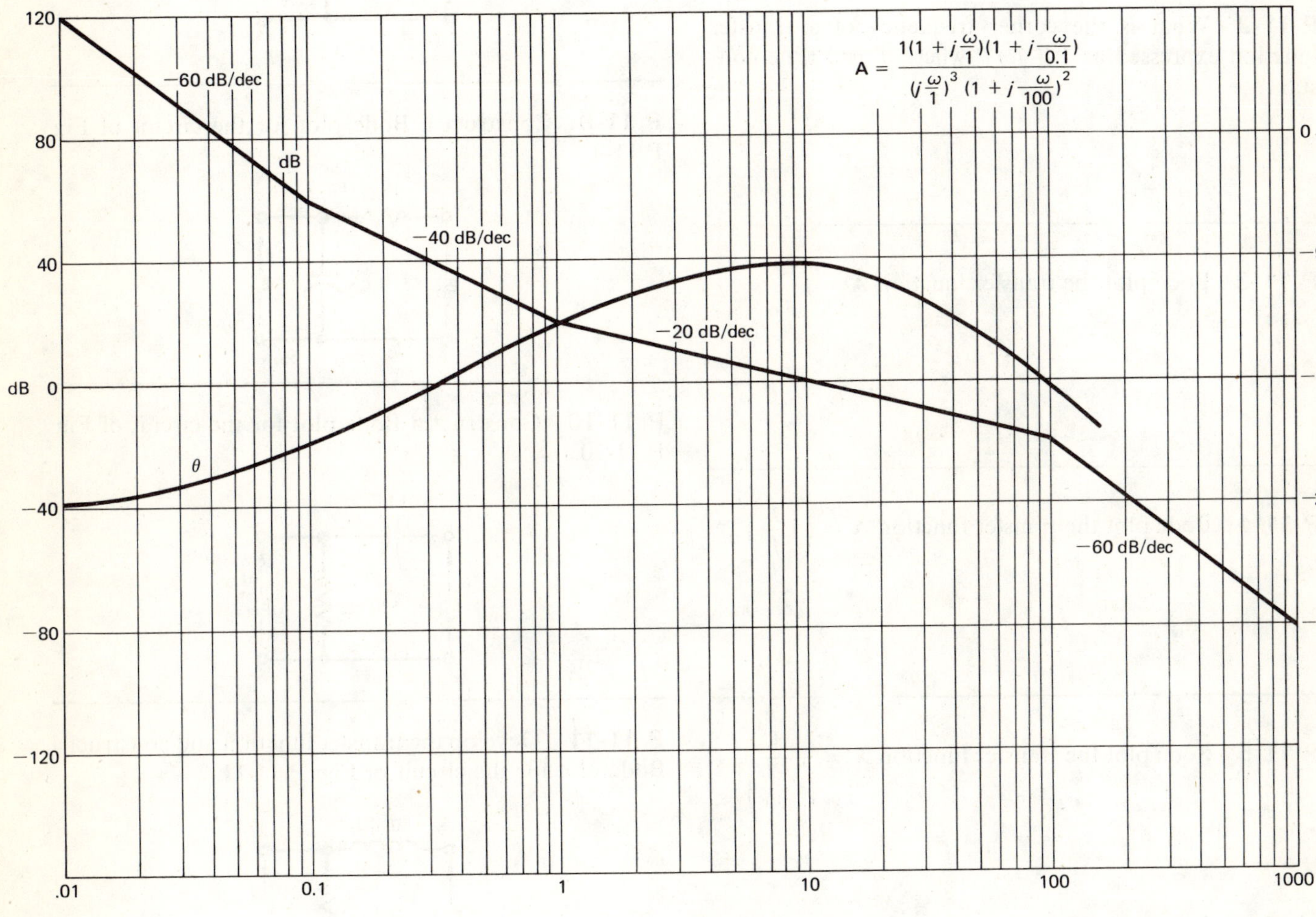

ANSWER See Fig. PA 11-10.

PROBLEMS WITHOUT ANSWERS

P 11-1 Bode plot the transfer function $\mathbf{A} = 28{,}000$.

P 11-2 What is the corner frequency of a transfer function expressed as $1 + j\omega T$ where T is a time constant?

P 11-3 Bode plot the transfer function $\mathbf{A} = j\dfrac{\omega}{\omega_c}$.

P 11-4 Bode plot the transfer function $\mathbf{A} = \left(j\dfrac{f}{f_c}\right)^3$.

P 11-5 Bode plot the transfer function $\mathbf{A} = \dfrac{1}{\left(j\dfrac{\omega}{\omega_c}\right)^2}$.

P 11-6 Bode plot the transfer function

$$\mathbf{A} = 1 + j\frac{f}{5\text{ kHz}}.$$

P 11-7 Bode plot the transfer function

$$\mathbf{A} = \frac{1}{\left(1 + j\dfrac{f}{80\text{ kHz}}\right)^2}.$$

P 11-8 Develop the transfer function and construct the Bode plot for the circuit of Fig. P 11-8.

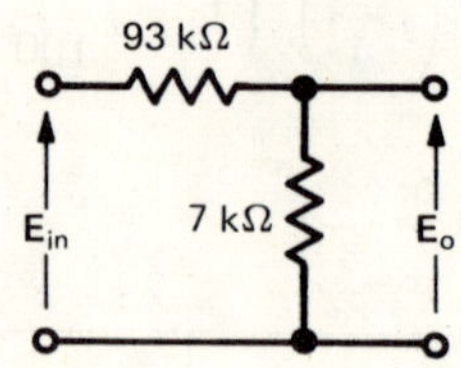

P 11-9 Construct a Bode plot for the circuit of Fig. P 11-9.

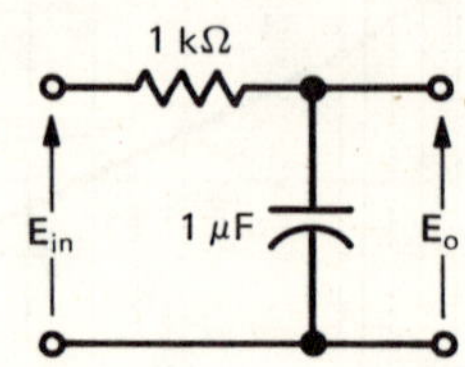

P 11-10 Construct a Bode plot for the circuit of Fig. P 11-10.

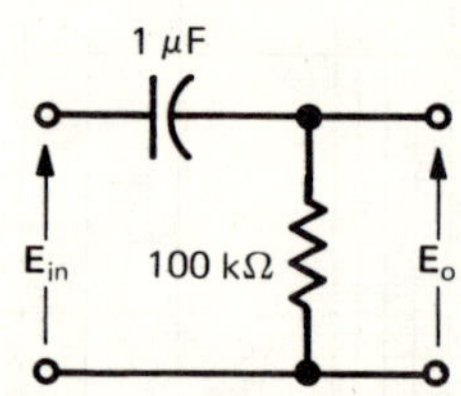

P 11-11 Develop the transfer function and construct a Bode plot for the circuit of Fig. P 11-11.

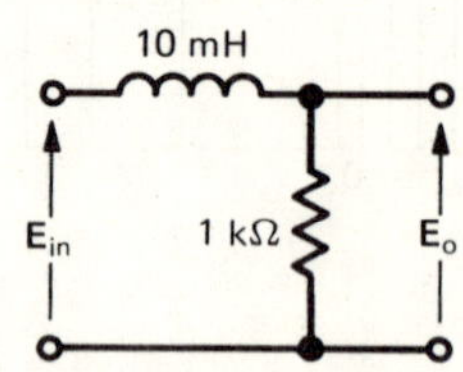

P 11-12 Develop the transfer function and construct a Bode plot for the circuit of Fig. P 11-12.

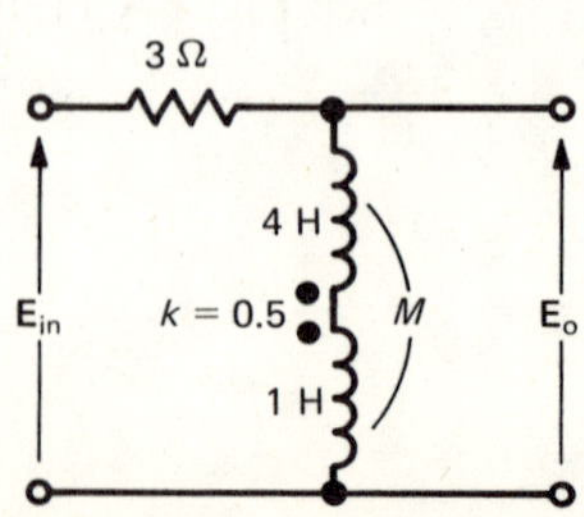

P 11-13 Sketch the Bode plot for the circuit of Fig. P 11-13.

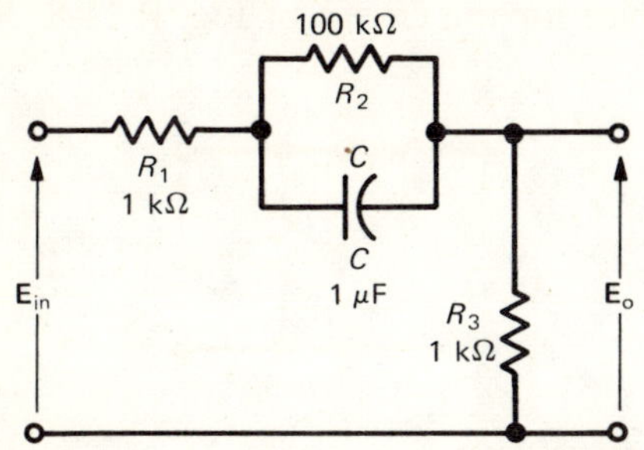

P 11-14 With regard to the transfer function

$$\mathbf{A} = \frac{100(1 + j\omega 1)}{(1 + j\omega 5)^2 (j\omega 1)}$$

fill in the blank spaces.

(*a*) The factor 100 in the numerator represents ________ dB of gain.
(*b*) The $1 + j\omega 1$ factor will break at $\omega =$ ________ and slope ________ dB/dec or ________ dB/oct. It will contribute ________ degrees of phase ________ at the corner frequency.
(*c*) The $(1 + j\omega 5)^2$ factor breaks at $\omega =$ ________ and slopes ________ dB/dec. It contributes ________ degrees of phase ________ at the corner frequency and the corresponding amplitude response is ________.
(*d*) The $j\omega 1$ factor contributes ________ degrees of phase ________ and slopes ________.

P 11-15 Estimate the transfer function of the Bode plot in Fig. P 11-15.

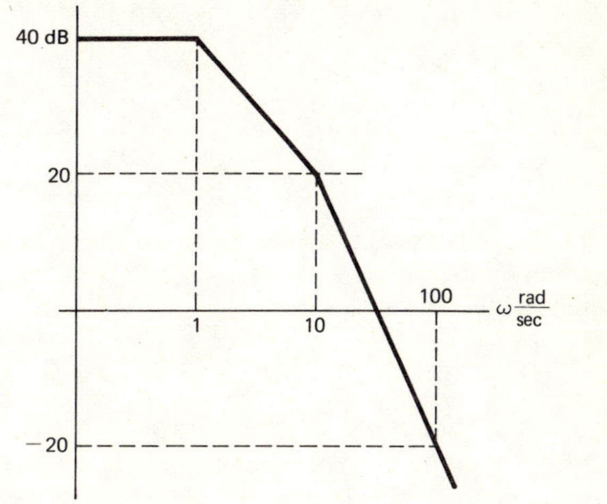

P 11-16 Estimate the transfer function of the Bode plot in Fig. P 11-16.

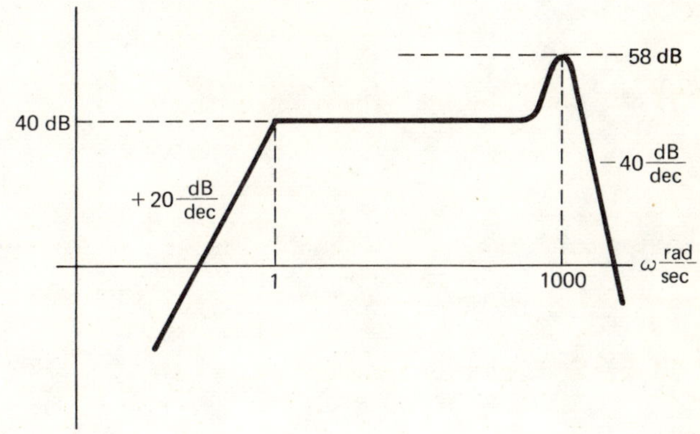

A current notation

The notation used in this text to represent an electric current is based upon the following considerations:

1) Electric current flow is designated with an upper- or lowercase symbol I and an appropriate subscript if required.
2) The assumed positive reference direction of current flow is indicated with an arrow as shown in Fig. A-1.

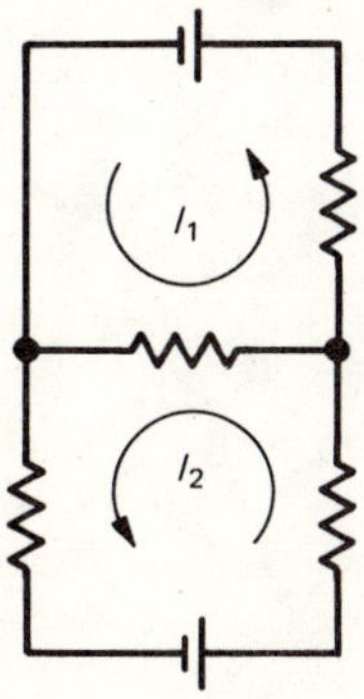

FIGURE A-1

3) Current flow in this text will, in general, imply electron or negative charge flow. Thus the current I in flowing through a resistor R polarizes the terminal which it first enters negative with respect to the terminal from which it emerges. This is illustrated in Fig. A-2.

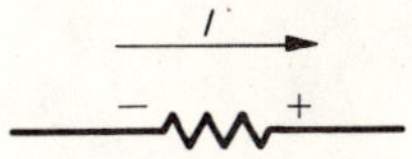

FIGURE A-2

If your previous training is such that you are more comfortable with so-called conventional current flow, which implies the transfer of positive rather than negative charge, you need only reverse the arrow associated with I in this text and proceed with the development presented or any problem you might be attempting to solve. As you can see from Fig. A-3, the resultant polarity of the voltage developed across an impedance is still the same, since the end of the impedance that the conventional current first enters is polarized positive with respect to the end from which it emerges.

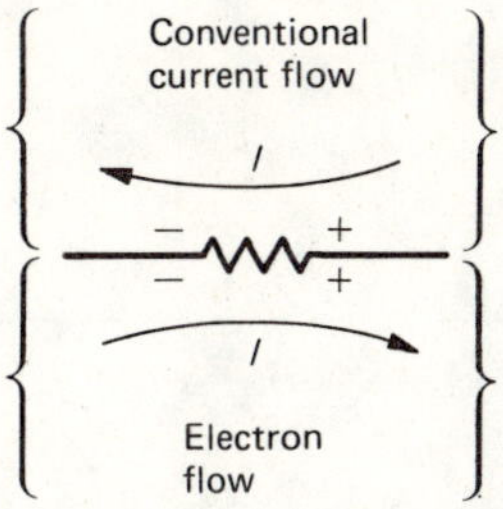

FIGURE A-3

For additional reference material regarding the notation used in this text, the reader is referred to Phillip Cutler, "Electronic Circuit Analysis," vol. 1, Chap. 1, McGraw-Hill Book Company, New York, 1960.

B voltage notation

The notation used in this text to describe a voltage is intended to avoid the confusion that sometimes arises in the student's mind with regard to a voltage drop or a voltage rise. Instead, this text merely refers to a voltage which is rigorously described according to the following convention:

1) The voltage between a pair of terminals is to be described by the symbol E or V in association with an arrow or double subscripts to indicate the terminals between which this voltage exists.
2) It will be agreed that the reference or starting point from which the voltage is reckoned is the tail of the arrow in arrow notation or the first subscript in double-subscript notation.
3) To evaluate V or E, you start at the tail of the arrow (or the first subscript) and follow some path through the circuit, while taking note of the various potentials encountered, until you arrive at the head of the arrow (or the second subscript). The resultant value of the expression for V or E will yield the voltage at the head of the arrow (or second subscript) relative to the reference (starting) point.
4) In evaluating (or expressing) V or E as you proceed from the reference point to the point in question, you will encounter various circuit elements. If in traversing this path you enter a circuit element at one terminal and emerge at the other terminal at a potential more positive than the point of entry, you have gone in the positive direction or up in potential, and this is indicated with a positive sign. On the other hand, if you emerge at a potential more negative than the point of entry, you have gone down in potential, and this is indicated with a negative sign.

Some illustrative examples are shown in Figs. B-1 and B-2. Remember, the current arrows indicate electron

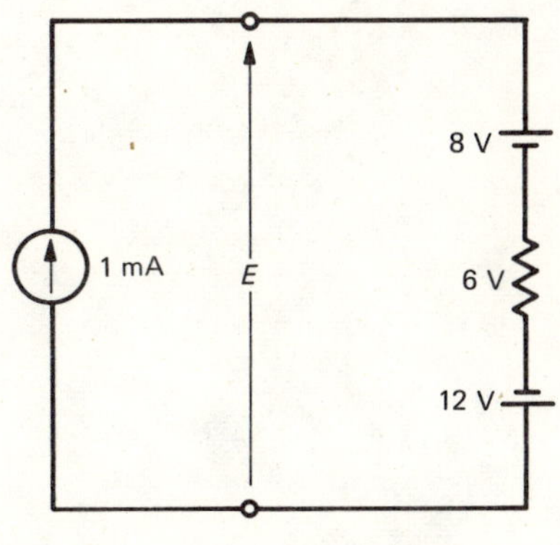

FIGURE B-1

$$E = -12\text{ V} - [6\text{ K}(1\text{ mA})] + 8\text{ V}$$
$$= -10\text{ V}$$

flow, and hence the end of the resistor that the current first enters is polarized negative with respect to the end from which it emerges.

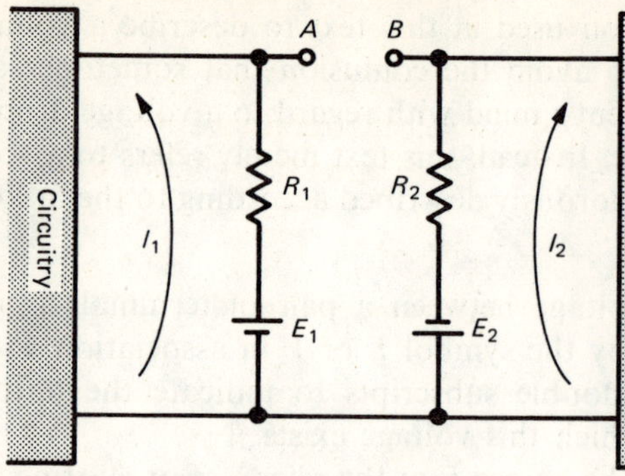

$$V_{AB} = -R_1 I_1 - E_1 - E_2 + R_2 I_2$$

FIGURE B-2

For additional reference material regarding this form of voltage notation, see Phillip Cutler, "Electronic Circuit Analysis," vol. 1, Chap. 1, McGraw-Hill Book Company, New York, 1960.

index